Cyberbiosecurity

Dov Greenbaum
Editor

Cyberbiosecurity

A New Field to Deal with Emerging Threats

Springer

Editor
Dov Greenbaum
Zvi Meitar Institute for Legal Implications
of Emerging Technologies
Reichman University
Herzliya, Israel

ISBN 978-3-031-26036-0 ISBN 978-3-031-26034-6 (eBook)
https://doi.org/10.1007/978-3-031-26034-6

This Springer imprint is published by the registered company Springer Nature Switzerland AG
The registered company address is: Gewerbestrasse 11, 6330 Cham, Switzerland

Preface

I initially became acquainted with the field of Cyberbiosecurity through a chance encounter with Oleg Brodt at Ben Gurion University in early 2020, before the full extent of COVID-19 appreciated.

Oleg was graciously introduced to me by Dr. Aviv Gaon at Reichman University. Oleg together with Rami Puzis, Dor Farbiash, Yuval Elovici and I eventually wrote an interesting paper on the subject of Cyberbiosecurity which we published during the 2020 lockdowns in *Nature Biotechnology*. I still recall having to quickly gather up all my research for the paper, and of course my oversized computer monitor, after the university announced that it was shutting down its physical campus due to the pandemic, and I was forced to decamp to my home.

Once completing and publishing the paper, the issues raised by our article further piqued my interest in the field of Cyberbiosecurity, especially given the general paucity of research on the subject.

My concerns related to the field were further exacerbated by the emerging long-term social, political, and economic effects of the COVID-19 virus: Although, as of this writing, the debate over the nature and source of the severe acute respiratory syndrome coronavirus 2 (SARS-CoV-2) that initially emerged in Wuhan China remains contentious, just the possibility that such a virus could have been created in a lab, intentionally, deliberately, mistakenly, maliciously, surreptitiously, or otherwise because of, or in part, due to failures of Cyberbiosecurity heightens the need for this book.

The field of Cyberbiosecurity has grown substantially over the past years. The research presented in this book should become a foundational component of the continued emerging research in Cyberbiosecurity and related areas.

This book is the result of substantial efforts by all of the chapter contributors who have graciously provided their time and experience to produce varied and excellent analyses of the developing field of Cyberbiosecurity.

In addition to the authors, I would like to thank the editorial staff at Springer Nature for all of their work in creating this book. I would also like to acknowledge my many academic mentors, the administration of Reichman University, the program manager at the Zvi Meitar Institute, the deans of the Harry Radzyner

Law School at Reichman University, and especially the funders of the of the Zvi Meitar Institute for Legal Implications of Emerging Technology, for their ongoing academic and financial backing. I want to further thank my parents for their constant love and support, and of course, my incredible wife and my loving family for always being there.

New Haven, CT, USA Dov Greenbaum

Contents

The Convergence of Biotechnology and Cybersecurity: A Primer on the Emerging Field of Cyberbiosecurity

Dov Greenbaum

Abstract This book represents the latest research and analysis in the nascent field of Cyberbiosecurity. It aims to provide early insights into the field and to aid in better describing and defining it. The papers herein discuss a range of topics related to this field, encompassing both the technical and the ethical, legal, and social implications of the many applications of the associated technologies.

Keywords Bioconvergence · Cyberneurosecurity · Bioeconomy · Digital Twins · MDIoT · Cyberbiosecurity

It is valuable to remember that Cyberbiosecurity is a new field, and much of it is still in formation and in flux. To wit, there even remain disagreements as to how to refer to the field itself, inconsistencies with regard to the exact metes and bounds of the area of study comprised by Cyberbiosecurity, and uncertainties as to how the field distinguishes itself from similar but different academic and practical pursuits such as cybersecurity and biosecurity.

Regardless as to how we eventually demarcate the ultimate characteristics of the field, this book represents a general consensus across a wide swath of the principle researchers and stakeholders in the field as to the growing necessity for the study and analysis of Cyberbiosecurity. The book also relates to the ongoing need to develop tools and methodologies to combat increasingly dangerous threats to the broader bioeconomy, – i.e., the part of the economy that includes economic activities relating to the fields of biology, biopharmaceuticals, health sciences, biotechnology, and agriculture, among others.

D. Greenbaum (✉)
Molecular Biophysics and Biochemistry, Yale University, New Haven, CT, USA

Zvi Meitar Institute for Legal Implications of Emerging Technologies, Reichman University, Herzliya, Israel

Harry Radzyner Law School, Reichman University, Herzliya, Israel
e-mail: dov.greenbaum@yale.edu

D. Greenbaum (ed.), *Cyberbiosecurity*, https://doi.org/10.1007/978-3-031-26034-6_1

To some degree, the field of Cyberbiosecurity is an example of the emerging concept of bioconvergence. Concisely, bioconvergence refers to the further expansion of the bioeconomy. More specifically, it refers to the idea that there are many opportunities for profitable synergy between the biosciences and other areas of technology, hi-tech in particular, that were formerly wholly independent of the biosciences. These areas include artificial intelligence (AI), big data, and cybersecurity.

Importantly, this synergy presents opportunities for newfound growth, discovery, and innovation in the biosciences, especially in the biomedical sciences. This growth includes but is not limited to the areas of diagnostics, drug delivery, and drug discovery. Relatedly, the incorporation of engineering methodologies within biology over the last couple of decades has led to progress within the field of synthetic biology [1, 2], albeit raising its own sets of ethical, legal, and social concerns [2, 3].

Another emerging area of bioconvergence relates to the concept of digital twins. Initially developed for mechanical and electrical engineering purposes by organizations such as NASA (National Aeronautics Space Administration) for use in the development of high-powered rocket engines and for rapidly and iteratively testing them in silico, the concept of creating approximating digital versions of biological systems has been suggested by many academics as a tool for rapidly advancing biomedical research without harming the human patient [4].

The creation and development of a digital twin of a biological process and biological system, or potentially of an entire human, requires substantial amounts of very personal medical and genomic data mediated by advanced artificial intelligence. The protection of the data and the control of its AI manipulation are subject to cybersecurity risks, and as such, Cyberbiosecurity will have an invaluable role in the near future when advanced digital twin technology comes online in the biosciences, and is employed by both academic labs and multinational biopharmaceutical firms for in silico development, and potentially even clinically trialing of new pharmaceutical agents.

In addition to the ideology of bioconvergence leading to greater interdisciplinary research in the biosciences, arguably some of the methodologies and tools promoted by the field of Cyberbiosecurity are themselves an outgrowth of this bioconvergence. However, in addition to these positive externalities, interdisciplinary approaches to the biosciences have also led to increasing cyber-related threats for biology, another area in the expanding purview of Cyberbiosecurity.

Notably, while bioconvergence is a relatively recent term and phenomenon, there have been long-standing efforts within the biosciences to incorporate research methodologies culled from other more computationally intensive fields, especially as the biosciences became, and continues to become, more data focused.

Relatively early examples of this shift to a more data-rich bioscience field include the completion of the human genome project and its early annotations, as well as the many follow-on omic's research endeavors that have subsequently evolved [5]. This shift has necessitated biologists to seek out more computational power and more digital storage to create large databases to support the deluge of diverse types of experimental data. Computational heft is also necessary for the tools to mine that data, as well as for the development and implementation of advanced biological system simulation and modeling tools. Many of the methodologies and ideologies emanating from this shift to a more data-intensive biology coalesced into the now-established fields of Bioinformatics [6] and Computational Biology.

Among many of its innovations, the successes of the field of Bioinformatics and a shift toward exploratory-driven biology and away from classical hypothesis-driven research lead to an even greater increase in the digitization of the biosciences in the search of more discoveries.

While Bioinformatics and the digitization of biological knowledge have resulted in incredible research advancements, the increasing reliance on complex computing power means that many areas within the biosciences are even more at risk from numerous cybersecurity concerns. These threats include various attempts to hack different components of the bioeconomy infrastructure, ranging from academic labs to commercial infrastructure, as well as the potential for new and novel pandemics resulting from security concerns in microbial gain-of-function research laboratories around the world [7].

These potential attacks within various components of the bioeconomy networks can be the result of both professional and amateur hackers. But regardless as to which malicious actors threaten the bioeconomy, every threat is a cause for concern given the ever-growing importance of biotechnology within our economies, and concomitant growth of in silico biological research that can be subject to complex and debilitating attacks. The potential damage that these threats can cause is compounded by the continued general lack of secure systems, particularly within academic bioscience laboratories. Attacks on academic labs can have real-world consequences: consider the possibility that unsecured databases relied upon for follow-on practical medical research can be corrupted through hacking and the like.

But it is not only academic infrastructure that is at risk. Authors within this edited volume will discuss the likelihood that external bioengineering and biomanufacturing infrastructure relied upon by both academic and clinical science labs can be sabotaged digitally, resulting in unreliable and even dangerously wrong academic and clinical results, or worse [6]. In addition to this sabotage, biotechnology infrastructure, both internal and external, can be at risk for corporate espionage attempts that aim to collect intellectual property and know-how, or data breaches resulting in the disclosure of proprietary and private information, as well as various other attacks on the infrastructure within the bioeconomy for fun or for profit.

Additional concerns of Cyberbiosecurity relate to the increasing number of implants and tools that fall within the scope of the medical device Internet of Things (MDIoT) [8]. These are devices such as sensors and diagnostic tools that often rely on unencrypted access to the Internet. Some of these devices employ complex edge computing, while others simply transmit data back to the cloud. Nevertheless, given their relatively simplicity and uncomplicated nature, they are often used as entry points for hackers seeking to attack valuable and vulnerable networks, such as those in hospitals and old-age homes [9].

As advances in healthcare create more MDIoT hardware and software solutions, each device within the MDIoT can be subject to Cyberbiosecurity threats ranging from the benign to the dire. For example, in the field of neurotechnologies, there are numerous devices, implanted or external, that, due in part to lax encryption and security standards, can be compromised at the expense of the patients' health and their rights or even at the expense of the security of an entire medical institution [10]. Arguably, this has already resulted in a subfield within Cyberbiosecurity: Cyberneurosecurity [11].

Many of the concerns raised in this book may seem to be limited to biology labs and other narrowly focused institutions within the biosciences and the biopharmaceutical fields. However, the field of Cyberbiosecurity is relevant for all citizens regardless of their role, or lack thereof, within scientific industries. For example, in a healthcare setting, patient data could be manipulated or misappropriated. Additionally, Cyberbiosecurity also relates to cyber concerns in the increasingly important area of food security where the roles of cybersecurity, biosecurity, and cyber-physical security coalesce to protect the underlying data, the intellectual property, and the agrotech infrastructure.

Worldwide supply chain disruptions were recently brought on in part by the COVID pandemic directing world leaders to reconsider their national food security policies. As our agriculture technologies evolve to include more complex systems, automations, and artificial intelligence, in both natural environments in the field and lab-based settings, there will be increasing security concerns. Security breaches can affect much of the associated technological infrastructure, ranging from global positioning systems to data management. The potential large-scale hacking of these systems, for example, by corrupting the data or data collection devices, such as the myriad of Internet of Things (IoT sensors) associated with any of the nodes along the food distribution pathway, could create significant food insecurity concerns.

Cyberbiosecurity plays an important role, both in hardening the various pathways associated with food security and stress testing, discovering areas of concern, and working toward quick solutions to emerging problems within the food supply chain.

This book comprises 14 chapters from various leaders in the field that assess and analyze numerous aspects of the Cyberbiosecurity field. They are summarized below:

The history of this emerging field and its initial mission is presented by Randall Murch who provides an intimate and personal account of the early development of the field. Alexander Titus and his coauthors present the national security implications that are relevant to the field. Titus et al. note how Cyberbiosecurity is distinct from both the fields of cybersecurity and biosecurity. Their paper further aims to cement the name of the field as Cyberbiosecurity, rather the similarly sounding biocybersecurity.

In contrast, Lucas Potter and Xavier-Lewis Palmer draw actionable distinctions between the two terms, biocybersecurity and Cyberbiosecurity, noting philosophical and practical research distinctions between these two often conflated terms.

In her chapter, Diane DiEuliis provides an introduction to the use of Cyberbiosecurity technologies for ensuring the security of an expanding range of digital tools in the biosciences and the bioeconomy, so that they are not misused for harmful purposes.

Eric Ni, Gamze Gursoy, and Mark Gerstein review the life cycle of human data within the biosciences sector, from the physical acquisition of a sample through the deposit of data into a database and the ultimate analysis of the data. Cyberbiosecurity threats lurk at each of the discrete steps along the way. The authors discuss how to best protect this data.

Peter Ney et al. similarly look at the data workflow process to highlight concerns in security including poor software security practices, insecure hardware, and the general lack of data integrity checks in genetic databases.

Rami Puzis and Isana Veksler-Lublinsky work toward developing a scoring system for systematically quantifying these and other risks emanating from weaknesses and vulnerabilities within biological systems ranging from biosensors to synthetic genes. In employing the Common Vulnerability Scoring System to assess Cyberbiosecurity attacks, they suggest that the system, with some adjustment, can be used to quantify Cyberbiosecurity risks.

In the paper authored by Mariam Elgabry, the author seeks to identify other types of concerns associated with Cyberbiosecurity, particularly criminal activities that are facilitated by synthetic biology or biotechnology. Employing a Delphi study, this chapter aims to demonstrate a framework for use in developing technologies that are secure by design from Cyberbiosecurity threats.

The pandemic has been a learning experience for many aspects of the bioeconomy but also specifically for the emerging field of Cyberbiosecurity. Siguna Mueller's chapter looks specifically to the COVID-19 pandemic as a lens to assess specific Cyberbiosecurity threats, including the use of cyber-tools to weaponize pathogens. The chapter suggests that researchers can use this as a tool to analyze gaps that could be exploited by malicious actors throughout the expanding bioeconomy.

Eleonore Pauwels notes how emerging cyberthreats against biological datasets can threaten the integrity of research. This is particularly disconcerting regarding medical databases and the algorithms derived therefrom. As such, Pauwels argues that there is a need to assess the current levels of regulatory oversight and national and international governance mechanisms in place to protect this valuable data.

These governance mechanisms can be employed to not only create standards for securing bioscience databases, but as Roba Abbas et al. discuss in their chapter, governance is also valuable in regulating the social justice aspects of Cyberbiosecurity, particularly areas at the interface of the biosciences and the digital world that can affect the equality and freedom of citizenry.

Daniel Sobien and coauthors look to a use case of applying Cyberbiosecurity solutions, particularly in the application of artificial intelligence and specifically with regard to high-stake areas such as water systems and agricultural technology.

Sterling Sawaya, Erin Kenneally, Demetrius Nelson, and Garrett Schumacher look specifically to the concerns related to access and abuse of genetic data, particularly the vulnerabilities associated with the exposure of identifying information and health and disease susceptibility data. The authors further explain how, with the growth of advanced artificial intelligence, this data could potentially be weaponized.

Finally, Naor Dalal and others similarly look to the use of Cyberbiosecurity in the protection of wet labs, particularly as they are increasingly relying on artificial intelligence and robotic automation.

In summary, this early foray into Cyberbiosecurity should provide the reader with a broad appreciation for the role of Cyberbiosecurity in many aspects of the bioeconomy and potentially even beyond. There are still many uncharted areas of research in the field of Cyberbiosecurity, particularly as new developments in biotechnology, such as human enhancement, biowarfare, and the increased acceptance of genetically modified organisms (GMOs), pose new challenges that need to be addressed. These and other areas of research remain fertile ground for further exploring new Cyberbiosecurity strategies and solutions.

References

1. H. Yu, X. Zhu, D. Greenbaum, J. Karro, M. Gerstein, TopNet: A tool for comparing biological sub-networks, correlating protein properties with topological statistics. Nucleic Acids Res. **32**(1), 328–337 (2004)
2. D. Greenbaum, An analysis of federal circuit discrimination: The evolution of the written description requirement vis-a-vis DNA and biotechnological inventions concerns for synthetic biology. Recent Patents DNA Gene Seq. (Discont.) **5**(3), 153–165 (2011)
3. D. Greenbaum, Biology's brave new world. Science **369**(6508), 1170–1170 (2020)
4. D. Greenbaum, Making compassionate use more useful: Using real-world data, real-world evidence and digital twins to supplement or supplant randomized controlled trials, in *Biocomputing 2021: Proceedings of the Pacific Symposium*, (2020), pp. 38–49
5. D. Greenbaum, N.M. Luscombe, R. Jansen, J. Qian, M. Gerstein, Interrelating different types of genomic data, from proteome to secretome: 'Oming in on function'. Genome Res. **11**(9), 1463–1468 (2001)
6. R. Puzis, D. Farbiash, O. Brodt, Y. Elovici, D. Greenbaum, Increased cyber-biosecurity for DNA synthesis. Nat. Biotechnol. **38**(12), 1379–1381 (2020)
7. F.M. Oeschger, U. Jenal, Addressing the misuse potential of life science research – Perspectives from a bottom-up initiative in Switzerland, in *Gain of Function Research of Concern*, ELSI in Science and Genetics Frontiers in Bioengineering and Biotechnology, ed. by D. Greenbaum, K. Berns, vol. 6, (2018), p. 38
8. D. Greenbaum, Avoiding overregulation in the medical internet of things, in *Big Data, Health Law and Bioethics*, (Cambridge University Press, 2018)
9. M. Sherman, Z. Idan, D. Greenbaum, Who watches the step-watchers: The ups and downs of turning anecdotal citizen science into actionable clinical data. Am. J. Bioeth. **19**(8), 44–46 (2019)
10. D. Greenbaum, Cyberbiosecurity: An emerging field that has ethical implications for clinical neuroscience. Camb. Q. Healthc. Ethics **30**(4), 662–668 (2021)
11. N. Liv, D. Greenbaum, Cyberneurosecurity, in *Policy, Identity and Neurotechnology: The Neuroethics of Brain-Computer-Interfaces*, ed. by V. Dubljevi, A. Coin, (2023)

Introduction: Origin and Intent for the New Field of Cyberbiosecurity

Randall Murch

Abstract While important, today's biosecurity is too narrowly focused on current and emerging threats. Most biological functions, operations, and outcomes rely on information technologies (IT). While extraordinarily beneficial, relying on cyber technologies also presents risks. Threats in cyberspace abound and could be focused on biological-based targets for an array of purposes and reasons all resulting in negative outcomes. Cyberbiosecurity was created to help the spectrum of the life sciences to begin to understand potential cyberthreats and develop defenses, recovery protocols, and resilience strategies.

Keywords Biosecurity · Cybersecurity · Information technologies · Cyberthreats · Emerging threats · Cyberbiosecurity

1 Cyberbiosecurity: How It Was Conceived and the Basis

Originally, cyberbiosecurity was the concept developed by three faculty members from Virginia Tech (two of the three have since moved to other universities). From their respective experiences and expertise, the three realized that cybersecurity by itself was insufficient to protect any biological activity or operation that relies on information technology applications, tools, or technologies. Thus, the concept was further developed and refined for further communications and marketing seeking visibility and funding. Following concept development, the originators undertook active engagement of the concept with commercial and academic colleagues as well as the US Federal agencies they were most familiar with. We had a number of agencies that provided audiences and eventually one which funded a large, three-university team to conduct a detailed system analysis of a bioprocess development facility which led to beginning to formalizing the new field. More on these are discussed below.

R. Murch (✉)
Virginia Tech, Virginia Tech Research Center, Arlington, VA, USA
e-mail: rmurch@vt.edu

D. Greenbaum (ed.), *Cyberbiosecurity*, https://doi.org/10.1007/978-3-031-26034-6_2

2 Why It Is Important

Cyberbiosecurity provides an alternative philosophy and approach to investigating postulated or encountered security gaps with biological and biosecurity-IT situations and scenarios. Many IT security, biopharma, agricultural, medical, biomedical, genomics, bio data analytics, and bioeconomy protection experts who were engaged in our process confirmed that cybersecurity alone was insufficient to detect, characterize, resolve, or attribute these threats. Most commented that cybersecurity experts would not be able to recognize or understand the significance of threats such as these. They all advised that a new approach was needed and cyberbiosecurity had great promise. One such IT security expert who had moved from a very large biopharma company to a smaller-scale, high-priority US military rapid response vaccine development program said it well, "I have a long IT security checklist I have developed for use here, inspired by my previous employment. But, at the end of the day, I have no idea what exactly I am supposed to protect here and how to recognize, detect or rehabilitate our program from deeper bio-cyber threats. Cybersecurity by itself is likely insufficient."

3 What Gap(s) Exist and Existed Prior

Our original team, and our rapidly increasing informal group catalyzed by the first workshop in 2017, all realized that no one was paying close attention to this problem set. Any official US Government program that should have had responsibility any of us reached out to all came back with "they think they have it covered." Experience had taught the team that, often, that attitude was a prescription for disaster. Following the original inquiries and further and deeper investigations below top-level agencies to achieve greater fidelity and precision, all agreed that something different had to be created, developed, and established. High-level cybersecurity prescriptions were not enough nor tailored for the enormous complexity and diversity of the dynamic cyber-bio interface. Now it was time to change attitudes and minds.

Beginning in 2014, a very powerful cascade of reports and articles were published on the risks associated with big data [1], initiating discussion of safeguarding the bioeconomy [2–5], and vulnerabilities with biolabs of the future [6–8] and more from a key champion from the US Federal Bureau of Investigation [5, 9]. All of these taken together provided a strong base for a new and big push to secure biological facilities and operations. While these were all crucial contributions, all of these stated the "problem" and made recommendations but none led to the next "next step," that is, an actionable path to an environment of deeper and tailored understanding of threats and risks and leading to developing, testing, and validating solution sets. These articles and reports strengthened the commitment to establishing a new way forward. Now it was time to find an opportunity to initiate a path to conceivable, or even implementable, solutions.

4 The First Project

Armed with a quad chart we could all agree upon, the original team set out to seek funding for a project that could lead us to demonstrating our philosophy and ideals for a new field. Selling this to a funding agency was quite an ordeal for over a year and then success. The funder (customer) made it quite clear that they wanted a product that was actionable and matched well to a problem set that they were facing. They also did not want us to tell them what to think or how to respond but rather provide an analysis that would allow them to adapt what was produced to their problem at hand, which we would not be privy to in detail. A random research project would not do. This had to be completed in a fiscal year or less. Our team came up with a prototype project, worked with the customer to develop the internal strategy for funding, and then formed the multi-university team and formulated the agreements and plan. The core theme for the project was a "system analysis" which was tightly coupled with customer needs, not a research project that had no understanding of what would be actionable.

"Systems analysis" can be defined as "the act, process, or profession of studying an activity (such as a procedure, a business, or a physiological function) typically by mathematical means in order to define its goals or purposes and to discover operations and procedures for accomplishing them most efficiently" (*Merriam Webster Dictionary*). The customer was not interested in a collection of mathematical models to support the analysis. Therefore, the analysis was performed in a descriptive manner seeking to understand the target system as a complex system of systems. The study was to include the human dimension, as well.

Thus, it was agreed that the Virginia Tech – University of Nebraska at Lincoln (UNL) – Colorado State University (CSU) team of experts would conduct a thorough systems analysis of the Biological Process Development Facility (BPDF). This 1-year effort was conducted in a system analysis project format with an outcome-focused approach which was actionable. The four themes of the project were the following:

Task 1: Characterize the information and physical ecosystems of biomanufacturing.
Task 2: Perform focused analysis of prioritized vulnerabilities (also known as "deep dives"), including the supply chain.
Task 3: Generate hypothetical yet plausible scenarios and concept of operation (CONOP, representation of an ideal state) incorporating both offensive (attack) and defensive (defend, protect) options.
Task 4: Final report (detailed US Government Only report).

One unique dimension was that we needed to ensure that all viable operational considerations and viability were integrated. The principal investigator (PI) from

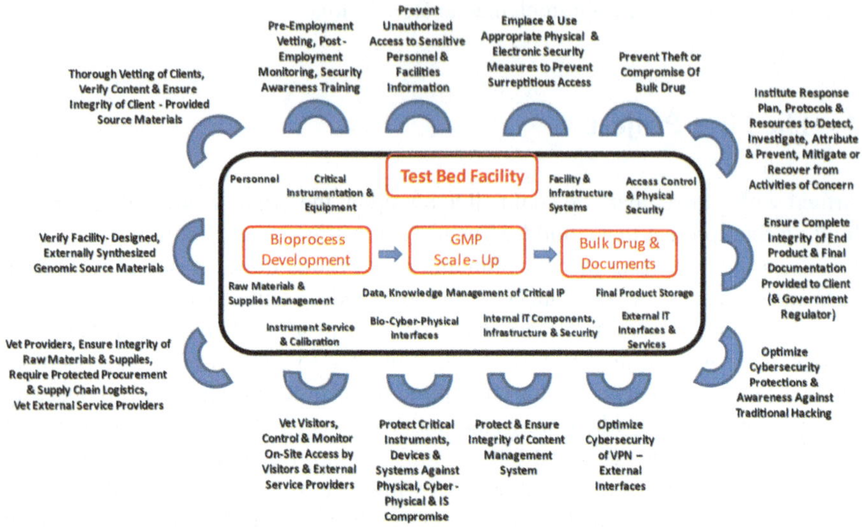

Fig. 1 A summary of results from this project is incorporated below (also see Ref. [11]). This constitutes a simplified pictorial summary of possible threats and defenses. While some results may seem mundane, both the team and customer realized that they must be incorporated for completeness and actionability. Success could not rely on singular or linear solutions but rather adaptive and combination solutions for various lines of attack and defense

Virginia Tech and the BPDF director ensured that the project was adaptive, dynamic, and focused on the project objectives and customer requirements. Continuing engagement with the customer throughout the project was absolutely necessary to ensure success, as were on-site (BPDF) interim program reviews (IPRs) with or without customer presence (Fig. 1).

5 Outcome and Encouragement by the Client to Extend and Expand Soonest

In September 2017 nearing the end of our project, the PI attended a final briefing with the customer which included many constituencies beyond the original office the team was working with. At the end of the briefing, the PI was advised that all were highly pleased with the outcome and benefit to internal groups which could actualize the team's work for their own purposes. Further, the leadership of the review team enthusiastically advised that following the end of our period of performance, "you should run as far and fast as you can with this . . . this is truly fantastic and groundbreaking work."

5.1 *Making Our Refined Ideas Public: "Cyberbiosecurity"*

Cyberbiosecurity seeks to bring attention to inherent and increasing insecurity of any biologically based activity or function that is supported by or interfaced with information systems. The need for cyberbiosecurity inquiries and developments has been effectively demonstrated over the past 4.5 years, since the publishing of the first two peer-reviewed papers and the Frontiers e-book and many national and global presentations. Though comparative studies have not been performed, as yet, the explorations of a variety of dimensions and applications have increased many times over. Thus, the worldwide community of investigators and commentators must believe that cyberbiosecurity provides a new avenue to examine and communicate vulnerabilities in cyberbiosecurity in their own scientific venues.

5.2 *The Rapid Expansion of Cyberbiosecurity*

5.2.1 The First Workshop

The very first workshop on cyberbiosecurity was held in October 2017, soon after the completion of the US Government-funded project. The purposes were threefold: rollout the concept of cyberbiosecurity, explore potential boundaries for this new field, and develop a core constituency including exactly the right representatives of ten academic institutions, including the participating universities in the first study, eight US Government agencies, two small companies which do biosafety and biosecurity training, and two key nonprofits. Creating this core paid off in ensuing years for several reasons. These included recruiting authors for the cyberbiosecurity e-book, see below; having a group of experts to call on to represent cyberbiosecurity in various fora; including pertinent US National Academies studies and presenters at high-level commissions; and continuing visibility for cyberbiosecurity wherever or whomever has been interested in learning more or showcasing this topic.

5.3 *The First Two Peer-Reviewed Articles*

Within 2 weeks after the workshop, our core project team moved to the next phase: the first-ever publications on cyberbiosecurity. In December 2017, the first article was published in *Trends in Biotechnology* [10]. It is primarily focused on the security of the genomics-bioinformatics laboratory and has significant impact (42 citations, 1 policy citation, 32 tweets, 11 news mentions, 70 reads, and 6 blog mentions). In April 2018, the second article was published in *Frontiers in Bioengineering and Biotechnology* (Section on Biosafety and Biosecurity) [11]. This paper took a more strategic approach, described the original systems analysis executed for the US Government, and laid out a path forward toward a new discipline. That article has received 14,380 views, 43 citations, and worldwide viewership (as of August 17, 2022). The above work also led to the first definition of cyberbiosecurity:

> Understanding the vulnerabilities to unwanted surveillance, intrusions, and malicious and harmful activities which can occur within or at the interfaces of comingled life and medical sciences, cyber, cyber-physical, supply chain and infrastructure systems, and developing and instituting measures to prevent, protect against, mitigate, investigate and attribute such threats as it pertains to security, competitiveness and resilience. [11]

While this definition seems unwieldy, it is sufficiently inclusive. Other definitions have emerged which have been tailored to specific communities, that is, the US Department of Defense.

5.4 *The e-Book*

The e-book was invited based on the principal investigator's role as an associate editor for the section in which the second article was published and its substantial success. The co-editor was a highly regarded expert in health sciences, biodefense, and biosecurity. Thus, an e-book [12] was formalized. The process of recruiting authors, editing contributions, and working with Frontiers through the entire process took approximately 1 year. This publication had 16 articles, 71 authors. Most of the authors participated in the first workshop and purposely covered a wide spectrum of topics and applications as the title suggested. It maximized the diversity of topics as the first of many hoped-for explorations. As of August 17, 2022, this e-book has achieved 115,162 views, 93,348 article views, 16,611 article downloads, and 5203 topic views. The number of views usually increases by ca. 1000 per week. Interest in this effort has spanned six continents, as evidenced by the demographics associated with this work. This is a very successful endeavor and one that should be on the reading list of all those interested in exploring and pursuing cyberbiosecurity or closely related topics.

5.5 *Outreach and Expansion: Many Presentations, Including at High-Level Fora*

The natural progression from this work has been presentations to a variety of audiences, more publications in both article and book form, and training and outreach. Alerts via ResearchGate have indicated that explorations of cyberbiosecurity continue to result in publications from across the globe. Personally, as just one example, I have 10 publications including the e-book and specific chapters and 20 public presentations both to audiences in the USA and internationally, several lectures at my university, and ca. 9 presentations within the US Government. These presentations include the Blue Ribbon Commission on Biodefense (www.biodefensecommission.org) and two panels at the US National Academies of Sciences, Engineering, and Medicine (NASEM). Other colleagues on the core team have made a host of presentations and issued related publications. As exciting, three undergraduate students from other universities within the USA have reached out to learn more and prepare to study this topic as graduate students. One of those three has worked with faculty at his campus as well as faculty from that university's main campus within the system to create a new, interdisciplinary program focusing on cyberbiosecurity. Reportedly, there are two other books in progress, in addition to this one, all with different foci. A recent report by a US National Academies study committee provides excellent context for investigations and developments in cyberbiosecurity [13].

5.6 Functional and Structural Expansion at a University

In response to the work of the author, and colleagues from the College of Agriculture and Life Sciences (CALS) at Virginia Tech (Virginia Polytechnic Institute and State University, Blacksburg, Virginia, USA), a new center has been created which has three main foci. Cyberbiosecurity is one of these. The other two are related: data analytics and the Smart Farm (agricultural operations heavily invested in information technology-supported systems). The center is called the Center for Advanced Innovation in Agriculture (CAIA) (see www.caia.cals.vt.edu). It is headed by a senior official of the Virginia Agricultural Experimentation Station system, who also holds a professorship in an agricultural discipline. Affiliate faculty come from departments and centers across the college as well as the Colleges of Engineering and Business. CAIA has a very strong relationship with the Commonwealth Cyber Initiative (CCI; https://cyberinitiative.org) which is a state-funded, multi-university program focused on highly advanced cyber research and education. Researchers from the Virginia Tech's Hume Center for National Security and Technology (https://hume.vt.edu) have been collaborating with CALS and CAIA faculty for some time. Colorado State University is another US university with strong interests in this arena. The author looks forward to learning about similar centers and collaborations from the USA and elsewhere in cyberbiosecurity.

5.7 The Future: Expansion to and Deepening of Cyberbiosecurity to Applicable Disciplines

Given the evolution and trajectory of cyberbiosecurity over the past 4.5 years, it is safe to say that this will continue, as well as deepen and broaden. Further, it is one thing to conduct research and experimental studies but quite another to implement cyberbiosecurity practices, programs, and policies. Transitioning from idea or concept to action could be a focus of near to midterm. Understanding how cyberbiosecurity investigators undertake, test, and validate a capability or application would be significantly beneficial to the discipline. Though knowing details may not be appropriate, understanding the process would be.

Though public health restrictions due to COVID-19 may be challenging, perhaps the time has come for direct communications between cyberbiosecurity investigators and observers. Options such as regional, national, or international webinars or conferences might provide significant value. Published summaries of presentations or abstracts would also be most valuable and likely increase the cyberbiosecurity constituency.

During the envisioned gatherings, one topic that might be explored could be "guidelines for practice and application" for cyberbiosecurity. Guidelines, and even standards, are known in many fields. The process of agreeing on guidelines which all investigators or trainers would follow would be incorporated when cyberbiosecurity methods are adopted or implemented. If such guidelines (eventually standards) could be developed, vetted, and adopted, confidence levels for quality, reliability, repeatability, and safety would be established or increased. If the cyberbiosecurity environment were to pursue and adopt such an approach, researchers would consider integrating such principles into anything they produce, whether knowledge or prototype technologies.

Policy development and implementation should be taken up for cyberbiosecurity, as it should be part of considerations of those who are concerned with biosecurity, biodefense, and safeguarding the bioeconomy (by sector, by country, by national networks, etc.). The big push in widening the view of biosecurity through publications on "big data" security, lab security of the future, and cyberbiosecurity should be fully incorporated into nonproliferation, counterproliferation, biosecurity, and biodefense discussions and actions.

The future of cyberbiosecurity is very bright. This book will widen the view and enhance its attraction and impact.

References

1. American Association for the Advancement of Science, Federal Bureau of Investigation and United Nations Interregional Crime and Justice Research Institute, *National and Transnational Implication of Security of Big Data in the Life Sciences* (American Association for the Advancement of Science, Washington, DC, 2014), p. 91
2. The National Academies of Sciences, Engineering and Medicine, *Meeting Recap, Workshop – Convergence: Safeguarding Technology in the Bioeconomy* (Organized by the Board on Chemical Sciences and Technology and the Board on Life Sciences, Washington, DC, 2014)
3. National Academies of Sciences, Engineering and Medicine, *Meeting Recap, Safeguarding the Bioeconomy: Applications and Implications of Emerging Science* (Organized by Board on Chemical Sciences and Technology, Washington, DC, 2015)
4. The National Academies of Sciences, Engineering and Medicine, *Meeting Recap, Safeguarding the Bioeconomy III: Securing Life Sciences Data* (Organized by the Board on Life Sciences

and Board on Chemical Sciences and Technology, Washington, DC. Weise, E. (2015, February 5), 2016)
5. K.G. Kozminski, D.G. Drubin, Biosecurity in the age of big data: A conversation with the FBI. Mol. Biol. Cell **26**(22), 3894–3897 (2015). https://doi.org/10.1091/mbc.E14-01-0027
6. E. Pauwels, A. Vidyarthi, *How Our Unhealthy Cybersecurity Infrastructure Is Hurting Biotechnology*. Wilson Briefs (The Wilson Center, Washington, DC, March 2016), p. 4
7. E. Pauwels, A. Vidyarthi, *Who Will Own the Secrets in Our Genes? A U.S. – China Race in Artificial Intelligence and Genomics*. Wilson Briefs (The Wilson Center, Washington, DC, February 2017), p. 14
8. E. Pauwels, G. Dunlap, *The Intelligent and Connected Bio-Labs of the Future: The Promise and Peril in the Fourth Industrial Revolution*. Wilson Briefs (The Wilson Center, Washington, DC, September 2017), p. 17
9. E.H. You, *Safeguarding the Bioeconomy: U.S. Opportunities and Challenges* (Testimony for the U.S. – China Economic and Security Review Commission, Washington, DC, March 2017)
10. J. Peccoud, J. Gallegos, R. Murch, W. Buchholz, S. Raman, Cyberbiosecurity: From naive trust to risk awareness. Trends Biotechnol. **36**(1), 4–7 (2017). https://doi.org/10.1016/j.tibtech.2017.10.012
11. R.S. Murch, K.L.W. So, S. Raman, W. Buchholz, J. Peccoud, Cyberbiosecurity: An emerging new discipline to help safeguard the bioeconomy. Front. Bioeng. Biotechnol. **6**, Article 39, 6 p (2018). https://doi.org/10.3389/fbioe.2018.00039
12. R.S. Murch, D. DiEuliis, Guest Editors. Mapping the Cyberbiosecurity Enterprise. Invited Research Topic Special Collection. Frontiers: Bioengineering and Biotechnology (Section on Biosafety and Biosecurity). Initiated Fall 2018, Articles Published in 2019 (16 Articles, 75 Authors) (2019), https://www.frontiersin.org/research-topics/8353/mapping-the-cyberbiosecurity-enterprise
13. The National Academies of Science, Engineering and Medicine, (Safeguarding the Bioeconomy, 2020), https://www.nap.edu/catalog/25525

Cyber and Information Security in the Bioeconomy

Alexander J. Titus, Kathryn E. Hamilton, and Michelle Holko

Abstract The expansion of digitally interconnected devices, laboratory equipment, and cloud computing is accelerating discovery and development across the bioeconomy. However, this interdependence simultaneously introduces a growing number of new cyber and information security concerns. As a result, new fields are emerging so that practitioners can better research, understand, and respond to these new threats. Biocybersecurity, cyberbiosecurity, and digital biosecurity are only a few of these new emerging fields that require definition and refinement as they grow. In this chapter, we discuss the growing need for cyber and information security in the bioeconomy and what each of these new fields mean and briefly highlight case studies where vulnerabilities are introduced, pre-existing infrastructure capable of response, and critical junctures where new technologies are required to mitigate these threats.

Keywords Biocybersecurity · Cyberbiosecurity · Digital biosecurity · Cybersecurity · Information security

A. J. Titus (✉)
Colossal Biosciences, Dallas, TX, USA

International Computer Science Institute, Berkeley, CA, USA

Bioeconomy.XYZ, Washington, DC, USA

Council on Strategic Risks, Washington, DC, USA
e-mail: publications@bioeconomy.xyz

K. E. Hamilton
Colossal Biosciences, Dallas, TX, USA

Bioeconomy.XYZ, Washington, DC, USA

M. Holko
International Computer Science Institute, Berkeley, CA, USA

Bioeconomy.XYZ, Washington, DC, USA

Defensive BioTech, LLC, Cabin John, MD, USA

D. Greenbaum (ed.), *Cyberbiosecurity*, https://doi.org/10.1007/978-3-031-26034-6_3

1 Introduction

Technology is accelerating at a pace faster than ever before, and as technology grows into new fields, industries, and markets, it is critical to stay ahead, or at least at pace, of the security threats. In 2011, Marc Andreessen of the renowned Andreessen Horowitz venture capital (a16z) firm famously made the claim that software is eating the world [1], recognizing that the digital revolution was changing everything from industrial manufacturing to healthcare via networks and Internet connectivity, increasing the cyberattack surface exponentially. In 2019, the firm released another technology manifesto, this time acknowledging an actively evolving paradigm shift as biology is now eating the world [2]. The life sciences and biotechnology are shaping our next industrial revolution, coevolving with technology and software; we must pay careful attention to the growth, development, and integration of these technologies into our biology and everyday lives. It is in the intersection of these two revolutionary domains that new threats and vulnerabilities are being discovered everyday and why the study of cyber and information security in the context of the bioeconomy is crucial to the long-lasting success of these industries.

The global bioeconomy is enormous, comprising large portions of national economies (>2% of the US GDP), as of 2017 [3] https://www.schmidtfutures.com/our-work/task-force-on-synthetic-biology-and-the-bioeconomy/. In 2020, the National Academies of Sciences, Engineering, and Medicine proposed the following definition for what comprises the United States' bioeconomy: "Economic activity that is driven by research and innovation in the life sciences and biotechnology, and that is enabled by technological advances in engineering and in computing and information services." This includes industries such as pharmaceuticals, biotechnology research and development, and medical diagnostics. It excludes industries such as beverages and tobacco, nature tourism, hunting, fishing, and paper products, among others. Some estimates see the bioeconomy being directly responsible for the infusion of $4 trillion per year over the next 10 years into the global economy [4]. And while an in-depth analysis of the bioeconomy is beyond the scope of this chapter, a brief review is warranted to level set why it is critical to safeguard the bioeconomy [5], particularly in the context of the cyber domain.

In the midst of the COVID-19 pandemic, the biotechnology industry experienced its largest year of financing in history [6]. In 2020, the global bioeconomy saw over 73 firms raising more than $22B dollars in initial public offerings as they moved to become publicly traded companies, and private funding grew 37% over the previous year. Additionally, global healthcare spending reached nearly $8T dollars in 2017 [7], and with a predicted growth trajectory for years to come.

In addition to financial gains, data in the bioeconomy is growing exponentially. In the field of genomics, the US National Human Genome Research Institute predicts that genomic projects will generate 40 exabytes of data in the next decade alone [8] and that genomic data is set to exceed the growth potential of Twitter, YouTube, and the entire field of astronomy [9].

The growth of the bioeconomy, both from financial and data perspectives, has been recognized across the world, and many countries are investing in efforts to strengthen their respective positions in the global bioeconomy. Within the USA, a number of proposed pieces of legislation have supported the investment of billions of dollars into the domestic bioeconomy [10–12]. These investments come with a growing concern about the impact of the wide array of potential cyber threats to the industry. In light of major cyber breaches impacting large portions of the global economy in the midst of the COVID-19 pandemic, there is an urgent need to improve the cyber posture of the entire bioeconomy in anticipation of the increasing cyber threat potential [13].

The COVID-19 pandemic has also placed a heightened emphasis on the bioeconomy as an essential component of domestic national security [14]. In an interconnected world of biomanufacturing, reagents, and biotechnologies, every country in the world is dependent on partner nations to provide critical components of their bioeconomy supply chains [15]. When supply chains are disrupted, as in the face of a global pandemic, costly and even deadly pressure is exerted on domestic healthcare and national security ecosystems [16]. An increased awareness of the global nature of the bioeconomy has led to interest in increasing onshore and nearshore options, as well as alternative suppliers, to improve resilience in the sector.

When considering the national security implications that lie at the intersection of the bioeconomy and cybersecurity in light of the COVID-19 pandemic, a summary of potential impacts was highlighted in a recent report by the US Cyberspace Solarium Commission, titled *Cybersecurity Lessons from the Pandemic*: [17]

1. Both a pandemic and a significant cyberattack can be, and likely will be, global in nature.
2. Both a pandemic and a significant cyberattack require a whole-of-nation response.
3. When no mitigations are readily available, innovations emerge slowly and thus require systems that are resilient, agile, and collaborative in nature, built between government and industry.
4. Investment in prevention and preparation will be far cheaper and more effective than relying solely on detection and response.

As bioeconomy organizations move to modernize their cyber posture, it is important to understand the relevant aspects of cyber and information security. In addition, an assessment of experience-tested knowledge and applications that the bioeconomy can pull from the well-established fields of cyber and information security will accelerate adoption and readiness. In areas where the bioeconomy has unique cybersecurity requirements, refining the understanding, communication, and translation of requirements across sectors is imperative to progress. This chapter intends to [1] describe what cybersecurity, information security, and biosecurity mean in the context of securing the bioeconomy, [2] provide a framing of how we can consider naming the field at the intersection of cybersecurity and the bioeconomy, [3] explore critical domains within the bioeconomy, and [4] highlight examples within these critical domains.

2 Setting the Stage with Definitions/Descriptions

To effectively discuss cyber and information security in the bioeconomy, a field increasingly referred to as cyberbiosecurity, we first propose working definitions of the independent fields of both cybersecurity and biosecurity. Both fields are well established and have a robust ecosystem of practitioners and academics, and as mentioned above, both have been thrust onto the global stage in the midst of the COVID-19 pandemic and global-scale cyberattacks.

2.1 What Is Cybersecurity?

Cybersecurity is the protection of the physical hardware, software, systems, and information that flow through networks. It is a broad and creative field, with a wide range of ever-growing techniques and technology-based solutions. Cybersecurity encompasses the protection of Internet-connected systems across domains and sectors, regardless of the application of these systems, networks, and information [18]. It is defensive in nature, as cybersecurity professionals are always trying to stay at least one step ahead of hackers and threat actors. As the world becomes more networked and interconnected, cyber vulnerabilities are introduced into nearly every aspect of our lives from our homes, cars, and appliances to large equipment and computer systems [19].

The wide and ever-evolving range of cybersecurity practices involves protecting networks, devices, and data from unauthorized access, criminal use, and tampering. One of the main tenets of cybersecurity is the CIA triad, the practice of ensuring confidentiality, integrity, and availability of information [20]. While there are some common strategies to achieve this, the methods depend on the system architecture as well as the range of individuals accessing the network. The National Institute of Standards and Technology (NIST) provides guidance and best practices for cybersecurity and risk management, including the Cybersecurity Framework [21]. As technology and system configurations, including hardware, virtualized environments, and software, continue to change, so must cybersecurity practices. The financial services industry was one of the leading sectors to develop robust cybersecurity practices. Many of these principles can be extrapolated to other industries, domains, and sectors, but it is important to realize that there are also sector- and industry-specific characteristics of networks, systems, and information that may warrant differences in cybersecurity approaches.

An example of how these practices have changed over time can be seen by reviewing different approaches for user authentication and network access/permissions. Technologies such as virtual private networks (VPNs) are very widely used and focus on authenticating a user or device upon entry to the network and restricting network access to only authenticated users, but the users generally have access to the full network within the VPN and are only authenticated upon entry [22]. In recent years, however, with the increasing pace of cyber breaches, security practices have evolved such that VPNs are not enough to provide adequate security to computer and network systems [23]. Two of these additional security practices include multifactor authentication (MFA) and the use of role-based profiles. Role-based profiles limit a user's access to the network to only the domains, programs, and information that are needed to perform the specific role. MFA requires users to have more than one source of authentication. These different sources can include (1) things you know, like a password or passphrase; (2) things you have, like a token or key; and (3) things you are, biometrics like facial recognition scanning, fingerprints, iris scans, and more.

A currently popular strategy to improve the verification of a user or devices beyond the VPN, is zero trust security. In the zero trust paradigm, users are only given access to the applications and information that they absolutely need, multifactor authentication is used, and reauthentication is required every time there is a request to access a new critical system, domain, or information [24]. The practice of constant reauthentication, in combination with MFA, is considered among the most secure paradigms for cyber and information security, as it subverts many types of cyberattacks, like brute force attacks. In recognition of the importance of zero trust security, a recent US executive order (EO) was issued mandating that federal agencies begin moving to a zero trust posture [25]. While this EO is focused on cyber modernization within the federal government, the practice of zero trust is growing among industries as well and will likely become the global standard for cyber and information security.

Other cybersecurity best practices include performing timely software updates and patching, regular system backups, requiring cybersecurity training for system users, developing data inventories, system maps, and incident response plans and network scanning and threat hunting to monitor for breeches or attempted breeches. Many of the ransomware attacks and other breeches could be avoided by timely and regular software updates and patching [26]. During the COVID-19 pandemic, a delay in regular updates was observed and may have contributed to the increase in ransomware activity seen especially in the healthcare industry [27]. Backups are also a critical component. An intact backup of the system and data, stored offsite or in the cloud, can eliminate the need to respond to ransomware attacks. But it is important to remember that the cybersecurity practices employed depend on the system architecture and technology used in your network and, as technology evolves, so will the threats, and the best security professionals will engineer cybersecurity in an effort to stay ahead of threat actors.

2.2 What Is Information Security?

Information security (InfoSec) is a critical subfield of cybersecurity focused specifically on data security and integrity [28].

Data is the foundation of much of the modern global economy [29]. Critical data ranges from financial, operational, and competitive information for a government or corporation to the personally identifiable information about individuals' lives, from their habits and preferences to their medical history. Keeping such information secure is paramount in the growing bioeconomy because more and more of the global economy relies on access to this information and these products, particularly as the global bioeconomy accelerates the use of precision and personalized medicine. In these cases, public health, national security, and an individual's ability to care for themselves and their families depend on this information. Large data breaches in recent years have unknowingly compromised the personal information of millions of global citizens [30], highlighting the urgent need for enhanced information security in the bioeconomy.

Within the bioeconomy, there is also a significant amount of "open data" research. Data sharing is paramount to discovery and innovation, and many organizations, including the National Institutes of Health (NIH), have endorsed "open data" initiatives in an effort to solve some of the greatest challenges in biomedical research [31]. This is obviously in tension with efforts to keep data secure and private. When considering this question of when it is okay to make data "open" vs secure and private, it is important to consider the data type, source, stakeholders, and national security impact. It may be easier to know the former three, whereas national security impact cannot always be understood at the time of decision-making, as data agglomeration and global circumstances can impact this evaluation and change over time.

There are several key regulations and policies that can help guide security decision-making for bioeconomy information. The Health Insurance Portability and Accountability Act (HIPAA) was enacted into law in 1996, modernized the flow of healthcare information, stipulated how personally identifiable information maintained by the healthcare and healthcare insurance industries should be protected from fraud and theft, and addressed some limitations on healthcare insurance coverage [32]. It generally prohibits healthcare providers and healthcare businesses from disclosing private information to anyone other than a patient and the patient's authorized representatives, while individuals are able to share their personal healthcare information voluntarily. The Genetic Nondiscrimination Information Act (GINA) is another piece of legislation, signed into law in 2008, to prohibit some types of genetic discrimination, another type of personal health information [33]. There are also laws that include but are not specific to bioeconomy data, like the General Data Protection Regulation (GDPR) in Europe [34] and California Consumer Privacy Act (CCPA) in California [35].

While there are some regulations that provide guidance about data security best practices, these are mostly focused on healthcare data and consumer information.

Not all bioeconomy data fall into these categories. There is also a tension with "open data" initiatives. All of this means that a thorough evaluation of the bioeconomy and national security implications of specific types and applications of data is needed to guide best practices. A conservative stance would be to treat all data as sensitive. There has also been significant development in algorithms to ensure secure storage and computation of sensitive data types, like genomic data [36]. This kind of innovation is increasingly needed in order to strike a balance between security and "open data" for discovery.

2.3 *What Is Biosecurity?*

Biosecurity is a broad field that includes practices and policies that aim to keep diseases away from animals, property, and people [37]. Best practices in biosecurity include laboratory practices, personnel security, escape containment, and information access protocols that limit who has access to critical information and physical samples that may cause damage or harm to the individual or at a national security level.

Policies and procedures are put in place to develop best practices for biosecurity and at a governmental and international level also include norms and agreements on what is appropriate and not acceptable to carry out [38]. Dual-use research of concern (DURC) is an example of a policy challenge that has plagued regulators and policy makers for decades. DURC refers to the research and analysis of potentially concerning features and functions of biological agents for the purpose of better understanding how to respond to such a threat. But given the precarious nature of verifying the intent of a specific research topic, it has often been a topic of contention.

Another program that has been implemented to help improve global biosecurity is the Biological Threat Reduction Program (BTRP) [39], funded by the Department of Defense and implemented by the Defense Threat Reduction Agency (DTRA). The goal of this program is to improve the global biosecurity posture by funding specific research projects in partner countries that help build safe infrastructure for biomedical research. This is an example of science and tech diplomacy, where allied countries work together to build infrastructure to ensure capabilities across the entire global allied network.

In principle, cybersecurity and biosecurity both aim to reduce the risk of harm and prevent threats in these distinct domains, both of which could be exploited to cause harm at the individual all the way to the national and global levels. Some of the ways biosecurity principles and practices are similar to cybersecurity practices are the practice of restricting DURC, research into vulnerabilities in an effort to understand how to mitigate them, and the rapidly changing nature of the technologies [40]. This overlap provides an opportunity to learn from cybersecurity to develop biosecurity best practices, and vice versa, all in support of securing the bioeconomy.

3 Defining Emerging Terminology

3.1 Cyberbiosecurity

If we examine the term "cyberbiosecurity" through the lens of interdisciplinary collaboration, the field would represent the act of applying biosecurity best practices to the cyber domain [41]. Computer network threats are commonly referred to as "viruses"; thus the overlap appears apparent and even reasonable, especially given the need to rapidly increase cyber literacy around the globe.

Many biosecurity best practices transcend the discipline silo and apply well to the practice of cybersecurity, where policies and best practices are intended to keep bad actors away from potentially harmful material and to mitigate the impact of harmful biological agents intentionally or unintentionally being released from a contained area. This includes personnel security, physical security, laboratory safety protocols, and escape containment protocols, among others [42]. Yet, while these practices may apply in certain situations to the cyber and information security needs of the bioeconomy, it does not encompass all that should be considered when building a robust focus on these topics. An example of where this field would apply well is when protecting critical software that, if released into the open Internet, would cause harm to cyber systems broadly. Much in the way that recent cyber viruses, such as WannaCry and Petya, have caused global challenges [43]. Thus, the field of cyberbiosecurity would be the focus of protecting offensive and defensive cyber capabilities.

3.2 Digital Biosecurity

The term "digital biosecurity" is also growing in popularity as a way of describing a broader set of concepts beyond cyber systems, to include biological and biomedical data and related algorithms [44]. In the case of digital biosecurity, the terminology implies the application of biosecurity best practices to the digital domain and would treat the breach, loss, and/or escape of critical information as a key focus. For example, the loss of personally identifiable information could compromise a patient's medical records, or the loss of a valuable algorithm designed for drug discovery would compromise the business success of a pharmaceutical company. Thus, the field of digital biosecurity would be the focus of protecting critical cyber capabilities, information, and algorithms in a life science context.

3.3 Biocybersecurity

The term "biocybersecurity" is also emerging into practice, albeit at a slower rate than both cybersecurity and digital biosecurity. The concept of biocybersecurity is focused on applying cybersecurity and information security best practices to the healthcare and life science domains [45]. In the cyber domain, these practices may include zero trust security protocols [46], software-defined network perimeters [47], properly segregated networks [48] for critical and noncritical systems, and more. In network segmentation, for example, DNA sequencing machines should not be connected to the same networks as sensitive personnel databases, creating threat vectors into business critical systems from less secure laboratory equipment. But in the case that there is only a single network, then zero trust security protocols would prevent unauthorized access from a DNA sequencer into such sensitive data systems. Thus, the field of biocybersecurity would be the field of applying cybersecurity methods to the bioeconomy.

4 What Do We Want These Emerging Fields to Mean?

There are a growing number of domains with different names, including cyber-biosecurity, digital biosecurity, biocybersecurity, and more, which are integrating the concepts and ideas from various disciplines into a unified effort. These fields are in fact subdisciplines within a broader context of cyber and information security within the bioeconomy and are the attempted articulation of an urgent need to bring a stronger security focus to industries across healthcare, agriculture, biotechnology, and the bioeconomy.

4.1 Why It's Important to Define Cyber and Information Security in the Bioeconomy

It's critical to define what we mean by cyber and information security in the bioeconomy because as new subdisciplines emerge, faster progress will be made by leveraging the great work already in practice and under development across both domains of cybersecurity and biosecurity, including their similarities and differences [40].

Within the field of cybersecurity, there are those practices previously listed, including zero trust security, software-defined networks, network segmentation as well as intrusion detection, monitoring, tracking, and mitigation. There is also a long-standing field of biosecurity work where best practices have been defined through decades of research and implementation, including laboratory, site, personnel, research, and information safety protocols.

Rarely before, however, have these two groups of practitioners interacted in a meaningful way; thus we need to level the way we communicate to reduce friction and redundancy in order to be effective. It is also important to define what we are not talking about or at least to subcategorize application domains, as one size does not fit all when it comes to the bioeconomy.

These subdomains, including scientific research and facilities, commercial operations and facilities, clinical care settings, and public health and government services, each have their own set of industry and government norms, standards, and protocols, and many differ among these subdomains. For example, research and manufacturing require precise and accurate results to ensure correct information is discovered, shared, and secured. Whereas in clinical care settings and government services, access to the personal information of patients and constituents is paramount to successful operations. Of course, there are many overlapping priorities, such as the accuracy of medical diagnoses in clinical care and the security of critical competitive intellectual property (IP) information in industry. Nevertheless, developing industry-specific and pan-industry foci within cyber and information security acknowledges and leverages the unique and varying nature of each subdomain.

5 Critical Domains for Enhanced Cyber and Information Security

There is no question that modern cybersecurity technologies should be the standard in the bioeconomy. But as with the distinction between definitions outlined above, there are distinctions that must be made between the fields in which we discuss cybersecurity. Just as the best cybersecurity practices from the financial services industry could be adopted to other industries but also need to be adapted to the industry-specific needs from a system, network, configuration, information, and user perspective, so do security practices in the cyber realm need to be applied in the context of biology. In many cases, the security paradigms must also be expanded or new paradigms developed to account for the industry- and domain-specific considerations.

The two primary domains, as outlined by cyber and information security, are just cybersecurity of systems and infrastructure and the information security of personally identifiable and non-identifiable data, both of which are critically important to various industries across healthcare, agriculture, biotechnology, and the bioeconomy.

5.1 Scientific Research and Facilities

5.1.1 Computer Networks

In the field of scientific research, there are a wide range of domains that are vectors for cyber vulnerabilities. The first, and most obvious from a cyber perspective, is the research computing network. Across the network infrastructure within universities, companies, and government research labs, the ability for a bad actor to tamper with scientific research is directly related to the ability to access these networks. In addition, in many organizations, research networks are not distinct from enterprise networks, and thus vulnerabilities to the network through research computing can put enterprise systems at risk and vice versa. In addition, since the bioeconomy, especially in academic research institutions, has a large number of trainees who connect to the network with their own devices that cannot be engineered for security at the organizational level, it is important to consider the security framework that provides access to these trainees while securing critical infrastructure and data.

5.1.2 Laboratory Equipment

Another vector for vulnerabilities in research facilities is via laboratory equipment. Most modern lab equipment is connected to the Internet, often to a networked environment, and vulnerabilities in the hardware and software that run the equipment can be exploited by threat actors and expose research networks to a wide range of threats. This leads to issues as described above. Additionally, within the equipment itself, a bad actor can disrupt experimental conditions and sensitive calibrations and damage data collection or analysis. Ideally, a thorough evaluation of the security settings and capabilities of lab equipment will be performed before purchase and integration with the network; understanding that this is not usually possible, a coordinated evaluation of the security posture and configuration of common lab equipment across the sector is needed to aid bioeconomy organizations in making secure decisions.

5.1.3 Personal Equipment

Personal equipment in a laboratory setting poses additional vulnerabilities if not properly protected. In academic research, personal computers are often used for research purposes and are directly connected to research networks (see computer networks section above). In industrial research settings, this may happen less, but the use of personal computers or the use of corporate computers for personal tasks is common. Similarly, the use of personal devices such as cell phones, thumb drives, and music players can create attack vectors in research settings. Cell phones are often plugged directly into laboratory or personal computers that are connected to the research network. Thumb drives are a common method of intrusion, and these are often handed out at conferences and in academic programs, creating opportunities—that the unsuspecting recipients of these devices do not even consider as a possible vulnerability— to seed infected drives into a pool of resources.

5.1.4 Open Source Software

Similar to "open data," there is also a large corpus of "open source software" relevant to the bioeconomy. Since a large part of the bioeconomy involves research, and research hinges on data and analysis sharing for discovery, it makes sense that this is prominent. It is also important to consider the security aspects of using open source software in your research, development, and production workflows. Open source software modules are not often maintained with the same security scrutiny as licensed software. It's important to consider the source of the open source software you are using and ideally create safe regions to use these tools within your network so that an intrusion wouldn't be able to penetrate the entire network. Also, when contributing open source software, consider evaluating the security posture regularly and posting updates to mitigate risks at scale.

5.1.5 Research Data

Research data is an additional domain of concern for cyber vulnerabilities. The risk posed via data loss ranges from nonpersonally identifiable information (PII) for research purposes and experimental conditions to critical health information and intellectual property.

There are substantial cyber and information security risks in the research setting, but these threat vectors can be mitigated with modern cybersecurity best practices. These include the same methods to protect networks as used in any industry, and leveraging existing technology and expertise is the most efficient way to build cyber resiliency in the bioeconomy.

5.2 Commercial Operations and Facilities

Commercial operations in the bioeconomy underpin biomanufacturing, storage, and distribution of life science technologies as well as the accompanying data. These operations range from small to large scale and also range with respect to overall security posture. Some of the smaller operations have fewer dedicated resources for security but may still be critical for the bioeconomy. It's important to note that tools are not always able to prevent cyberattacks and promote an overall security posture. Commercial operations and facilities would be well served to invest in security professionals to guide the overall architecture of the system, data infrastructure, and tool monitoring.

5.2.1 Computer Networks

Computer networks are a common threat vector across subdomains, but where research networks focus on connected infrastructure in laboratories, operational networks in commercial settings pose an additional scope of risk due to the expanded access across business and mission-critical systems. In a commercial setting, customer information, product orders, personnel records, and financial statements are only a small subset of the information involved in commercial operations. In addition to the information risk, these networks connect and control critical industrial equipment and facilities.

5.2.2 Manufacturing

In a manufacturing setting, most modern processes are networked and automated. In the bioeconomy, these could be manufacturing facilities producing medical devices, pharmaceutical ingredients, finished pharmaceutical products, or biotechnologies, all of which require precise operational steps for success. Disrupting these manufacturing processes would, at best, force product recalls and, at worst, lead to loss of life from the failure of medical devices or drugs. One way to protect this process and flow is to containerize the system and not connect it to the Internet. This has implications for modifying the process and making changes in real time but is actually a sound option for some types of manufacturing scenarios. Post-manufacturing, these products must be stored and distributed, often in climate-controlled settings, and disruption to these steps in the industrial process, such as a dethaw of vaccine doses or a mis-shipment of medication, could result in direct loss of life. During the COVID-19 pandemic, there were batches of vaccine doses that had to be discarded due to a lack of continuity of conditions. It is important not only to maintain the conditions but also to make sure that there is an accurate accounting of these conditions for regulators. Technology is almost always used for this and is also subject to cyber manipulation as other systems.

5.3 Clinical Care Settings

5.3.1 Computer Networks and Connected Medical Devices

Clinical care settings are among the most sensitive to disruption within the bioeconomy due to the direct nature of injury and disease management. As with other subdomains, network security is critical to successful cyber and information security in healthcare but poses additional direct consequences on patient well-being. Modern healthcare is highly networked, both through medical equipment at the bedside and Internet of Things (IoT)-connected personal medical devices. So while network intrusions in research or industrial settings may have adverse secondary and tertiary effects on customers and patients, network intrusions into critical medical equipment and devices may have direct adverse effects on patients.

5.3.2 Diagnostics

As diagnostics move from the hospital to the clinic and the home, many of these home-based platforms rely on technology and networking to operate. A fully distributed diagnostic network means that there are not only data integrity considerations but also possibly distributed infiltration issues. Building these systems so that a breach will be local and not global is critical. If each distributed site requires access to a local network, a globally network system, if breached, would mean infiltration into each distributed network system. Thankfully, there are many ways to achieve the security to prevent this possibility, but it's important for developers and innovators to consider the security implications and best practices to avoid major breaches, IP loss, and questions around data integrity.

5.3.3 Virtual Care

In addition to the direct effects on medical care from network intrusions, there is a growing movement toward virtual care and telehealth services. Through these services, patients are sharing personal health information with providers over phone- or Internet-based appointments from personal devices, thus introducing an expanding domain of responsibility for secure medical care. End-to-end security is challenging when a medical system has no control over a patient's personal devices; therefore special attention needs to be paid to developing cyber and information security best practices in this domain, which may include cloud-based services with zero trust protocols in place to mitigate the impact of personal device or endpoint security risks.

5.4 Public Health and Government Services

Public health and those services provided by local, state/territory, and national governments to care for their citizens have many of the same vulnerabilities as those outlined in industrial and clinical settings. However, due to the scale at which these services are deployed, they require particular consideration. For example, while a breach to a healthcare provider network can have devastating consequences, the breach of a nationalized healthcare system can have widespread consequences beyond that of any single institutional breach [49].

6 A Representative Set of Examples Are Examined Below

6.1 COVID-19 Pandemic

The COVID-19 pandemic was the largest-scale public health crisis in over a century. Amidst the crisis, a series of widespread cyberattacks demonstrated the challenges that large-scale public health crises can create for the world [50], including both primary, secondary, and tertiary impacts and beyond.

During the COVID-19 pandemic, all of the previously mentioned subdomains posed threat vectors. In a race to develop effective vaccines to combat SARS-CoV-2, vaccine and therapeutic research data became critical on a global scale. Once the vaccines were developed, the pace at which manufacturing was required meant that disruptions to the supply chain would cause widespread effects as well. Post-manufacturing, the vaccines required distribution to public health sites and storage across a respective country, and then logistics management for people to sign up for vaccination appointments was required. Each of these steps in the process posed significant challenges and risks if disrupted through a cyberattack and a loss of information security.

In the midst of returning economies to work, a healthy workforce is critical to success. If any of the primary effects of cyberattacks and information loss occurred, then citizens may be required to quarantine, risking becoming ill, or possibly die from COVID-19. The secondary effects, and beyond, of such attacks may include job and economic loss, the closure of businesses, and recessions that require prolonged periods of recovery. The public health impacts of such secondary effects will only be fully realized in time but are undoubtedly present [51].

6.2 Precision and Personalized Medicine

Both precision and personalized medicine aim to use the specific information about an individual (e.g., race, gender, age, genetics) in order to guide a more individualized approach to clinical care. An individual's identifiable genetic information is commonly used to create individual-specific care regimes. This is a particularly poignant concern given that once digital genetic information is inappropriately disclosed or stolen in a cyberattack, it cannot be amended or retrieved as with other types of information like social security number, bank accounts, passports, etc. [52] The gravity of potential breaches to the systems of consumer-facing organizations—like those providing genetic counseling, food sensitivity and allergy testing, and ancestral information—as once genetic information has been co-opted is impossible to rescue, the highest form of identity theft.

Biorepositories, along with the collection, storage, and supply chain distribution, hallmarks of the precision and personalized medicine development, and implementation process all face distinctive vulnerabilities from bad actors [53]. This information can be used in nefarious ways without an individual's consent, including profiling, public distribution of medical conditions, and even person-specific bio-warfare.

6.3 Organ Donor Registration

Organ donor registration is another representative example of public health services that have direct impacts on lives if disrupted. While the impact is more localized to individuals, the magnitude of damage done if a disruption were to occur is likely greater at the individual level. In these public services, supply chain resiliency is key, as well as information security for those registered to receive a transplant and those registered as donors.

6.4 Additional Impacts

The two examples above illustrate a small subset of public health domains and services that are critical to protect. The cyber and information security requirement of public health services includes every person, in every community and in every country. Leveraging best practices across well-established cybersecurity domains is the fastest route to provide critical services to global citizens.

7 Emerging Technologies at the Intersection of Biology and Cyber Domains

There is a subset of novel threats developing out of the intersection of biology and computer science, driven by advances in DNA data storage technologies and DNA information systems [54]. It has been shown that malware can be synthesized into artificial DNA, and upon sequencing (reading) the DNA, the sequencing machine may be compromised with the embedded malware [55]. Current applications of such technologies are in a nascent stage of development, with demonstrations of the effects derived in carefully crafted experimental systems. But nonetheless, these are growing vectors of concern that we should meet in an anticipatory posture, rather than a reactive posture.

As such, the storage of malware into biological systems may be a novel vector, requiring new domains of security to be developed. However, in the face of such techniques, it is even more imperative that good cyber and information security practices be in place. For example, if a DNA sequencer is segregated from business and mission critical systems, then the effectiveness of such hacks can be localized and contained. Thus, focus should be placed on studying this growing capability (and mitigations), but we must refrain from overstating the unique threats caused by such capabilities.

Another novel vector is the emerging potential of DNA and other molecules to serve as a high-density storage vessel. The production of data by humans today is occurring at an unprecedented rate. DNA offers a solution to this accumulation of data. Unlike current models of data storage which exhibit limitations in the longevity of storage, DNA—when kept in optimal conditions—can go thousands of years without degradation. Storing information in DNA was first exhibited in 1988, and in 2012, George M. Church, Yuan Gao, and Sriram Kosuri successfully demonstrated the ability to convert an html-coded book including more than 50,000 words, 11 JPG images, and 1 JavaScript program into a 5.27 megabit bitstream [56]. The major bottlenecks in this methodology—cost and reliance on polymerase chain reaction (PCR)— are being resolved by advancements in photolithographic synthesis addressing cost and the development of systems like the Dynamic Operations and Reusable Information Storage (DORIS), which do not rely on PCR [57, 58].

As DNA-based information storage increases in financial and practical viability, the potential uses for biomolecular cryptology practices proportionally increase. Cryptography is a scientific technique that secures information via transformation from a readable message into something indecipherable so that only those possessing the mathematical or logic cipher can access its original meaning [59]. DNA-based cryptography is not grounded in mathematical coding so it makes it substantially more difficult to hack.

Given the rapid pace of innovation in both the storage of data in DNA and its potential for next-generation encryption of existing data, there should be increased focus on the development of tools and techniques at the intersection of biology and computer science. A holistically defensive and preventative posture is achievable by prioritizing granular advancements in the form of domain-specific tools and techniques over increasing siloed approaches to poignant vulnerabilities.

Undoubtedly, the pace of novel vulnerabilities will accelerate as biotechnology advances and cyber operations become more accessible to less sophisticated actors—all the more reason to build up a strong practice of cyber and information security in the bioeconomy, along with a creative mindset, in advance.

8 Closing the Cyber and Information Security Gap in the Bioeconomy

It is without debate that there is an urgent need to improve both cybersecurity broadly and information security more specifically, in the bioeconomy. There are a growing number of ways to describe these practices (e.g., cyberbiosecurity, digital biosecurity, biocybersecurity), but no matter which subdomain is emphasized, there are vulnerabilities that must be addressed in both the health- and non-health-related bioeconomy domains, and it will take a holistic collaboration between practitioners in cybersecurity, biotechnology, biomanufacturing, biosecurity, healthcare, and more to drive substantive impact.

The strongest posture we can create within the bioeconomy is built on the techniques and practices that exist today in robust fields of traditional cybersecurity as well as biosecurity and healthcare. In these areas, new subdomains are not likely to be necessary, and healthy collaboration and information exchange across industries is the strongest position. In the emerging cases where biotechnology and cyber vulnerabilities truly intersect, such as the examples of malware written into DNA [46], new subdisciplines will be necessary.

However, we must recognize that there are true differences between industries within the bioeconomy, and a one-size-fits-all approach to cybersecurity will be inadequate. A matrix approach to viewing the impact of technology across industries would help identify differences and similarities. Example technology domains that may require a specific emphasis include DNA synthesis technologies [60], biological databases [61, 62], and biomanufacturing [63], among many others [64]. These technologies, and how they impact across industries such as pharmaceuticals, healthcare, agriculture, biotechnology, and energy, will drive the growth of the bioeconomy over the next decade and beyond [65]. As such, each will introduce new threat vectors for bad actors to target, and a critical approach to protecting these areas is essential.

References

1. M. Andreessen. Why Software is Eating the World (2011)
2. J. Conde, V. Pande, J. Yoo. Biology is Eating the World: A Manifesto (2019)
3. R. Carleson. Congressional testimony of managing director Dr. Rob Carleson (2019)
4. Congressional Research Service. The bioeconomy: a primer (2021)
5. National Academies of Sciences, Engineering, and Medicine. Safeguarding The Bioeconomy (2020)
6. L. DeFrancesco. Financing breaks all records in 2020 (2021)
7. World Health Organization. Global spending on health: a world in transition (2019)
8. NHGRI. Genomic data science (2021), Accessed 22 Aug 2021
9. Z. D. Stephens et al. Big Data: Astronomical Or Genomical? PLoS Biology (2015)
10. U.S. House of Representatives. Engineering biology research and development act of 2019 (2019)

11. U.S. Senate. Bioeconomy research and development act of 2021 (2021)
12. U.S. Senate. United States innovation and competition act of 2021 (2021)
13. K. B. Alexander, et al. Covid-19 and the cyber challenge. The Cyber Defense Review (2021)
14. A. J. Titus, et al. Biotechnology in defense of economic and national security (2020)
15. Advanced Regenerative Manufacturing Institute. National technology roadmap for pandemic response and recovery (2021)
16. A. J. Titus. A national bioeconomy manufacturing and innovation initiative (2020)
17. Senator Angus King and Representative Mike Gallagher. Cybersecurity lessons from the pandemic: CSC white paper #1 (2020)
18. Cybersecurity & Infrastructure Security Agency. Cybersecurity (2021), Accessed 29 Aug 2021
19. Nick Galletto, et al. Cyber, cyber everywhere: Is your cyber strategy everywhere too? (2019)
20. Cybersecurity & Infrastructure Security Agency. Meeting confidentiality, integrity, and availability requirements (2021) Accessed 29 Aug 2021
21. National Institute of Standards and Technology. Cybersecurity framework (2021), Accessed 12 Sept 2021)
22. P. Ferguson, et al. What is a VPN? (1998)
23. S. Rahimi, et al. Analysis of the security of VPN configurations in industrial control environments (2012)
24. S. Rose, et al. Zero trust architecture (2020)
25. U.S. White House. Executive order on improving the nation's cybersecurity (2021)
26. S. Furnell, et al. The ABC of ransomware protection (2017)
27. S. Bradley. 2020: a look back at patching and the pandemic (2020)
28. Cisco. What is information security (2021), Accessed 29 Aug 2021)
29. J. Manyika, et al. Big Data: The next frontier for innovation, competition, and productivity (2011)
30. Y. B. Choi. Organizational cyber data breach analysis of facebook, equifax, and uber cases (2021)
31. U.S. National Library of Medicine. NIH data sharing policies (2021), Accessed 3 Oct 2021
32. U.S. Health and Human Services. The HIPAA privacy rule (2021), Accessed 3 Oct 2021
33. U.S. Health and Human Services. Genetic information (2021), Accessed 3 Oct 2021
34. GDPR.EU. Complete guide to gdpr compliance (2021), Accessed 3 Oct 2021
35. California Office of the Attorney General. California consumer privacy act (2021), Accessed 3 Oct 2021
36. A. Titus, et al. SIG-DB: Leveraging homomorphic encryption to securely interrogate privately held genomic databases (2018)
37. USDA Animal and Plant Health Inspection Service. Defend the flock – biosecurity 101 (2021), Accessed 29 Aug 2021
38. C. Wilson. Cybersecurity and cyber weapons: is nonproliferation possible? (2013)
39. National Academies. The biological threat reduction program of the department of defense: from foreign assistance to sustainable partnerships (2007)
40. K. Hoyt. What biosecurity and cybersecurity research have in common (2017)
41. L. C. Richardson, et al. Cyberbiosecurity: a call for cooperation in a new threat landscape (2019)
42. S. Hinchliffe, et al. Security life: the emerging practices of biosecurity (2008)
43. R. Badhwar. The advanced malware prevention playbook (2021)
44. D. DiEuliis. Parsing the digital biosecurity landscape (2020)
45. R. Riggs. The growing importance of bio-cybersecurity (2019)
46. J. Warner. What is zero trust security? (2021), Accessed 28 Aug 2021)
47. B. Lantz, et al. A network in a laptop: rapid prototyping for software-defined networks (2010)
48. N. Wagner, et al. Towards automated cyber decision support: A case study on network segmentation for security (2016)
49. Cybersecurity & Infrastructure Security Agency. Measuring the COVID-19 pandemic's effect on the national critical function provide medical care (2021)
50. B. Pranggono, et al. COVID-19 pandemic cybersecurity issues (2021)

51. D. DiEuliis, et al. Beyond 1918: bringing pandemic response into the present, and future (2020)
52. G. J. Schumacher, et al. Genetic information security as state of the art (2020)
53. S. Sawaya, et al. Artificial intelligence and the weaponization of genetic data (2020)
54. P. M. Nay. Securing the future of biotechnology: a study of emerging bio-cyber security threats to DNA-information systems (2019)
55. P. M. Nay. Computer security, privacy, And DNA Sequencing: compromising computers with synthesized DNA, privacy Leaks, And more (2017)
56. G. M. Church, et al. Next generation digital information storage in DNA (2012)
57. P. L. Antkowiak, et al. Low cost DNA data storage using photolithographic synthesis and advanced information reconstruction and error correction (2020)
58. K. N. Lin et al. Dynamic and scalable DNA-based information storage (2020)
59. J. Fruhlinger. What is cryptography? How algorithms keep information secret and safe (2020)
60. R. Puzis, et al. Increased cyber-biosecurity for DNA synthesis (2020)
61. A. J. Titus, et al. SIG-DB: Leveraging homomorphic encryption to security interrogate genomics databases (2018)
62. S. Arshad, et al. Analysis of security and privacy challenges for DNA-genomics applications and databases (2021)
63. J. C. Reed, et al. Cyberbiosecurity implications for the laboratory of the future (2019)
64. A. M. George. The National security implications of cyberbiosecurity (2019)
65. M. Walsh, et al. Security measures for safeguarding the bioeconomy (2020)

Mission-Aware Differences in Cyberbiosecurity and Biocybersecurity Policies: Prevention, Detection, and Elimination

Lucas Potter and Xavier-Lewis Palmer

Abstract The fields biocybersecurity (BCS) and cyberbiosecurity (CBS) are terms that are sometimes used by members of the public interchangeably. In some respects, this is logical – the fields are both from the nexus of biosecurity and cybersecurity. However, BCS and CBS are different in several key ways. To facilitate this delineation, a comparative literature and source review was completed. Foremost, the philosophical difference (or focus) of CBS is mainly to secure biological resources with cyber-enabled methods. The primary focus of BCS is to use cyber-enabled technologies to create biological threats. Secondarily, there are differences in core research interest and in the ability of CBS and BCS to leverage various kinds of threats. Finally, the policies used to predict, prevent, mitigate, or respond to CBS and BCS threats vary significantly. These topics will be discussed in the context of several scenarios.

Keywords Biocybersecurity · Cyberbiosecurity · Policy · Mission · Prevention

1 Introduction

Since the mid-2010s, the terms cyberbiosecurity (CBS) and biocybersecurity (BCS) have emerged, focusing on the intersection of the domains of cybersecurity, biosecurity, and cyber-physical security [80, 83]. These terms have found their

Authors "Lucas Potter" and "Xavier-Lewis Palmer" have equally contributed to this chapter.

L. Potter (✉)
School of Cybersecurity, Old Dominion University Norfolk, Norfolk, VA, USA

Biomedical Engineering Institute, Department of Engineering and Technology, Old Dominion University, Norfolk, VA, USA
e-mail: lpott005@odu.edu

X.-L. Palmer
Biomedical Engineering Institute, Department of Engineering and Technology, Old Dominion University, Norfolk, VA, USA

D. Greenbaum (ed.), *Cyberbiosecurity*, https://doi.org/10.1007/978-3-031-26034-6_4

popularity considering heightened technological innovations within each subdomain, in addition to limited but significant demonstrations of exploits to be found where these subdomains meet, and the use of these terms has only grown since. CBS and BCS have been often packaged in two-word terms, such as "cyberbiosecurity" and "biocybersecurity," perhaps out of caution, but as time proceeded, BCS and CBS increasingly have been noticed to be packaged as single words. Further, these words have been used in media interchangeably, especially by some authors as the two terms involved the same domains and differences appeared initially trivial. However, further reflection has motivated a change in this treatment of the terms while the intersection in which they reside is still new in both literature and mainstream conversation. We believe that building nuance between CBS and BCS now will help the organization and coordination of responses toward exploitations in the intersections spoken of.

This paper's mission is to open discussion to properly branch the two terms and does so in multiple sections. First, we start with offering working definitions of both CBS and BCS, sampling rich, initial offerings from the literature provided. We then map the growth of papers carrying the terms, in reasonable capacity, from 2017 to 2020, showing the reader key trends in publication type, number, and country institutions involved. Following, we discuss funding of CBS and BCS threat analysis and reduction, wherein funding concerns are made clearer. Given how funding differences may exist based on the route of technology use, this section provides a financial foundation for specificity between CBS and BCS. Next, we discuss policy differences for CBS and BCS and how these translate to operational effects when such policy may be enforced. This discussion is then followed by discussion of legislation of policy in this new age dubbed the "Cyber-Bio" Age, wherein increased focus on new vulnerabilities and exploits at the intersections of CBS and BCS are expected to find an increase in directed policy. We then conclude with reflection on all that has been covered. This said, we will now move into discussion of working definitions.

1.1 Working Definitions of CBS and BCS

Cyberbiosecurity (CBS) and biocybersecurity (BCS) are fields born of similar Nexi, and both face toward a contemporary explosion in research [66]. For the purposes of this developing work and discussion, we propose differing definitions that may prove helpful. The first, CBS, is the more frequently discussed of the two. CBS focuses primarily on how cyber assets (for instance, computer networks) can affect biosecurity, which is "... generally associated with travel, supply chains, terrorist activities, and defense," though it is also of marked importance in academic settings with high amounts of trust [80]. CBS applies the methodologies of cybersecurity to conventional life science settings. Policies that seek to alleviate CBS threats tend to stay within domains that are conventionally known to biosecurity – such as the agricultural industry [14, 32] and pharmaceutical production [62]. This utilizes the well-defined fields of cybersecurity and biosecurity and allows for threats that have become conventional to counter with well-proved methods. The latest development of CBS is the enabling of a "specialized techno-surveillance network" [50]. This is not meant to limit the purview of CBS as a field but to allow for the use of doctrine and gathered knowledge in both the fields of cybersecurity and biosecurity to reach their full use within the context of that study. This has made CBS well enough established to begin grappling with the finer aspects, for instance, training [93] or in laboratory safety [91].

BCS can instead be the study of how biological assets (either nominally offensive or not) can be leveraged with cyber assets (for instance, machine learning, artificial intelligence (AI), advanced logistical networks, or biometric systems) to perform acts that are not condoned by legal or ethical means. One previously (and regrettably overly expansive) working definition was "Any cybersecurity system where a biological component, target, or interlock is involved in the terminal or intermediate stages" [72]. A narrower and more serviceable work simply noted that "Thus, in a biocybersecurity context, biological phenomena can act as interlocks, and even as facilitatory steps in a [bio]cybersecurity system" [83]. Another example would be the utilization of DNA as a means of information transmission of malware [48].

In short, CBS is better reserved for when one can look at a threat and understand where the line demarcating cyber and bio and the linkage between cyber and bio are clear. BCS is better reserved for when the melding of the two threats is unclear, or the target is a resource that is not well characterized by either biosecurity or cybersecurity. A simple method of delineation is that where CBS seeks to be ever vigilant to defend conventional assets from cyber and biological threats, BCS constantly morphs with the available biomedical information technology of the time to produce novel threats that may not be well characterized within the strictures of the CBS field or act in conjunction with other threat modalities. So, for instance, whereas a CBS threat may be the utilization of trusted collaborator networks to insert dangerous or non-functional organisms into a research setting [80], a BCS threat may be using AI methods to create false articles, reducing the clearance throughput of an academic journal [97] or malfeasance in precision medicine [29].

The differences are not at all meant to separate the goals of defense but to maximize the effectiveness of both fields at doing what they do best – another call for cooperation [94], this time between specialists in the same field maximizing their talents. The difference between the two is likely to merge with advances in data science and in biomedical accessibility. Once advances in programmatic intelligence can be immediately applied to biomedical resources, the fields will likely become indistinguishable and could be added to the burgeoning monolith of cybersecurity. This mentality is foreshadowed by proponents of the US Department of Defense's notion of "Defending Forward" [71, 108]. For the moment, it is important to have a field led not by the conventional awareness and policies that surround cybersecurity but by the infinite possibilities promised by biological sciences and disruptive computational resources.

As a final word on the topic of a formal separation between the two fields, a list of contemporary threat citations has been included below to facilitate discussion between readers. Ransomware attacks on healthcare systems and medical devices have been recorded [3, 10, 17, 46, 95] along with in-person cyber-enabled healthcare denial of service [4] and a rise in tracking and "DNA barcoding [which] refers to the use of DNA sequences to the use of DNA sequences from a signature region of the genome to make species-level identifications" [47, 52, 113]. Data storage using nucleic acids is now available [58, 64, 106], with all of the security implications this holds [68].

Now that the functional differences between CBS and BCS have been established, it is important to map differences between the two in publication intensity, to establish interest and potential future needs for both.

2 Mapping the Dynamic State of CBS and BCS from 2017 to 2020 and Methods

Prior work in submission [97] notes both the under exploration of the Earth's resources especially in the global south, along with the threat of malfeasance by actors wishing to exploit those resources. These resources include not just standard agricultural stores but also those carrying either components for new drugs or those carrying the next pandemic. Though current readers in literature will likely not need a reminder, future readers need only to research the impacts of the COVID-19 pandemic to learn the need to discover the microbiological denizens of the globe. Vital to this is quality record creation and keeping. All considered, with biosecurity and cybersecurity engaged through the concern of record integrity, this matter falls under the broad domains of CBS and BCS as it meets matters of biosecurity and digitalization in the form of bio-data [16, 30, 84, 98]. Here, we find importance in mapping vital work within the intersection that BCS and CBS address.

The method by which this research was conducted was as follows. Papers from December 8, 2017, to December 8, 2020, were searched under two batches, via Google Scholar, one for "biocybersecurity" and one for "cyberbiosecurity." In addition to relevant ecological papers for discussion in the broader scope (see Appendix: Analyzed CBS/BCS Papers). Only the papers within the selected data range were analyzed. The BCS publication was searched under the terms "biocybersecurity," "biocybersecurity," and "biocybersecurity." The BCS batch of publications was searched under the terms "cyberbiosecurity" and "cyberbiosecurity." Articles were added to a spreadsheet and followed until the last pages. For the purposes of visualizations and simple statistics, counts were made according to publications per year, the number of institutions, types of institutions, countries represented by affiliation, and impact factor, where accessible. Affiliations were counted one to a researcher and counted to represent unique countries participating. Regions were mapped using Microsoft Excel to produce a basic world map that shows areas of institutional participation or penetrance. The main issue is that only publicly available works could be utilized – classified reports or documents could be the predominant output of a nation, but these would not be counted.

2.1 Research Methods Limitations

Not all factors documented were useful, and of them, impact factors were not taken deeply into account given the novelty of the field. An emergent issue with the use of Google Scholar is that duplications were found under titles appearing in Chinese and English. Graphics reflect the count for someone who compiled under the original terms, minus removal of duplicates. Duplicates found were minimal. Further, articles found under the search terms, but not engaging CBS or BCS, were removed from final graphic tallies. A few articles were picked up by Google since they cited at least one CBS or BCS paper. It is possible that papers from 2021 and beyond picked up by indexing tools will continue to pick these up, without adaptive searches employed. These were nonetheless listed and counted for note. Additionally, the research conducted did not pick up the increased interest in the field caused by the COVID-19 pandemic and explosion of medical misinformation.

2.2 CBS and BCS Research Differences Visualized

Whether research conducted was termed under the title of BCS or CBS, the intersection of biosecurity, cybersecurity, and cyber-physical security has received tremendous input from institutions across the world. Between both, CBS is the dominant term. Matching with the majority focus and emphasis on exploits from the cyber and cyber-physical realms, this makes sense. Threaded within and between articles, select authors have opted for the term, BCS, to emphasize the realm of biocybersecurity. Neither choice is incorrect, and thus many of the graphics generated will discuss the field from the context of both BCS and BCS batch lists combined. More than 100 articles have been put forth representing more than 200 and 10 institutions across at least 25 countries. These institutions are commercial, academia, and military in basis, which communicates that a considerable degree of interest has been gathered across multiple spheres of research. Papers within this section have investigated questions such as the following:

1. The generation and implications of biocentric data's accessibility
2. Issues in manipulation, masking, exploitation, and revocability
3. Volume and variety of such data, along with affected parties
4. Moral questions and those relating to evolutions of technologies at the intersection of BCS/CBS, evolutions of their interplay, and hegemonies of power

Plotting the results above, we get the following graph, showing the nearly exponential rise in both CBS and BCS publications – with the CBS type being consistently an order of magnitude more popular (Fig. 1).

This pattern seems to be continuing along the same path since the original research was completed. A cursory analysis of Google Scholar results will show that 2021 saw 80 CBS and 11 BCS publications. Journals are the primary source of documentation, which points to an over-localization of knowledge held.

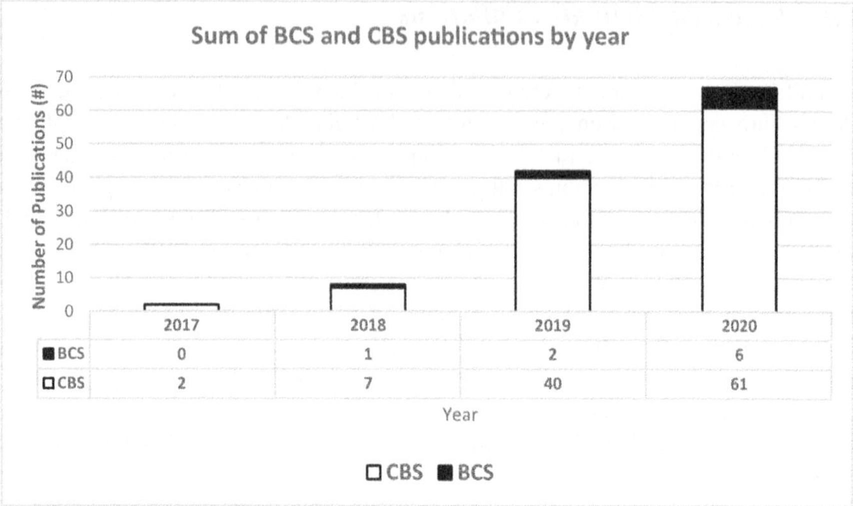

Fig. 1 Differential growth of CBS and BCS (2018–2020)

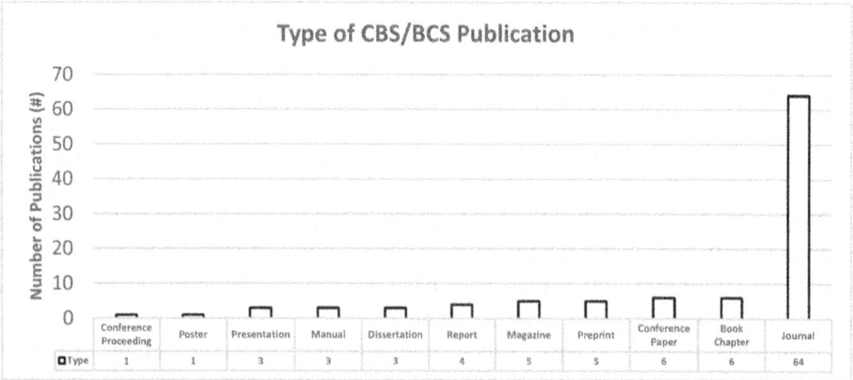

Fig. 2 Mapping of communications means of BCS and CBS by medium (2017–2020)

A more fascinating demonstration of the CBS/BCS field is the type of documentation being produced. Most works were classified as journal articles. This is an indication that CBS and BCS threats are for now mostly academic in nature (Fig. 2).

As may be expected, the Global North is the leader in producing both CBS and BCS documentations – particularly the USA was the undisputed global leader (Fig. 3).

By the numbers and mapping shown, multiple trends can be observed. One is that developed countries are highly represented, and the USA leads by far, which should be of little surprise. Its educational, economical, technological, and military superiority grants it advantages in achieving this reach. Additionally, it takes no small amount of responsibility for military stability in many of the nations of the

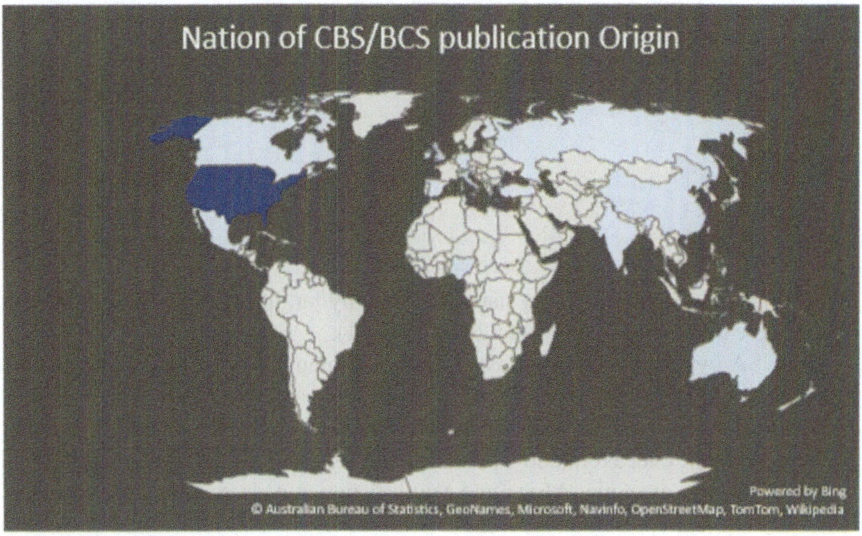

Fig. 3 Mapping of CBS and BCS publications by nation (2017–2020)

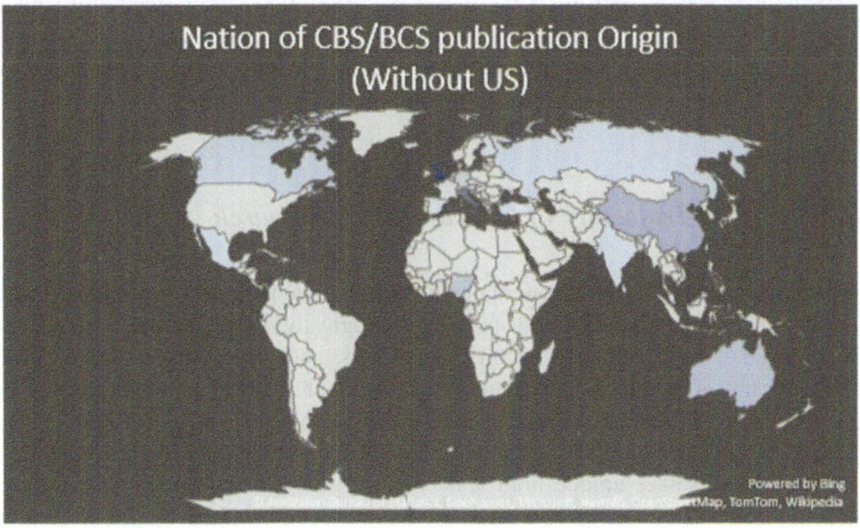

Fig. 4 Mapping CBS and BCS publications (2017–2020) – without US contributions

world. The USA has openly admitted to around 600 facilities in non-US states or territories [11, 112] (Fig. 4).

For a simpler analysis, the table below lists the top seven countries by publications via institutional authors counted. To note, these do not include researchers

Table 1 Institutional country of authors in CBS and BCS publishing between 2017 and 2020

USA	215
UK	33
Italy	20
Israel	10
China	9
Nigeria	6
Australia	6

listed in acknowledgments. That said, the West, by far, has performed the majority of CBS research between 2017 and 2020 (Table 1).

The USA by far holds those most institutional members who have published on cyberbiosecurity. The top five non-US countries are the UK, Italy, Israel, China, and Australia and Nigeria (tied for fifth). The Global South is largely shut out of these research endeavors and vulnerable through its lack of representation and participation. For example, Africa and South America possess almost no countries with institutional representation. Nations within the Global South would do well to prioritize partnerships with nations ahead and to increase budgets toward education and technological adaptation. Such can be seen in the economic partnerships of global south nations with the USA or (in the case of Africa, particularly) China [39], but such leaning can have complications. This circumstance gives reasons for countries to continue developing their biological resources and their ability to manage them.

2.3 *Future Trajectory of CBS and BCS Research*

Our best treatments and worst weapons lie ahead at the intersection of biology and computing. Thus, the means of cataloging biologically diverse species, the means for doing so, the creation of secure repositories, and protecting the integrity of those means, is incredibly important. It is not just the Global South that needs to be considering biocybersecurity, but the Global South remains the most vulnerable by lacking infrastructure and manpower to combat its vulnerabilities. Each country and region possesses unique flora and fauna at different scales and will require varying resources to take advantage of for protection and investigation, but a common bottleneck exists in repositories [97]. Protection from false information and flooding of records is paramount for their growth as biotechnology increasingly takes prominence.

Future work can be imagined as projects to realize more broadly what biocybersecurity means for each country and region and how that may change the nature of politics for citizens within. Each country could benefit from using such future work as a springboard for consideration of their biocybersecurity and biosecurity deterrence platforms and what that means for their growth specifically. Their unique biological resources, owed to novel habitats that each country has, which can be taken advantage of from the individual to the state level, will vary but need to be adequately mapped for thoughtful consideration to be given. Untold prosperity and security await down the line for each country that can endeavor to pursue matters in this sphere.

What can be observed is that the twin fields of CBS and BCS are quickly growing, and an increasing number of nations are taking notice through industry and university participation. No major surprises were held on hegemony in representation and expertise, but what does show cause for intrigue is the wide inequalities witnessed. These can expect to play ongoing roles in power imbalances and exploits of the Global South for decades to come if policy and funding are not focused to address BCS. Overall, it was found that articles introduced under the earlier mentioned search terms increased rapidly over 2019 and 2021. It is the firm expectation that publications would be on the increase in 2021 and beyond as more countries seek insight into the intersections occurring within BCS and CBS. Nonetheless, it is hoped that this brief mapping is helpful as a guide to the trajectory of CBS and BCS and how international security within it can be safeguarded.

3 Funding of CBS and BCS Threat Analysis and Reduction

The popular idea of the primary opponent of appropriate defensive stature in either CBS or BCS threats (for those outside of the field) is a shadowy, government-sponsored figure in a well-appointed and secure location. This makes for interesting stories. But the primary threat to secure CBS and BCS operations on the scale of a national government is not a nation-scale enemy at parity, internal political factions, or internecine strife. It is money.

The total cost in 2008 of nuclear weapon-related appropriations for the federal government of the USA was in total 52.4 billion dollars [101], with nuclear threat reduction and nuclear incident management being 5.865 billion (approximately 11%). The Congressional Budget Office (CBO) predicts the cost of all appropriations from 2019 to 2028 to rise to 432 billion over the decade (slightly less than the actual cost in FY 2008, 43.2 billion per annum) [13]. This evens with the noted degradation of the deterrent effect of nuclear weapons [18].

The entirety of DHS-CISA (Department of Homeland Defense-Cybersecurity and Infrastructure Security Agency, of which CBS and BCS threat prevention will hopefully be a part) has a planned (FY 2023) annual operating budget of approximately 2.5 billion to prevent cyber threats of all types [21, 26]. Making the generous assumption that all CISA funds go to preventing disruption of CBS and BCS threat, this would give the national leader in world's biodefense spending less than 10% of the money used to reduce nuclear threats. To be fair, being a newer field, a more apt comparison may be to look at the amount of financial support given toward nuclear security spending at its founding. Additionally, the

US government has given additional support for disinformation campaigns (some linked to BCS methods) via other means (see Sect. 4.1 "Cyber-Enabled Biological Misinformation: A Case Study" concerning the utilization of misinformation to augment public health emergencies).

This disparity in financial prioritization is not at all illogical – nuclear weapons are a singularly terrible threat. The low probability belies an unfathomably high impact [56]. However, as the rest of this text will outline, the nation-scale threats of CBS and BCS are far more insidious and may have a higher likelihood of occurring than nuclear threats. There will be two repeating mantras underpinning the findings of this work. One of them will be the relative cost in CBS and BCS. For comparison, relative cost of casualties in terms of land area is "$2,000 per square kilometer with conventional weapons, $800 with nuclear weapons, $600 with nerve-gas weapons, and $1 with biological weapons" [31]. These bioweapons are also much cheaper and easier to produce than ever [67] with a much wider variety of options [103]. For a cyber threat, this cost may be harder to quantify, but it could very well be much less. This is not, however, to say that biological defense programs would be without cost at all [42]. Cyber threats that have impacted biological operations include attacks on critical infrastructure such as freshwater ecosystems [100], potable water treatment systems [40], and healthcare systems [19]. The second mantra is the ability for a CBS or BCS offensive to be obfuscated to the level of absolute deniability.

The methods of assessing this infrastructure surrounding CBS resources have been analyzed [99], including the ever-present human factors of any system, irrespective of CBS or BCS differences [36, 81, 83].

4 Policy Differences for CBS and BCS: Operational Effects

The execution of policy is subject to many operational effects. There will be issues of compliance to policy (if one is unfamiliar with this concept, feel free to ask your local information technology staff how many people are using appropriate and sanitary password procedures) and issues of interpretation of policy – as in any set of rules that make the migration from ideation to concrete, actual practice.

The foremost question remaining to the responsible CBS and BCS defender is then how strategic, operational, and tactical decisions can be aligned with policy to prevent attacks from both modalities. Here the strategic goals of both are similar – to preserve and protect what is nominally the status quo in day-to-day operations and in terms of trust in conventional resources (i.e., government authority, academic integrity, medical service, etc.) or to inhibit the ability of a rival organization from accessing their CBS or BCS assets. The operational differences are widely varied, and the tactical decisions will most likely be highly dependent on the setting and available resources. Here the focus will be on the operational differences.

In this work, strategy is defined as the overarching goal of an organization. Operations are defined as the regulations, procedures, leadership, and institutional knowledge used to enact this strategy. Tactics would be the day-to-day operations undertaken to carry out that strategy as enabled by the operations. These definitions are largely borrowed from previous work [43, 82].

To better delineate the differences between CBS and BCS, let us start with an analysis of what types of offending attacks would fall under each category. Both tend to use emerging or disruptive technologies [45]. For instance, in the more conventional CBS field, an attack may be a cyber-based advanced persistent threat (APT) against a medical conglomerate researching a needed vaccine [77]. There the lines are clear – a cyber threat (the APT) is being used to penetrate or otherwise inhibit a biological resource (the medical conglomerate).

A BCS threat, by contrast, could be the use of biological data collection to characterize the staff of a medical conglomerate and then the generation of an artificial intelligence/machine learning (AI/ML) model to create the ideal combination of biological agents to infect most of those staff members, in essence, a data-driven approach to biological warfare. Another example could be base-level infections of laboratory equipment [68] or the general supply of nucleic acids [88].

An ideal policy for offensive CBS research is the maintenance or creation of "red team" approaches for one's own networks or resources. This could be the following:

1. Maintaining libraries of CBS and BCS attacks that have been demonstrated and are theoretically possible.
2. Setting numerous goals of BCS and BCS defense, after updating known vulnerabilities and exploits. It is important that Global North countries understand that vulnerabilities allowed to fester and be perfectly exploited in the Global South can eventually pose problems for the Global North.
3. Consistent reconnaissance through which new networks, employee types, and applications relevant to BCS and CBS emerge.
4. Engaging in the analogue of pen testing and phishing exercises for simple testing of regional supply chains, followed by exercises that test institutional internal resilience and responses to activity escalation in participating countries.
5. Having a global committee review responses and update guidance on proper BCS and CBS defense.

An ideal policy for offensive BCS research relies on one of its more unique aspects as an offensive tool. Where the presence of a gun or a mushroom cloud undoubtedly carries the connotation of being in a state of war, BCS threats are more diaphanous and unobservable. A war fought with only BCS attacks, which could, if waged competently enough, be nothing but a series of increasingly unfortunate coincidences. Thus, the ideal offensive posture would be to have a constellation of Man in the Middle threats located in an area of operations (AO), with attacks planned (for instance, the abovementioned listing of which biological agents would be most effective compared to genomic analysis of those subjects and via which medicines were not regularly stocked or able to be embargoed to that AO) and revised based on new intelligence.

Defensive posture is the field in which CBS excels – the constant innovations of cybersecurity and the applications to biological resources have been in a race against potential threats for the entirety of its existence.

However, capability can be blunted through oppositional control of biological data. Thus, it is important to discuss cyber-enabled biological misinformation. The next section will discuss this more in depth.

4.1 Cyber-Enabled Biological Misinformation: A Case Study

At this point, a small digression into the field of dis- and misinformation is required. The utilization of disinformation to augment biowarfare has already been noted [5, 73, 74]. One issue that has become more readily apparent in the last half-century, and especially since the onset of what has become known as the Global War on Terror (GWAT), is the use of asymmetric warfare. Though some have noted that "Arguably, it [asymmetric warfare] meant so many different things that it became a useless, ambiguous term," it is more rigorously defined as follows:

> Asymmetric warfare is population-centric nontraditional warfare waged between a militarily superior power and one or more inferior powers which encompasses all the following aspects: evaluating and defeating asymmetric threat, conducting asymmetric operations, understanding cultural asymmetry and evaluating asymmetric cost. [12]

Previous works have codified some of its methods fairly well [115]. This is logical as what is commonly referred to as "The West," more specifically referring to NATO (North Atlantic Treaty Organization) nations, with the primary agent being the US defense apparatus being a rarely disputed champion of conventional warfare.

This has reached new heights in recent years, especially with the fairly evident onset of concentrated dis/misinformation campaigns. The most well-acknowledged campaigns being the ongoing Russian Federation campaigns for which a US congressional hearing was held 5 years ago in the acknowledgment of the emergence of hybridized warfare [20]. This is likely rooted in propaganda efforts on the part of the Russian Federation as far back as 2008 in the invasion of Georgia [79], with roots in the Soviet era medical disinformation campaign being claims of the US government's entirely fictitious role in creating the virus that causes AIDS [8].

Bertoli in 2015 characterized the style of Russian misinformation with a trinary goal: it "entertains, confuses, and overwhelms the audience" [6]. It is also noted to be "rapid, continuous, and repetitive, and it lacks commitment to consistency" [79]. This has also been linked to a phenomenon named "Truth Decay" which is related to the acceptance of new disinformation or rejection of provable information [90]. Thus, disinformation campaigns are not necessarily a strictly operational offensive tool but can be seen as an investment in future misinformation advances.

The use of misinformation in war is hardly a new development, though its dissemination directly to citizens through the laissez-faire apparatus of the Internet was a novel concept. It was not under the purview of BCS, however, until the augmentation of this misinformation with a medical goal – the denial and propagandizing of COVID-19 via foreign state interference [1, 2]. The targeting of a population to inhibit willingness to participate in infection-prevention methods would be an exceptional goal. However, the methods demonstrated so far, especially the use of "burner" accounts (accounts made on a site and abandoned quickly after their original use), show that verifiable individuals do not often either see or share such information [2]. More effective means of spreading misinformation tend to come from sources that are already well established to give medical advice (whether that advice is accurate or helpful being an open question). The "Disinformation Dozen" is the name given to a group of people responsible for either generating or sharing the majority of COVID-19 vaccine misinformation, many of them being associated with "alternative" medicine [9, 75]. The networks that do end up exposing citizens to misinformation are commonly in networks that are entirely unrelated to health, medicine, or established government sources: "relatively small-size but very well-distributed organized communities of distrust that have embedded themselves with just everyday other communities, such as pet lovers or parents' associations" [54]. Interestingly, the "Deepfake" methodology of using AI to fake a known individual's face and voice to generate false statements seems to have not been ruled an effective measure [37].

Earlier analysis has shown that the cost for such a campaign is now microscopic in comparison to the cost of resources for conventional conflict, such as mobile armor [76]. And efforts to create methods to combat misinformation for civilian intake are currently ongoing [27, 28, 102] though not without detractors [7, 51].

4.2 *Advantages of Cyber-Enabled Asymmetric Biological Information Warfare*

One issue of note in the arena of CBS and BCS conflict is the asymmetry of information available. To put it one way, not many people, no matter their level of education or experience, will look at the mushroom cloud and refuse the reality of an extant nuclear attack. Yet propaganda could very well be used in the case of a BCS or CBS attack to obfuscate even the most basic existence of said attack [2, 76].

In the terms of a nation–nation interaction, formal declarations of conflict have not been the norm for several decades. The USA, for instance, has not formally declared war since 1942 [109]. Other euphemisms are generally preferred – for instance, the Russia-Ukraine conflict is, according to Russian sources, termed a "Special Military Operation" [70]. While several nations have agreed to report requirements for certain weapons systems [107] with BCS or CBS threats, there is practically a guarantee that there will be no formal declarations of beginning a campaign to access secured resources, misinform a public, or eliminate adversarial cyber or biological capabilities.

However, the nature of a BCS attack could be to reduce trust in academic institutions (as explored above). This is especially notable in conflicts that could utilize NGO (nongovernment organization) structures to obfuscate their purpose. The ability of a corporate authority to legitimize the flow of information has long been noted [34]. This power has lately been circumscribed by a handful of either malicious actors using the open architecture of the Internet to supplant the conventional sources of information [1, 90]. This is especially pertinent in the case of the USA and its non-cleared citizens [49], though not necessarily the official channels used for military or national defense information.

Implementation of policies to protect against biodiversity loss and potential biological agents can benefit common citizens when being undertaken by as many countries as possible, and this can be strengthened by international agreements. Implementation would require stronger biosecurity practices, but also cybersecurity practices across the continent, given that biological databases remain key to tracking resources [97]. Investments to improve connectivity across less serviced regions but also stronger biosecurity investments paired with appropriate mirror investments in infrastructure and practices for housing biological resources and exchanging data upon them are crucial. Arguably, the creation of more universities with suitable biology labs can address this. An attractive benefit is that deepened BCS investments can strengthen the growing efforts in combating epidemics from country to country and more efficiently help contain epidemics. However, massive investments, political unity, and strong anti-corruption measures would be required as not all countries of the Global South are stable. Individual and incremental steps, perhaps through additional foreign aid or through regional partnerships, might be the path forward as each country finds its path. Assuming that this could be, each country will need to consider its unique biospheres and biological resources to protect that which can be exploited. That requires a means of increased scouting, public learning, and accounting beyond mere conservatorships of wildlife preserves and zoos. However, this is a matter yet unanswered as the Global South develops in all of its technological and economic capacities.

One issue in combating the rising threat of CBS/BCS activities is how easily available the equipment is to acquire. This is evidenced by the monotonic rise of identity fraud and other cyber-enabled crimes [22, 65]. At some point, nation-state sponsored cybercrimes could even be seen as a methodology by which to interfere with a country's interest at a national level [71], especially since the investment in destroying a national trust in authority (such as medical practitioners) could be seen as a valuable investment as an auxiliary method to conducting conventional operations [84]. No analysis of cost will be offered, as the decrease in cost of Internet-enabled devices will likely decrease in the time it takes to publish this work, and the cost will no doubt be incredibly diminished in comparison to the relatively steady cost of investing in conventional forces or CBRN (Chemical, Biological, Radiological, Nuclear) weapons systems.

A threat which is unique to BCS would be the utilization of algorithms that usually are reserved for use in medical procedures or research for bioweapons research (so-called dual-use tools). For instance, work in conceptually simple codebases such as physiological simulators [86], when appended to AI interfaces, could be used to create unique combinations of biological agents to maximize fatalities of a given population [85]. This has already been reported to create combinations of molecules that are similar to VX nerve gas [110]. Such use of AI/ML methods to create novel threats is why the differentiation between CBS and BCS research ought to exist.

5 Legislation of Policy in the Cyber-Bio Age

In terms of disinformation campaigns, which could affect nation-level resources, legislation to protect from that subset of BCS threats is critical. The Sars-Cov-2 pandemic, in addition to the usual waves of flu and other colds, has been accompanied by numerous misinformation campaigns that have affected healthcare and put a strain on economies worldwide. Therein, the benefit of facilitating positive and truthful communication is key. This is not without difficulty as the fallibility of leading health organizations, coupled with overleveraging of power, has eroded public trust. We propose a considerable rethinking of laws designed to punish malicious messaging (disinformation) and misuse of public data, combined with incentives for beneficial science communication. This will not be easy but pressing for such remains important as our societies delve deeper into the Cyber Bio-Age.

A more pressing concern is the rise of technologies that blend the biologically active and cyber components, such as brain-computer interface technology. Herein, researchers have noted concerns about corporate malfeasance that could occur with access to curated customer data from the BCIs used [59]. An indirect possibility that exists comes from the potential of malicious users exploiting openings exposed or created by corporate failings to protect customer biological assets, for example, confidential patient data [17] or even biological implants [104]. At this intersection, the lack of a consistent set of legislated cybersecurity approaches for biologically active technology is a large target for potential cyber threats. As previously noted, extant legislation "imperfect deterrents in the biological arena. Deterrents and laws preventing malevolent Cyber-Bio activity have not been legislated in many countries" [41]. This hinges on the development of biofabrication models that would enable the long-term functional implantation of electrically active pieces of computing hardware in a biological environment [44].

5.1 Bio-Cyber Bill of Rights

Yet even as the fighting of nations dictates the media diet of typical citizenry, the possibility of over-encroachment of nations to methods of conflict with NGO within their own territories, some methods may be in conflict with legal precedent or violate strongly held personal beliefs.

One acclaimed that US institution is the clear delineation of personal, unin-fringeable rights placed into the founding constitutional document [111]. Workplace safety laws were founded in the USA by the labor movement at the turn of the twentieth century (though the official founding of OSHA was not until 1970 [69]). In the 1940s, there were rumblings of an economic bill of rights that would delineate the labor right of citizens [96]. This is especially pertinent with using government interventions in ways of preventing disease spread (for instance, sick days and government-sponsored healthcare), and calls for renewed interest in an economic bill of rights are returning [105]. Even the methods by which those regulations could be put in place are of current interest [38]. This use of organized labor to prevent health impacts has been shown to carry over to the present day [78], with the further implication that the proper organization of labor could aid in the prevention of disease and a more tightly integrated series of national defense.

The cyber-enabled economy has found many ways to invade the reasonable expectation of privacy. And the digital storage of medical information implies that even what previously was guaranteed privacy (under Health Insurance Portability and Accountability Act (HIPAA)) may require updating, considering this. Already, such data is now a target for CBS/BCS attacks [16]. The European Union (EU) has passed the GDPR [35] which has advanced the cause of data protection of citizenry on the easily accessible surface Internet. In the USA, data regulations have not had an equivalent amount of consideration, to the point that US sold voice-controlled speaker systems have been linked to advertisement databases [53]. In a similar vein, the USA has grappled with the four Internet freedoms proposal [87] which, when combined with timely, accurate, and timely vetting of informational sources, could create a more informed populace via Internet connections.

A previous call for national CBS measures included calls to "Set forth clear consequences for individuals or countries that undertake such [CBS threats] actions without imperiling the legitimate sharing of scientific data and information" and "Allow for the establishment of voluntary standards in partnership with the private sector" [41]. The first point is urgently needed. The voluntary nature of private sector standards will be explored in Sect. 4, and the failures of a voluntary framework in terms of information veracity will be discussed. The exact nature of the US right to privacy is now more of an open question than ever, with the actions of the CDC in tracing compliance to COVID-19 precautions via purchased user geo-location data – though this report has not been fully verified [23].

With a delineation of rights comes the question of how much transparency is required between citizens, industry, and government. The answers provided so far are blurry at best [55]. Another point of contention between citizens and industry, especially as the lines between biological and cyber tools become blurred, is the responsibility of a company to support their biologically linked products in terms of cybersecurity [104]. This lack of support is logical from a strictly business perspective – after all, why would a commercial enterprise spend money it did not have to on a product they deem economically untenable? But for those that depend on these technologies to function, this is likely not a comfort.

While the Internet may aspire to the ideals of free and open information access for all connected to it [60], it is rapidly showing that the ideal of liberty on the Web has rapidly declined to a state of informational anarchy [33, 63]. The link from misinformation in general to medical misinformation has previously been established [90]. Not only are many social media sites repositories of misinformation, in some cases, the spread of misinformation was encouraged by an abject lack of corporate responsibility and efforts to obscure their part in the spread of misinformation [116]. This lack of action to stem misinformation was admitted by a volunteer dealing with misinformation on one of the world's largest social media sites [89]. Hopefully in the coming years, this misinformation will be effectively combated while keeping the open framework users have come to expect from the Internet.

To that end, a Bio-Cyber Bill of rights ought to include such things as a reasonable expectation of persistent privacy from commercial interests, the right to erasure of bio-data (both direct and indirect) [114], the right to reasonable support for purchased biomedical devices (including open access for discontinued devices), freedom from ownership of one's person by commercial interests, freedom from an internationally recognized corporate equipment (not technique) from assessing one's biology, freedom to practice biotechnological work in so far as it does not infringe on another's biology or property, freedom toward the inclusion of DNA evidence and other biotechnological assays in the defense of defendants in criminal trials, freedom of bodily autonomy in augmentation, and freedom of bio-digital representation.

5.2 Supergovernmental Framework for Academic Research

As established in earlier work, academic settings are rife with the possibility of BCS and CBS threats [80]. This is aided by the divergent evolutions in academic disciplines – for evidence of that look no further than the myriad styles of citations in each discipline, which sometimes even change across journals in the same field [24]. This is not to mention the lack of any coherent data sharing policy expectation in journals, which makes it easier to fabricate results.

Nominally, the solution could be to empower the national or regional government to inspect journal offices and university labs to verify their quality and their statements to confirm their veracity. Yet that runs into a scaling issue – some national governments may not have a person on staff that could adequately assay statements from a lab and know what to ask in case some statements are fabricated. If they do, there is no telling those individuals would not be in some way connected to that lab, especially in smaller regional governments. The ideal solution would be to empower a supergovernmental agency for two primary tasks: first, to establish data sharing guidelines to which an academic journal would need to adhere to for accreditation as a reputable journal, and second, to audit academic research centers in the case of suspicious findings.

The first task is already noted as a necessary change – not just in the case of a malicious actor actively planting false or fake information as has been previously reported [54]. In one case, social science journals have notably lacked a coherent data policy and a singular ideology of research misconduct [25, 92]. However, as far as security and the most likely publication avenues of data relevant to BCS/CBS, physical and life science journals are more likely to be used. In this case, when 709 journals were inspected, less than 40% of journals analyzed had a "strong" data sharing policy [57].

The immediate drawback of this approach is that the highest output journals are located in the Global North or more specifically the "Western World" (nebulously defined as Europe and North America). Therefore, the burden of sending auditors to the Global South to inspect relatively newer publications may be untenable and create a scientific caste system where outputs from the Global North are seen as more accurate or pertinent. This is already noted in the case of English being the accepted language of international scientific publishing which limits the rate of output of non-native speakers [61]. To combat this, the ideal solution would be for an office to be an extension of a pre-existing organization (such as the United Nations) which could fund all activities without regard to geographic location. Another (though not mutually exclusive solution) is to have the costs of inspections and audits be a fee structure that allows for more well-funded universities to subsidize the costs of audits of less-funded universities and journals [97]. The resources of the Global North in the context of CBS have already been undertaken [15, 16].

Thus, the administration of such a hypothetical organization could be similar to the UN Security council. A handful of representatives from the most trusted journals or scientific accreditation organizations and a rotation board of all members – and perhaps the most pressing decisions – could be put to a plebiscite of members, who presumably would either be or have been active contributors and understand how to verify claims. This model could then be utilized to offer scientific clarification on mass-market publications such as newspapers or television programs, from which most of the population is informed.

6 Conclusion

Potential and demonstrated exploits at the intersection of domains of biocybersecurity, cyber-physical security, and cybersecurity have meaningful differences. Particularly, differences in direction and application of technologies within the intersection of these domains justify the decision to split biocybersecurity and cyberbiosecurity from this point on. It is the hope of the authors that conversations at this intersection increase in nuance, for the purposes of planning and coordinating the conducting of sensible security policy.

Appendix: Analyzed CBS and BCS Papers

1. Ficetola, G. F., Canedoli, C., & Stoch, F. (2019). The Racovitzan impediment and the hidden biodiversity of unexplored environments. *Conservation Biology*, *33*(1), 214–216.
2. Forrest, A. (2020). The Panthalassa project: The future of ocean research for conservation. *Conservation Letters*, e12743. https://efile.fara.gov/docs/3634-Informational-Materials-20201002-20.pdf
3. Kristensen, N. P., Seah, W. W., Chong, K. Y., Yeoh, Y. S., Fung, T., Berman, L. M., ... & Chisholm, R. A. (2020). Extinction rate of discovered and undiscovered plants in Singapore. *Conservation Biology*. https://doi.org/10.1038/nature18315.
4. Mora, C., Tittensor, D. P., Adl, S., Simpson, A. G., & Worm, B. (2011). How many species are there on Earth and in the ocean?. *PLoS Biol*, *9*(8), e1001127.
5. Martin, C. (2011). A global view of funding for the plant sciences. https://doi.org/10.1016/j.cub.2011.05.027
6. Bromham, L., Dinnage, R., & Hua, X. (2016). Interdisciplinary research has consistently lower funding success. *Nature*, *534*(7609), 684–687. https://www.nature.com/articles/s41559-018-0561-z
7. Simon J. Anthony, Christine K. Johnson, Denise J. Greig, Sarah Kramer, Xiaoyu Che, Heather Wells, Allison L. Hicks, Damien O. Joly, Nathan D. Wolfe, Peter Daszak, William Karesh, W. I. Lipkin, Stephen S. Morse, PREDICT Consortium, Jonna A. K. Mazet, Tracey Goldstein, Global patterns in coronavirus diversity, Virus Evolution, Volume 3, Issue 1, January 2017, vex012, https://doi.org/10.1093/ve/vex012
8. Su, S., Wong, G., Shi, W., Liu, J., Lai, A. C., Zhou, J., ... & Gao, G. F. (2016). Epidemiology, genetic recombination, and pathogenesis of coronaviruses. *Trends in microbiology*, *24*(6), 490–502. https://doi.org/10.1016/j.tim.2016.03.003
9. George, A. M. (2019). The national security implications of cyberbiosecurity. Frontiers in Bioengineering and Biotechnology, 7, 51.
10. Tyler, V. E. (2001). The future of botanical drugs and the rainforest. *Journal Of Herbal Pharmacotherapy*, *1*(1), 5–12.
11. Leite, M. J., Silva, V. F., Silva, M. A., e Silva, A. C. L., Silva, G. H., Aguiar, M. M., ... & Rodal, M. J. (2019). Ecological Variability Prediction Based on Functional Characteristics of an Urban Rainforest. *Journal of Experimental Agriculture International*, 1–12.
12. Stropp, J., Umbelino, B., Correia, R. A., Campos-Silva, J. V., Ladle, R. J., & Malhado, A. C. M. (2020). The ghosts of forests past and future: deforestation and botanical sampling in the Brazilian Amazon. Ecography. https://doi.org/10.1111/ecog.05026
13. Bohannon, J. (2015). Hoax-detecting software spots fake papers.
14. Cyril Labbé, Dominique Labbé. Duplicate and fake publications in the scientific literature: how many SCIgen papers in computer science?.

Scientometrics, Springer Verlag, 2012, pp.10.1007/s11192-012- 0781-y. ff10.1007/s11192–012-0781-yff. ffhal-00641906v2f.
15. DiEuliis, D., Lutes, C. D., & Giordano, J. (2018). Biodata Risks and Synthetic Biology: A Critical Juncture. Journal of Bioterrorism & Biodefense, 9(01).
16. Murch, R. S., So, W. K., Buchholz, W. G., Raman, S., & Peccoud, J. (2018). Cyberbiosecurity: an emerging new discipline to help safeguard the bioeconomy. Frontiers in bioengineering and biotechnology, 6, 39.
17. MILLETT, K. K., dos Santos, E., & MILLETT, P. D. (2019). Cyber-Biosecurity Risk Perceptions in the Biotech Sector. Frontiers in bioengineering and biotechnology, 7, 136.
18. Potter, L., Ayala, O., & Palmer, X. L. (2020). Biocybersecurity–A Converging Threat as an Auxiliary to War. arXiv preprint arXiv:2010.00624.
19. DiEuliis, D., Lutes, C. D., & Giordano, J. (2018). Biodata Risks and Synthetic Biology: A Critical Juncture. Journal of Bioterrorism & Biodefense, 9(01).
20. Ney, P. M. (2019). Securing the future of biotechnology: A study of emerging bio-cyber security threats to DNA-information systems (Doctoral dissertation).
21. DeFranco, J., DiEuliis, D., & Giordano, J. (2019). Redefining Neuroweapons. PRISM, 8(3), 48–63.
22. Potter, L., & Palmer, X. L. (2020). Human Factors in Biocybersecurity Wargames. arXiv preprint arXiv:2005.02135.
23. Palmer, X. L., & Karahan, S. (2020, March). Defending Forward: An Exploration through the Lens of Biocybersecurity. In ICCWS 2020 15th International Conference on Cyber Warfare and Security (p. 373). Academic Conferences and publishing limited.
24. Palmer, X. L., Potter, L., & Karahan, S. (2020, March). On the Emerging Area of Biocybersecurity and Relevant Considerations. In Future of Information and Communication Conference (pp. 873–881). Springer, Cham.
25. Olofinbiyi, S. A., & Singh, S. B. (2020). The Role and Place of Covid-19: An Opportunistic Avenue for Exponential World's Upsurge in Cyber Crime. International Journal of Criminology and Sociology, 221–230.
26. Giordano, J. J. (2020). Catholic Ethics and the Challenge of COVID-19-Part 5: Pandemic Public Surveillance.
27. Wintle, B. C., Boehm, C. R., Rhodes, C., Molloy, J. C., Millett, P., Adam, L., ... & Doubleday, R. (2017). Point of View: A transatlantic perspective on 20 emerging issues in biological engineering. Elife, 6, e30247.
28. Ruppel, E. K., Gross, C., Stoll, A., Peck, B. S., Allen, M., & Kim, S. Y. (2017). Reflecting on connecting: Meta-analysis of differences between computer-mediated and face-to-face self-disclosure. Journal of Computer-Mediated Communication, 22(1), 18–34.
29. Peccoud, J., Gallegos, J. E., Murch, R., Buchholz, W. G., & Raman, S. (2018). Cyberbiosecurity: from naive trust to risk awareness. Trends in biotechnology, 36(1), 4–7.
30. 肖. (2018). 美土安全家提出"网生物安全" 新概念. 科技中, (5), 106–106.

31. Gillum, D., Carrera, L. A. O., Mendoza, I. A., Bates, P., Bowens, D., Jetson, Z., ... & O'Donnell, M. (2018). The 2017 Arizona Biosecurity Workshop: An Open Dialogue About Biosecurity. Applied Biosafety, 23(4), 233–241.
32. Liao, W., Bazarova, N. N., & Yuan, Y. C. (2018). Unpacking medium effects on social psychological processes in computer-mediated communication using the social relations model. Journal of Computer-Mediated Communication, 23(2), 90–106.
33. Pauwels, E., & Denton, S. W. (2018). The internet of bodies: life and death in the age of AI. Cal. WL Rev., 55, 221.
34. Pauwels, E., & Denton, S. W. (2018). Searching for privacy in the Internet of Bodies. The Wilson Quarterly, 42(2).
35. Schabacker, D. S., Levy, L. A., Evans, N. J., Fowler, J. M., & Dickey, E. A. (2019). Assessing cyberbiosecurity vulnerabilities and infrastructure resilience. Frontiers in bioengineering and biotechnology, 7, 61.
36. Mantle, J. L., Rammohan, J., Romantseva, E. F., Welch, J. T., Kauffman, L. R., McCarthy, J., ... & Lee, K. H. (2019). Cyberbiosecurity for biopharmaceutical products. Frontiers in Bioengineering and Biotechnology, 7, 116.
37. Reed, J. C., & Dunaway, N. (2019). Cyberbiosecurity Implications for the Laboratory of the Future. Frontiers in bioengineering and biotechnology, 7, 182.
38. Murch, R., & DiEuliis, D. (Eds.). (2019). Mapping the cyberbiosecurity enterprise. Frontiers Media SA.
39. Turner, G. (2019, May). The Growing Need for Cyberbiosecurity. In InSITE 2019: Informing Science+ IT Education Conferences: Jerusalem (pp. 207–215).
40. Jordan, S. B., Fenn, S. L., & Shannon, B. B. (2020). Transparency as Threat at the Intersection of Artificial Intelligence and Cyberbiosecurity. Computer, 53(10), 59–68.
41. Vinatzer, B. A., Heath, L. S., Almohri, H. M., Stulberg, M. J., Lowe, C., & Li, S. (2019). Cyberbiosecurity Challenges of Pathogen Genome Databases. Frontiers in bioengineering and biotechnology, 7, 106.
42. Guttieres, D., Stewart, S., Wolfrum, J., & Springs, S. (2019). Cyberbiosecurity in Advanced Manufacturing Models. Frontiers in Bioengineering and Biotechnology, 7, 210.
43. Richardson, L. C., Lewis, S. M., & Burnette, R. N. (2019). Building Capacity for Cyberbiosecurity Training. Frontiers in Bioengineering and Biotechnology, 7, 112.
44. Duncan, S. E., Reinhard, R., Williams, R. C., Ramsey, F., Thomason, W., Lee, K., ... & Murch, R. (2019). Cyberbiosecurity: A new perspective on protecting us food and agricultural system. Frontiers in bioengineering and biotechnology, 7, 63.
45. Carmi, G., & Bouhnik, D. (2019, May). The Effect of Rational Based Beliefs and Awareness on Employee Compliance with Information Security Procedures: Case Study of a Financial Firm. In InSITE 2019: Informing Science+ IT Education Conferences: Jerusalem (pp. 001–002).

46. Hudson, C. (2019). Cybersecurity in DNA Design and Verification Tools: Risks and Solutions (No. SAND2019-11338PE). Sandia National Lab.(SNL-CA), Livermore, CA (United States).
47. Walsh, M., & Streilein, W. (2020). Security Measures for Safeguarding the Bioeconomy. Health security, 18(4), 313–317.
48. Hudson, C., & Oehman, C. (2019). Blue Ribbon Panel Remarks (No. SAND2019-11058C). Sandia National Lab.(SNL-CA), Livermore, CA (United States).
49. Hudson, C. (2019). Modeling realistic genomic and synthetic biology facilities at scale (No. SAND2019-0817C). Sandia National Lab.(SNL-CA), Livermore, CA (United States).
50. Motta, M. Political Scientists: A Profile of Congressional Candidates with STEM Backgrounds. PS: Political Science & Politics, 1–6.
51. Hudson, C. (2019). Genomic and Synthetic Biology Cybersecurity (No. SAND2019-4923PE). Sandia National Lab.(SNL-CA), Livermore, CA (United States).
52. Adames, N. R., Gallegos, J. E., Hunt, S. Y., So, W. K., & Peccoud, J. (2019). Hands-On Introduction to Synthetic Biology for Security Professionals. Trends in biotechnology, 37(11), 1143–1146.
53. Diggans, J., & Leproust, E. (2019). Next steps for access to safe, secure DNA synthesis. Frontiers in bioengineering and biotechnology, 7, 86.
54. DeFranco, J., DiEuliis, D., & Giordano, J. (2019). Redefining Neuroweapons. PRISM, 8(3), 48–63.
55. Caswell, J., Gans, J. D., Generous, N., Hudson, C. M., Merkley, E., Johnson, C., … & Ting, C. L. (2019). Defending our public biological databases as a global critical infrastructure. Frontiers in Bioengineering and Biotechnology, 7, 58.
56. Mueller, S. (2019). On DNA Signatures, Their Dual-Use Potential for GMO Counterfeiting, and a Cyber-Based Security Solution. Frontiers in bioengineering and biotechnology, 7, 189.
57. Berger, K. M., & Schneck, P. A. (2019). National and transnational security implications of asymmetric access to and use of biological data. Frontiers in bioengineering and biotechnology, 7, 21.
58. Richardson, D. B., Langholz, B., & Kelly-Reif, K. (2019). General Relative Rate Models for the Analysis of Studies Using Case-Cohort Designs. American journal of epidemiology, 188(2), 444–450.
59. Duncan, S., & Ford, R. (2019). SmartFarm Innovation Network.
60. Mueller, S. (2019). Are Market GM plants an unrecognized platform for bioterrorism and biocrime?. Frontiers in bioengineering and biotechnology, 7, 121.
61. Easttom, C., & Mei, N. (2019, October). Mitigating Implanted Medical Device Cybersecurity Risks. In 2019 IEEE 10th Annual Ubiquitous Computing, Electronics & Mobile Communication Conference (UEMCON) (pp. 0145–0148). IEEE.

62. Edwards, B. (2019). Synthetic Biology and Dilemmas of Insecurity. In Insecurity and Emerging Biotechnology (pp. 79–87). Palgrave Pivot, Cham.
63. Mondal, M., & Ray, K. S. (2019). Review on DNA Cryptography. arXiv preprint arXiv:1904.05528.
64. Carter, S. R., & Warner, C. M. (2018). Trends in synthetic biology applications, tools, industry, and oversight and their security implications. Health security, 16(5), 320–333.
65. DiEuliis, D. (2019). Key National Security Questions for the Future of Synthetic Biology. Fletcher F. World Aff., 43, 127.
66. Hu, Y., Green, G. S., Milgate, A. W., Stone, E. A., Rathjen, J. P., & Schwessinger, B. (2019). Pathogen detection and microbiome analysis of infected wheat using a portable DNA sequencer. Phytobiomes Journal, 3(2), 92–101.
67. Рассохина, И. И., Коткова, Д. Н., & Платонов, А. В. (2019). Анализ мировой публикационной активности по направлению "биоэкономика". Проблемы развития территории, (3 (101)).
68. Misuse Potential of Systems Biology – New Challenges for Biological Arms Control? KATHRYN NIXDORFF Science Peace Security '19, 125.
69. Barosy, W. (2019). Successful Operational Cyber Security Strategies for Small Businesses.
70. Edwards, B. (2019). Insecurity and Emerging Biotechnology: Governing Misuse Potential. Springer.
71. Arevalo-Barzallo, J. R., & Flores Pusda, E. V. (2019). Manual De Bioseguridad Dirigido a Estudiantes De La Escuela De Cosmiatria De La Universidad Iberoamericana Del Ecuador (Doctoral dissertation, Unibe).
72. Pauwels, E. (2019). The New Geopolitics of Converging Risks.
73. Fiorella, R. (2019). International Seminars On Nuclear War And Planetary Emergencies-49th Session. World Scientific.
74. Mueller, S. (2020). Facing the 2020 Pandemic: What does Cyberbiosecurity want us to know to safeguard the future?. Biosafety and Health.
75. Farbiash, D., & Puzis, R. (2020). Cyberbiosecurity: DNA Injection Attack in Synthetic Biology. arXiv preprint arXiv:2011.14224.
76. Schmid, D. C. (2020). Risks and Consequences of Hazard Agents to Human Health. In Toxic Chemical and Biological Agents (pp. 129–141). Springer, Dordrecht.
77. Berger, K. M. (2020). Addressing Cyber Threats in Biology. IEEE Security & Privacy, 18(3), 58–61.
78. Özkoç, E. E., & Mannion, M. (2020). Siberbiyogüvenlik Uygulamalarında DNA Dizilimleri için Özet Algoritmaları Karşılaştırılması. Avrupa Bilim ve Teknoloji Dergisi, (18), 656–663.
79. DiEuliis, D. (2020). Parsing the Digital Biosecurity Landscape. Georgetown Journal of International Affairs, 21, 166–172.
80. Hester, R. J. (2020). Bioveillance: A Techno-security Infrastructure to Preempt the Dangers of Informationalised Biology. Science as Culture, 29(1), 153–176.

81. MAJID, M. A. (2020). A Gene Synthesis Regime for Malaysia to Emulate in Securing Future Bioprinted Vaccines. MALIM: JURNAL PENGAJIAN UMUM ASIA TENGGARA (SEA Journal of General Studies), 21.
82. van der Linden, D., Michalec, O. A., & Zamansky, A. (2020). Cybersecurity for smart farming: socio-cultural context matters. IEEE Technology and Society Magazine.
83. Bernal, S. L., Martins, D. P., & Celdrán, A. H. (2020, June). Distributed Denial of Service Cyberbioattack Affecting Bacteria-based Biosensing Systems. In 2020 17th International Conference on Electrical Engineering/Electronics, Computer, Telecommunications and Information Technology (ECTI-CON) (pp. 279–282). IEEE.
84. Ibrahim, M., Liang, T. C., Scott, K., Chakrabarty, K., & Karri, R. (2020). Molecular Barcoding as a Defense against Benchtop Biochemical Attacks on DNA Fingerprinting and Information Forensics. IEEE Transactions on Information Forensics and Security.
85. Glaz, B., Wang, C., Hurley, M., & Kott, A. (2020). Artificial Intelligence in Synthetic Biology, Cyber Defense, and Aeromechanical Design. CCDC Army Research Laboratory Adelphi United States.
86. McCartney, A. T., Yeo, J. Y., Blomquist, T. M., & Huntley, J. F. (2020). Whole-Genome Sequencing of Bacterial Isolates That Degrade the Cyanobacterial Toxin Microcystin-LR. Microbiology Resource Announcements, 9(40).
87. Gallegos, J. E., Kar, D. M., Ray, I., Ray, I., & Peccoud, J. (2020). Securing the exchange of synthetic genetic constructs using digital signatures. ACS Synthetic Biology, 9(10), 2656–2664.
88. Elgabry, M., Nesbeth, D., & Johnson, S. D. (2020). A Systematic Review of the Criminogenic Potential of Synthetic Biology and Routes to Future Crime Prevention. Frontiers in bioengineering and biotechnology, 8.
89. Cunningham, M. A., & Geis, J. P. (2020). A National Strategy for Synthetic Biology. Strategic Studies Quarterly, 14(3), 49–80.
90. López, S. B., Martins, D. P., & Huertas, A. C. (2020). Distributed Denial of Service Cyberbioattack Affecting Bacteria-based Biosensing Systems.
91. Spanakis, E. G., Bonomi, S., Sfakianakis, S., Santucci, G., Lenti, S., Sorella, M., ... & Magalini, S. (2020, July). Cyber-attacks and threats for healthcare–a multi-layer thread analysis. In 2020 42nd Annual International Conference of the IEEE Engineering in Medicine & Biology Society (EMBC) (pp. 5705–5708). IEEE.
92. WANG, X. L. (2020). The Age of Biosecurity: New Biotechnology Revolution and National Biosecurity Governance. China Biotechnology, 40(9), 95–109.
93. O'Brien, J. T., & Nelson, C. (2020). Assessing the Risks Posed by the Convergence of Artificial Intelligence and Biotechnology. Health security, 18(3), 219–227.
94. DeFranco, J., Rhemann, M., & Giordano, J. (2020). The Emerging Neurobioeconomy: Implications for National Security. Health security, 18(4), 267–277.

95. Gallegos, J. E., Rogers, M. F., Cialek, C. A., & Peccoud, J. (2020). Rapid, robust plasmid verification by de novo assembly of short sequencing reads. Nucleic acids research, 48(18), e106-e106.
96. Bartley, B. A., Beal, J., Karr, J. R., & Strychalski, E. A. (2020). Organizing genome engineering for the gigabase scale. Nature Communications, 11(1), 1–9.
97. Frye-Mason, G. C. (2020). Laboratory Directed Research and Development 2019 Annual Report (No. SAND2020-3752R). Sandia National Lab.(SNL-NM), Albuquerque, NM (United States).
98. Weber, J., & Kämpf, K. M. (2020). Technosecurity Cultures: Introduction.
99. Firestone, J. (2020). The need for soft law to regulate synthetic biology. Jurimetrics, 60(2), 139–173.
100. 王小理. (2020). 生物安全代: 新生物科技革与家安全治理. 中生物工程志, 40(9), 95–109.
101. Nakamoto, I., Wang, S., Guo, Y., & Zhuang, W. (2020). A QR Code–Based Contact Tracing Framework for Sustainable Containment of COVID-19: Evaluation of an Approach to Assist the Return to Normal Activity. JMIR mHealth and uHealth, 8(9), e22321.
102. National Academies of Sciences, Engineering, and Medicine. (2020). HORIZON SCANNING AND FORESIGHT METHODS. In Safeguarding the Bioeconomy. National Academies Press (US).
103. National Academies of Sciences, Engineering, and Medicine. (2020). ECONOMIC AND NATIONAL SECURITY RISKS PERTAINING TO THE BIOECONOMY. In Safeguarding the Bioeconomy. National Academies Press (US).
104. Nave Jr., G. K., Hall, N., Somers, K., Davis, B., Gruszewski, H., Powers, C., ... & Ross, S. D. (2020). Wind dispersal of natural and biomimetic maple samaras. arXiv preprint arXiv:2010.12553.
105. Gallegos, J. E., Hayrynen, S., Adames, N. R., & Peccoud, J. (2020). Challenges and opportunities for strain verification by whole-genome sequencing. Scientific reports, 10(1), 1–9.
106. Serrà, A., Pip, P., Gómez, E., & Philippe, L. (2020). Efficient magnetic hybrid ZnO-based photocatalysts for visible-light-driven removal of toxic cyanobacteria blooms and cyanotoxins. Applied Catalysis B: Environmental, 268, 118,745.
107. Wang, Q., Ding, J., Xie, H., Hao, D., Du, Y., Zhao, C., ... & Wang, B. (2020). Phosphorus removal performance of microbial-enhanced constructed wetlands that treat saline wastewater. Journal of Cleaner Production, 125119.
108. Olson, N. E., Cooke, M. E., Shi, J. H., Birbeck, J. A., Westrick, J. A., & Ault, A. P. (2020). Harmful Algal Bloom Toxins in Aerosol Generated from Inland Lake Water. Environmental Science & Technology, 54(8), 4769–4780.
109. Tran, N. H., Li, Y., Reinhard, M., He, Y., & Gin, K. Y. H. (2020). A sensitive and accurate method for simultaneous analysis of algal toxins in freshwater using UPLC-MS/MS and 15 N-microcystins as isotopically labelled internal standards. Science of The Total Environment, 139727.

110. Arango, J. G., & Nairn, R. W. (2020). Prediction of Optical and Non-Optical Water Quality Parameters in Oligotrophic and Eutrophic Aquatic Systems Using a Small Unmanned Aerial System. Drones, 4(1), 1.
111. Bruynseels, K. (2020). Responsible innovation in synthetic biology in response to COVID-19: the role of data positionality. Ethics and Information Technology, 1–9.
112. Sardi, A., Rizzi, A., Sorano, E., & Guerrieri, A. Cyber Risk in Health Facilities: A Systematic.
113. 彭耀. (2020). 合成生物代: 生物安全, 生物安保与治理. 安全研究, 38(5), 29–57.
114. Bruynseels, K. Responsible Innovation in Synthetic Biology in response to Covid-19.
115. Richardson, M. D., Lemoine, P. A., Stephens, W. E., & Waller, R. E. (2020). Planning for Cyber Security in Schools: The Human Factor. Educational Planning, 27(2), 23–39.
116. Demestichas, K., Peppes, N., & Alexakis, T. (2020). Survey on Security Threats in Agricultural IoT and Smart Farming. Sensors, 20(22), 6458.
117. Mian, I. S., Twisleton, D., & Timm, D. A. (2020). What is the resource footprint of a computer science department? Place, people, and Pedagogy. Data & Policy, 2.
118. Ugya, A. Y., Hasan, D. U. B., Ari, H. A., Ajibade, F. O., Imam, T. S., Abba, A., & Hua, X. (2020). Natural freshwater microalgae biofilm as a tool for the clean-up of water resulting from mining activities. All Life, 13(1), 644–657.
119. Cano, J. J. (2018). Seguridad y ciberseguridad en los dispositivos médicos. Sistemas, (149), 55–66.
120. Fiorentino, F., Lasinio, G. J., Careddu, G., Caputi, S. S., Rossi, L., Calizza, E., & Costantini, M. L. (2020). New epilithic δ15N-based analytical protocol for classifying Nitrogen impact in Lake Bracciano. Ecological Indicators, 117, 106663.
121. Firestone, J. (2020). Application of Software Engineering Principles to Synthetic Biology and Emerging Regulatory Concerns.
122. Philp, J. (2020). 6 Digitalisation in the bioeconomy: Convergence for the bio-based industries. The Digitalisation of Science, Technology and Innovation Key Developments and Policies: Key Developments and Policies, 143.
123. Kar, D. M., Ray, I., Gallegos, J., & Peccoud, J. (2018, August). Digital signatures to ensure the authenticity and integrity of synthetic DNA Molecules. In Proceedings of the New Security Paradigms Workshop (pp. 110–122).
124. Smyth, S. J., Macall, D. M., Phillips, P. W., & de Beer, J. (2020). Implications of biological information digitization: Access and benefit sharing of plant genetic resources. The Journal of World Intellectual Property.
125. Udayanga, D., Miriyagalla, S. D., Herath, I. S., Castlebury, L. A., Ferdinandez, H. S., & Manamgoda, D. S. (2020). Foliar pathogenic fungi: growing threats to global food security and ecosystem health. Ceylon Journal of Science, 49(5).

126. Field, R., Quach, T. T., & Ting, C. (2020). Efficient Generalized Boundary Detection Using a Sliding Information Distance. IEEE Transactions on Signal Processing, 68, 6394–6401.
127. Phillipson, J. A. (2020). Active Defense and Industrial Control System Security in Precision Agriculture (Doctoral dissertation, Utica College).
128. Kar, D. M., Ray, I., Gallegos, J., Peccoud, J., & Ray, I. (2020). Synthesizing DNA molecules with identity-based digital signatures to prevent malicious tampering and enabling source attribution. Journal of Computer Security, (Preprint), 1–31.
129. Faezi, S., Chhetri, S. R., Malawade, A. V., Chaput, J. C., Grover, W., Brisk, P., & Al Faruque, M. A. (2019, January). Oligo-Snoop: a non-invasive side channel attack against DNA synthesis machines. In Network and Distributed Systems Security (NDSS) Symposium 2019.
130. Puzis, R., Farbiash, D., Brodt, O., Elovici, Y., & Greenbaum, D. (2020). Increased cyber-biosecurity for DNA synthesis. Nature Biotechnology, 1–2.
131. Wang, X. (2020). COVID-19 Epidemic and Enhancing China's National Biosecurity System. Journal of biosafety and biosecurity.
132. Lentzos, F., Goodman, M. S., & Wilson, J. M. (2020). Health security intelligence: engaging across disciplines and sectors.
133. Schumacher, G. J., Sawaya, S., Nelson, D., & Hansen, A. J. (2020). Genetic information insecurity as state of the art. bioRxiv.
134. El-Fatyany, A., Wang, H., Abd El-atty, S. M., & Khan, M. Biocyber Interface-Based Privacy for Internet of Bio-nano Things.
135. Farbiash, D., & Puzis, R. (2020). Cyberbiosecurity: DNA Injection Attack in Synthetic Biology. arXiv preprint arXiv:2011.14224.
136. Mueller, S. (2019). On DNA Signatures, Their Dual-Use Potential for GMO Counterfeiting, and a Cyber-Based Security Solution. Frontiers in bioengineering and biotechnology, 7, 189.
137. DiEuliis, D., Lutes, C. D., & Giordano, J. (2018). Biodata Risks and Synthetic Biology: A Critical Juncture. Journal of Bioterrorism & Biodefense, 9(01).
138. Frazar, S. L., Hund, G. E., Bonheyo, G. T., Diggans, J., Bartholomew, R. A., Gehrig, L., & Greaves, M. (2017). Defining the synthetic biology supply chain. Health security, 15(4), 392–400.
139. Ney, P., Koscher, K., Organick, L., Ceze, L., & Kohno, T. (2017). Computer Security, Privacy, and {DNA} Sequencing: Compromising Computers with Synthesized {DNA}, Privacy Leaks, and More. In 26th {USENIX} Security Symposium ({USENIX} Security 17) (pp. 765–779).
140. Mirsky, Y., Mahler, T., Shelef, I., & Elovici, Y. (2019). CT-GAN: Malicious tampering of 3D medical imagery using deep learning. In 28th {USENIX} Security Symposium ({USENIX} Security 19) (pp. 461–478).

References

1. I. Ahmed, *The Disinformation Dozen* (Center for Countering Digital Hate, London, 2021)
2. B. Allyn, *Study Exposes Russia Disinformation Campaign That Operated in the Shadows for 6 Years* (NPR.org, 2020, June), p. 16, Retrieved from https://www.npr.org/2020/06/16/878169027/study-exposes-russia-disinformation-campaign-that-operated-in-the-shadows-for-6-
3. S. Banerjee, T. Hemphill, P. Longstreet, Is IOT a threat to consumer consent? The Perils of wearable devices' health data exposure, in *The Perils of Wearable Devices' Health Data Exposure*, (2017)
4. S.L. Bernal, D.P. Martins, A.H. Celdrán, Distributed denial of service cyberbioattack affecting bacteria-based biosensing systems, in *17th International Conference on Electrical Engineering/Electronics, Computer, Telecommunications and Information Technology (ECTI-CON)*, (IEEE, Phuket, 2020), pp. 279–282
5. R. Bernard, G. Bowsher, R. Sullivan, F. Gibson-Fall, Disinformation and epidemics: Anticipating the next phase of biowarfare. Health Secur. **19**(1) (2021). https://doi.org/10.1089/hs.2020.0038
6. G. Bertolin, *Conceptualizing Russian Information Operations: Info-War and Infiltration in the Context of Hybrid Warfare* (IO Sphere, 2015)
7. A. Blake, *The Tempest Over DHS's Disinformation Governance Board* (The Washington Post, 2022, May 2), Retrieved from https://www.washingtonpost.com/politics/2022/04/29/disinformation-governance-board-dhs/
8. T. Boghardt, Soviet Bloc intelligence and its AIDS disinformation campaign. Stud. Intell. **53**(4), 1–24 (2009), Retrieved from https://web.archive.org/web/20100324175917/https://www.cia.gov/library/center-for-the-study-of-intelligence/csi-publications/csi-studies/studies/vol53no4/pdf/U-%20Boghardt-AIDS-Made%20in%20the%20USA-17Dec.pdf
9. S. Bond, *Just 12 People Are Behind Most Vaccine Hoaxes on Social Media, Research Shows* (NPR.org, 2021, May 14), Retrieved from https://www.npr.org/2021/05/13/996570855/disinformation-dozen-test-facebooks-twitters-ability-to-curb-vaccine-hoaxes
10. S. Bond, V. Romo, L. Wamsley, *U.S. Hospitals Targeted in Rising Wave of Ransomware Attacks, Federal Agencies Say* (NPR.org, 2020, October 29), Retrieved from https://www.npr.org/2020/10/29/928979988/u-s-hospitals-targeted-in-rising-wave-of-ransomware-attacks-federal-agencies-say
11. R. Brooks, *How Everything Became War and the Military Became Everything: Tales from the Pentagon (ISBN: 9781476777870)* (Simon & Schuster, New York, 2016)
12. D.L. Buffaloe, *Defining Asymmetric Warfare (No. 58)* (The Institute of Land Warfare, Arlington, 2006)
13. C.B. Office, *Projected Costs of U.S. Nuclear Forces, 2019–2028* (CBO, Washington, DC, 2019), Retrieved from https://www.cbo.gov/system/files/2019-01/54914-NuclearForces.pdf
14. R. Carneiro, S. Duncan, F. Ramsey, H. Seyyedhasani, R. Murch, *Cyber Attacks in Agriculture: Protecting Your Farm and Small Business with Cyberbiosecurity*, FST-387 (Virginia Cooperative Extension, Blacksburg, 2021)
15. J. Caswell, J.D. Gans, N. Generous, C.M. Hudson, E. Merkley, C. Johnson, et al., Defending our public biological databases as a global critical infrastructure. Front. Bioeng. Biotechnol. (2019). https://doi.org/10.3389/fbioe.2019.00058
16. Centers for Disease Control and Prevention, *Health Insurance Portability and Accountability Act of 1996 (HIPAA)* (Centers for Disease Control and Prevention, Atlanta, 2018), Retrieved from https://www.cdc.gov/phlp/publications/topic/hipa
17. I. Cherenko, G. Williams, *Reducing the Incidents of Phishing, Protecting the Confidentiality of HIPAA Data and, Ensuring the Availability of Critical Systems Vital to the Success of the Healthcare System Using a Layered-Defense Approach* (Metropolitan State University of Denver-Department of Computer Sciences, Denver, 2021)

18. T.C. Coglitore, *The Erosion of US Nuclear Deterrence Credibility in the 21st Century* (Air War College, Air University, Maxwell AFB United States, 2010), Retrieved from https://apps.dtic.mil/sti/citations/AD1018583
19. R. Collier, NHS ransomware attack spreads worldwide. CMAJ **189**(22), E786–E787 (2017)
20. Commission on Security and Cooperation in Europe, The Scourge of Russian Disinformation, *The Scourge of Russian Disinformation* (U.S. Helsinki Commission, Washington, DC, 2017, September 14), Retrieved from https://www.csce.gov/international-impact/events/scourge-russian-disinformation
21. Cong., 1. U., *Cybersecurity and Infrastructure Security Agency Act of 2018 H.R. 3359* (2018), Retrieved from https://www.congress.gov/bill/115th-congress/house-bill/3359
22. N.Y. Conteh, Global, M. D, The unprecedented rise in cybercrime and the role of the human vulnerability factor, in *Ethical Hacking Techniques and Countermeasures for Cybercrime Prevention*, (2021), pp. 32–43
23. J. Cox, *CDC Tracked Millions of Phones to See If Americans Followed COVID Lockdown Orders* (VICE.com, 2022, May 3), Retrieved from https://www.vice.com/en/article/m7vymn/cdc-tracked-phones-location-data-curfews
24. M. Crosas, The evolution of data citation: From principles to implementation. IASSIST Q. **37**(1–4), 62–62 (2014)
25. M. Crosas, J. Gautier, S. Karcher, D. Kirilova, G. Otalora, A. Schwartz, Data policies of highly-ranked social science journals. SocArXiv (2018)
26. D.O. Defense, *DHS-CISA Budget Overview Fiscal Year 2023 Congressional Justification* (Department of Homeland Security, Washington, DC, 2021), Retrieved from https://www.dhs.gov/sites/default/files/2022-03/Cybersecurity%20and%20Infrastructure%20Security%20Agency%20%28CISA%29_Remediated.pdf
27. Department of Homeland Security, *Fact Sheet: DHS Internal Working Group Protects Free Speech and Other Fundamental Rights When Addressing Disinformation That Threatens the Security of the United States* (DHS, Washington, DC, 2022), Retrieved from https://www.dhs.gov/news/2022/05/02/fact-sheet-dhs-internal-working-group-protects-free-speech-other-fundamental-rights
28. Department of State – Office of the Spokesperson, *Fact vs. Fiction: Russian Disinformation on Ukraine* (Department of State, Washington, DC, 2022), Retrieved from https://www.state.gov/fact-vs-fiction-russian-disinformation-on-ukraine/
29. D. DiEuliis, J. Giordano, Balancing act: Precision medicine and national security. Mil. Med. **187**(1), 32–35 (2022). https://doi.org/10.1093/milmed/usab017
30. D. DiEuliis, C.D. Lutes, J. Giordano, Biodata risks and synthetic biology: A critical juncture. J. Bioterror. Biodefense **9**(1), 2–14 (2018)
31. J.D. Douglass, N.C. Livingston, *America the Vulnerable: The Threat of Chemical and Biological Warfare. ISBN: 9780669120806* (Lexington Books, Lexington, 1990)
32. S.E. Duncan, R. Reinhard, R.C. Williams, F. Ramsey, W. Thomason, K. Lee, R. Murch, Cyberbiosecurity: A new perspective on protecting the U.S. food and agricultural system. Front. Bioeng. Biotechnol. (2019). https://doi.org/10.3389/fbioe.201900
33. E. Dwoskin, *Misinformation on Facebook Got Six Times More Clicks Than Factual News During the 2020 Election, Study Says* (Washington Post, 2021, September 4)
34. A. Einstein, Why socialism? Mon. Rev. (1949)
35. European Council, *Regulation (EU) 2016/679 of the European Parliament and of the Council of 27 April 2016 on the Protection of Natural Persons with Regard to the Processing of Personal Data and on the Free Movement of Such Data. GDPR* (European Union, Brussels, 2016, April 27), Retrieved from https://eur-lex.europa.eu/eli/reg/2016/679/oj
36. N. Evans, *Information Technology Social Engineering: An Academic Definition and Study of Social Engineering – Analyzing the Human Firewall* (Iowa State University Digital Repository, Ames, 2009). https://doi.org/10.31274/etd-190810-436

37. P. Ewing, *Why Fake Video, Audio May Not Be as Powerful in Spreading Disinformation as Feared* (NPR.org, 2020, May 7), Retrieved from https://www.npr.org/2020/05/07/851689645/why-fake-video-audio-may-not-be-as-powerful-in-spreading-disinformation-as-feare
38. B. Farley, Blending powers: Hamilton, FDR, and the backlash that shaped modern congress. J. Pol. Hist. **33**(1), 60–92 (2021)
39. H. French, *China's Second Continent (ISBN: 9780307946652)* (Random House, New York, 2015)
40. B. Fung, A. Marquardt, *Hacked Florida Water Plant Reused Passwords and Had Aging Windows Installations* (CNN, 2021, February 11), Retrieved from https://www.cnn.com/2021/02/11/us/florida-water-plant-hack/index.html
41. A.M. George, The national security implications of cyberbiosecurity. Front. Bioeng. Biotechnol. (2019). https://doi.org/10.3389/fbioe.2019.00051
42. L. Goodman, Biodefense cost and consequence. J. Clin. Investig. **114**(1), 2–3 (2004), Retrieved from https://www.ncbi.nlm.nih.gov/pmc/articles/PMC437981/
43. J. Gow, The NEw Clausewitz? War, force, art and utility – Rupert Smith on 21st century strategy, operations and tactics in a comprehensive tactics. J. Strateg. Stud. **29**(6), 1151–1170 (2006)
44. S.G. Gray, *The Dawn of a 3D Biofabrication & Biomanufacturing Metaverse* (2021, October 29), Retrieved from 3DPRINTINGMEDIA.NETWORK https://www.3dprintingmedia.network/the-dawn-of-a-3d-biofabrication-biomanufacturing-metaverse/
45. A. Grynkewich, M. Clark, G. Fogg, G. Harrigan, G. Ackerman, R.E. Burentt, et al., *On the Horizon: Security Challenges at the Nexus of State and Non-State Actors and Emerging/Disruptive Technologies* (NSI Boston United States, Boston, 2019), Retrieved from https://apps.dtic.mil/sti/citations/AD1094006
46. B. Harris, *FDA Issues New Alert on Medtronic Insulin Pump Security* (HEALTHCARE-ITNEWS.COM, 2019, July 1), Retrieved from https://www.healthcareitnews.com/news/fda-issues-new-alert-medtronic-insulin-pump-security
47. P.D. Hebert, A. Cywinska, S.L. Ball, J.R. DeWaard, Biological identifications through DNA barcodes. Proc. R. Soc. Lond. Ser. B Biol. Sci. **270**(1512), 313–321 (2003)
48. N. Hemsouth, *Study: AI Detects Backdoor-Unlocking DNA Samples* (THEREGISTER.com, 2022, February 25), Retrieved from https://www.theregister.com/2022/02/25/dna-security-healthcare/
49. E.S. Herman, N. Chomsky, *Manufacturing Consent: The Political Economy of the Mass Media* (Pantheon Books, New York, 1988)
50. R.J. Hester, Bioveillance: A techno-security infrastructure to preempt the dangers of informationalised biology. Sci. Cult. **153–176** (2020). https://doi.org/10.1080/09505431.2019.1705270
51. K. Hooper, *Mayorkas Cites Misinformation About Homeland Security's Disinformation Board* (Politico, 2022, May 01), Retrieved from https://www.politico.com/news/2022/05/01/mayorkas-defends-dhs-disinformation-board-00029182
52. M. Ibrahim, T.C. Liang, K. Scott, K. Chakrabarty, R.I. Karri, Molecular barcoding as a defense against benchtop biochemical attacks on DNA fingerprinting and information forensics. IEEE Trans. Inf. Forensics Secur. **15**, 3595–3609 (2020)
53. J.P. Tuohy, *Researchers Find Amazon Uses Alexa Voice Data to Target You with Ads* (THEVERGE.com, 2022, April 28), Retrieved from https://www.theverge.com/2022/4/28/23047026/amazon-alexa-voice-data-targeted-ads-research-report
54. N. Johnson, *A New Study Explores the Spread of Misinformation About Coronavirus on Facebook* (A. Chang, Interviewer) (NPR.org, 2020, May 13), Retrieved from https://www.npr.org/2020/05/13/855611088/a-new-study-explores-the-spread-of-misinformation-about-coronavirus-on-facebook
55. S.B. Jordan, S.L. Fenn, B.B. Shannon, Transparency as threat at the intersection of artificial intelligence and cyberbiosecurity. Computer **53**(10), 59–68 (2020)

56. E.K. Keller, D.E. Warren, N.K. Hayden, H.D. Passell, L.A. Malczynski, G.A. Backus, *Nuclear Security Futures Scenarios* (Sandia National Labs, Albuquerque, 2017), Retrieved from https://web.archive.org/web/20201106231727id_/https://prod-ng.sandia.gov/techlib-noauth/access-control.cgi/2017/170913.pdf
57. J. Kim, S. Kim, H.-M. Cho, J.H. Chang, S.Y. Kim, Data sharing policies of journals in life, health, and physical sciences indexed in Journal Citation Reports. PeerJ **eCollection**, e9924 (2020). https://doi.org/10.7717/peerj.9924
58. A. Kumar, K. Sharma, H. Singh, S.G. Naugriya, S.S. Gil, R. Buyya, A drone-based networked system and methods for combating coronavirus disease (COVID-19) pandemic. Futur. Gener. Comput. Syst. **115**, 1–19 (2021)
59. S. Lesaja, X.L. Palmer, Brain-computer interfaces and the dangers of neurocapitalism. arXiv Preprint, arXiv:2009.07951 (2020)
60. L. Lessig, *Free Culture (9781594200069)* (Penguin Books, New York, 2004)
61. L. Liverpool, Researchers from Global South under-represented in development research. Nature (2021). https://doi.org/10.1038/d41586-021-02549-9
62. J.L. Mantle, J. Rammohan, E.F. Romantseva, J.T. Welch, L. Kauffman, J. McCarthy, K.H. Products, Cyberbiosecurity for biopharmaceuticals. Front. Bioeng. Biotechnol. (2019). https://doi.org/10.3389/fbioe.2019.00116
63. M. Martin, *Far-Right Misinformation Is Thriving on Facebook. A New Study Shows Just How Much* (NPR.org, 2021, March 6)
64. L.C. Meiser, P.L. Antkowiak, J. Koch, W.D. Chen, A.X. Kohll, W. Stark, R.N. Grass, Reading and writing digital data in DNA. Nat. Protoc. **15**(1), 86–101 (2020)
65. S. Monteith, M. Bauer, M. Alda, J. Geddes, P. Whybrow, T. Glenn, Increasing cybercrime since the pandemic: Concerns for psychiatry. Curr. Psychiatry Rep. **23**(4), 1–9 (2021)
66. S. Mueller, Facing the 2020 pandemic: What does cyberbiosecurity want us to know to safeguard the future? Biosaf. Health, 11–21 (2021). https://doi.org/10.1016/j.bsheal.2020.09.007
67. V. Nathanson, Bioweapons: Usable weapons are technically easier to produce now, but we lack legal protection against them. BMJ **325**(7367), 727–728 (2002)
68. P. Ney, K. Koscher, L. Organick, L. Ceze, T. Kohno, Computer security, privacy, and DNA sequencing: Compromising computers with synthesized dna, privacy leaks, and more. *26th Usenix Security Symposium* (Usenix Association, 2017), Retrieved from https://www.usenix.org/conference/usenixsecurity17/technical-sessions/presentation/ney
69. Occupational Safety and Health Act of 1970, *Public Law 91–596 [S. 2193], (29 U.S.C., Chapter 15)* (U.S., Washington, DC, 1970, December 29)
70. A. Osborn, P. Nikolsaya, *Russia's Putin Authorises 'Special Military Operation' Against Ukraine* (Reuters, 2022, February 24), Retrieved from https://www.reuters.com/world/europe/russias-putin-authorises-military-operations-donbass-domestic-media-2022-02-24/
71. X. Palmer, L. Potter, Biocyberwarfare and crime: A juncture of rethought, in *ECCWS 2021 20th European Conference on Cyber Warfare and Security*, ed. by T. Eze, (Academic Conferences International, 2021)
72. X.L. Palmer, L. Potter, S. Karahan, On the emerging area of biocybersecurity and relevant considerations, in *Future of Information and Communication Conference*, ed. by K. Arai, S. Kapoor, R. Bhatia, (Springer, Cham, San Francisco, 2020), pp. 873–881
73. X.L. Palmer, E. Powell, L. Potter, Matters of biocybersecurity with consideration to propaganda outlets and biological agents, in *European Conference on Cyber Warfare and Security* Academic Conferences International Limited, pp. 525–XIV (2021)
74. X.L. Palmer, E. Powell, L. Potter, Biocyberwarfare and crime: A juncture of rethought, in *European Conference on Cyber Warfare and Security*. Academic Conferences International Limited, pp. 517–XIV (2021)
75. X. Palmer, L.N. Potter, S. Karahan, COVID-19 and biocybersecurity's increasing role on defending forward. Int. J. Cyber Warf. Terror **11**(3), 15–29 (2021)
76. X.-L. Palmer, E. Powell, L. Potter, Matters of biocybersecurity with consideration to propaganda outlets and biological agents, in *20th European Conference on Cyber Warfare and Security (ECCWS 2021)*, ed. by T. Eze, L. Speakman, C. Onwubiko, (Academic Conferences International, Chester, 2021), pp. 525–533

77. X.L. Palmer, L. Potter, S. Karahan, An exploration on APTs in biocybersecurity and cyberbiosecurity. Int. Conf. Cyber Warf. Secur. **17**(1), 532–535 (2022)
78. A. Pattani, *For Health Care Workers, the Pandemic Is Fueling Renewed Interest in Unions* (NPR.org, 2021, January 11), Retrieved from for Health Care Workers, the Pandemic Is Fueling Renewed Interest in Unions
79. C. Paul, M. Matthews, *The Russian "Firehose of Falsehood" Propaganda Model* (RAND Corporation, Santa Monica, 2016), Retrieved from https://www.rand.org/pubs/perspectives/PE198.html
80. J. Peccoud, J. Gallegos, R. Murch, W. Buchholz, S. Raman, Cyberbiosecurity: From naive trust to risk awareness. Trends Biotechnol. **36**(1), 4–7 (2018). https://doi.org/10.1016/j.tibtech.2017.10.012
81. D. Perkins, E. Fabregas, *Mitigating Insider Threats Through Strengthening Organizations' Culture of Biosafety, Biosecurity, and Responsible Conduct* (Department of Health and Human Services, Washington, DC, 2017)
82. M. Ploumis, AI weapon systems in future war operations: Strategy, operations, and tactics. Comp. Strateg. **41**(1), 1–18 (2022)
83. L. Potter, X.L. Palmer, Human factors in biocybersecurity wargames, in *Future of Information and Communication Conference*, ed. by K. Arai, (Springer, Cham, Vancouver, 2021), pp. 666–673
84. L. Potter, O. Ayala, X.L. Palmer, Biocybersecurity – A converging threat as an auxiliary to war. arXiv, arXiv:2010.00624 (2020)
85. L. Potter, X. Palmer, K. Saltuk, *Biocybersecurity and Applications of Predictive Physiological Modeling*. Manuscript Submitted (2022)
86. L. Potter, S. Shetty, S. Karahan, X.–L. Palmer, Biocybersecurity and applications of predictive physiological modeling. J. Syst. Eng., Submitted Publication
87. M.K. Powell, *The Digital Broadband Migration: Toward a Regulatory Regime for the Internet Age. Remarks of Michael K. Powell at the Silicon Flatirons Symposium* (University of Colorado School of Law, Bolder, 2004), Retrieved from https://docs.fcc.gov/public/attachments/DOC-243556A1.pdf
88. R. Puzis, D. Farbiash, O. Brodt, Y. Elovici, D. Greenbaum, Increased cyber-biosecurity for DNA synthesis. Nat. Biotechnol. **38**, 1379–1381 (2020). https://doi.org/10.1038/s41587-020-00761-y
89. Qu1nlan, *Managing Misinformation on Reddit* (L. Garcia-Navarro, Interviewer, 2019, December 8)
90. RAND Corporation, *Truth Decay* (2022, May 1), Retrieved from RAND.org https://www.rand.org/research/projects/truth-decay/about-truth-decay.html
91. J.C. Reed, N. Dunaway, Cyberbiosecurity implications for the laboratory of the future. Front. Bioeng. Biotechnol. **7** (2019). https://doi.org/10.3389/fbioe.2019.00182
92. D.B. Resnik, D. Patrone, S. Peddada, Research misconduct policies of social science journals and impact factor. Account. Res. Pol. Qual. Assur. **17**(2), 79–84 (2010). https://doi.org/10.1080/08989621003641181
93. L.C. Richardson, S.M. Lewis, R.N. Burnette, Building capacity for cyberbiosecurity training. Front. Bioeng. Biotechnol. **7** (2019a). https://doi.org/10.3389/fbioe.2019.00112
94. L.C. Richardson, N.D. Connell, S.M. Lewis, E. Pauwels, R.S. Murch, Cyberbiosecurity: A call for cooperation in a new threat landscape. Front. Bioeng. Biotechnol. (2019b). https://doi.org/10.3389/fbioe.2019.00099
95. J. Riggi, *Ransomware Attacks on Hospitals Have Changed* (2021, October 9), Retrieved from AHA.org https://www.aha.org/center/cybersecurity-and-risk-advisory-services/ransomware-attacks-hospitals-have-changed
96. F.D. Roosevelt, *A Second Bill of Rights* (U.S., Washington, DC, 1944, January 11), Retrieved from https://fdrfoundation.org/a-second-bill-of-rights-video/
97. I.A. Samori, X.L. Palmer, L. Potter, S. Karahan, Commentary on biological assets cataloging and AI in the Global South, in *Intelligent Systems and Applications: Proceedings of the 2022 Intelligent Systems Conference (IntelliSys) Volume 3*, (Springer, Cham, 2022), pp. 734–744

98. K.K. Santos, P.D. Santos, Cyber-biosecurity risk perceptions in the biotech sector. Front. Bioeng. Biotechnol. **7**, 136 (2019)
99. D.S. Schabacker, L.A. Levy, N.J. Evans, J.M. Fowler, E.A. Dickey, Assessing cyberbiosecurity vulnerabilities and infrastructure resilience. Front. Bioeng. Biotechnol. (2019). https://doi.org/10.3389/fbioe.2019/00128
100. D.G. Schmale, A.P. Ault, W. Saad, D.T. Scott, J.A. Westrick, Perspective on harmful algal blooms (HABs) and the cyberbiosecurity of freshwater systems. Front. Bioeng. Biotechnol. (2019). https://doi.org/10.3389/fbioe.2019/00128
101. S.I. Schwartz, D. Choubey, *Nuclear Security Spending: Assessing Costs, Examining Priorities* (Carnegie Endowment for International Peace, Washington, DC, 2009), Retrieved from https://carnegieendowment.org/files/nuclear_security_spending_low.pdf
102. A. Seitz, *Disinformation Board to Tackle Russia, Migrant Smugglers* (AP News, 2022, April 28), Retrieved from https://apnews.com/article/russia-ukraine-immigration-media-europe-misinformation-4e873389889bb1d9e2ad8659d9975e9d
103. S. Sivagnanam, Bioweapons. BMJ **325** (2002). https://doi.org/10.1136/bmj.325.7367.727
104. E. Strickland, M. Harris, *Their Bionic Eyes Are Now Obsolete and Unsupported* (IEEE Spectrum, 2022, February 15), Retrieved from https://spectrum.ieee.org/bionic-eye-obsolete
105. C.R. Sunstein, We need to reclaim the second bill of rights. Chron. High. Educ. **50**(40), B9–B10 (2004)
106. S.K. Tabatabaei, B. Wang, N.B. Athreya, B. Enghiad, A.G. Hernandez, C.J. Fields, O. Milenkovic, DNA Punch Cards for storing data on native DNA sequences via enzymatic nicking. Nat. Commun. **11**(1), 1–10 (2020)
107. United Nations General Assembly, Treaty on the prohibition of nuclear weapons, in *Treaty on the Prohibition of Nuclear Weapons*, (United Nations, New York, 2017)
108. United States Department of Defense, *US DOD 2018 Cyber Strategy Summary* (United States Department of Defense, Washington, DC, 2018), Retrieved from https://media.defense.gov/2018/Sep/18/2002041658/-1/-1/1/CYBER_STRATEGY_SUMMARY_FINAL.PDF
109. United States Senate, *About Declarations of War by Congress* (2022), Retrieved from SENATE.gov https://www.senate.gov/about/powers-procedures/declarations-of-war.htm
110. F. Urbina, F. Lentzos, C. Invernizzi, S. Ekins, Dual use of artificial-intelligence-powered drug discovery. Nat. Mach. Intell. **4**, 189–191 (2022). https://doi.org/10.1038/s42256-022-00465-9
111. US Const (Amend I-X), *Constitution* (U.S. Congress, Philadelphia, 1787)
112. D. Vine, *The Pentagon's New Generation of Secret Military Bases* (2012, July 16), Retrieved from Mother Jones.com https://www.motherjones.com/politics/2012/07/pentagon-new-generation-military-bases-tom-dispatch/
113. K.M. Westfall, T.W. Therriault, C.L. Abbott, A new approach to molecular biosurveillance of invasive species using DNA metabarcoding. Glob. Chang. Biol. **26**(2), 4240448 (2020)
114. B. Wolford, *Everything You Need to Know About the "Right to Be Forgotten"* (2022), Retrieved from GDPR.eu https://gdpr.eu/right-to-be-forgotten/
115. M. Zedong, J. DeFrancis, *Annotated Quotations from Chairman Mao* (Yale University Press, New Haven, 1975)
116. E. Zuckerman, *Facebook Is Blocking Access to Data About How Much Misinformation It Spreads and Who Is Affected* (NIEMANLAB.org, 2021, November 2)

Revisiting the Digital Biosecurity Landscape

Diane DiEuliis

Abstract The world is currently experiencing a revolution in the biological sciences, fostered not only by advances in the fundamental understanding of biology but in the tools of the technologies that enable such understanding. Advances in biotechnology have been iterative for several decades, but in the past decade, a convergence of biotechnology and computational and information technologies has created a unique convergence. Information, in the form of biological data (or "biodata"), now has the potential to drive real, physical biological outcomes – in the form of novel organisms, tools for biological manufacturing, and the creation of novel products made by biology or possessing unique biological characteristics. This rich landscape of innovation presents a special challenge for traditional biosecurity – how best to secure and ensure that biodata and the tools for digital manipulation of biology are not used for harm? This chapter revisits a discussion of "digital biosecurity" as a novel form of biosecurity and is a prelude to a set of examinations throughout this volume of how best to secure the digital future of biotechnology.

Keywords Digital biosecurity · Bioeconomy · Synthetic biology · Digitization · Biodata

1 Introduction

The world is currently experiencing a true revolution in the biological sciences; in fact some have described innovations in emerging biotechnologies as part of the fourth industrial revolution [15]. The ability to make groundbreaking research discoveries and use biology to create sustainable products for the economy and environment is made possible by the digitization of biology [1]; as genetic code is embedded into digital ones and zeros, it enables the application of the classic

D. DiEuliis (✉)
National Defense University, Washington, DC, USA
e-mail: diane.dieuliis.civ@ndu.edu

D. Greenbaum (ed.), *Cyberbiosecurity*, https://doi.org/10.1007/978-3-031-26034-6_5

"design, build, and test" cycle of engineering to be applied directly to the function of biological organisms. Through in silico design, scientists can alter genetic code and gene expression for health and medicine, food, and agriculture or construct genetic circuits that can create high-value compounds for varied industrial sectors, including novel materials, chemicals with exceptional properties, sustainable food products, and biological fuels, to name a few [2].

The emergence of such a capability has clear dual-use implications, and as a novel component of biosecurity – commonly described as digital biosecurity or cyberbiosecurity [18] – it is currently not well covered by existing biosecurity policies or controls. Over the past several years, attempts to define [17] this aspect of biosecurity have been initiated [5], generating much-needed awareness [24] to this blind spot in the biosecurity realm. As the bioeconomy grows, so does the awareness of the magnitude of risks and vulnerabilities associated with biodata [9] and its usage. But the sheer scope and complexity of digital biosecurity have made structured policy and governance action difficult, for several reasons: the data is essential for benefits to be realized, and the types of data vary, as do the institutions which gather such data across academia, industry, nonprofits, and government. Another complication is the tremendous variability in the biodata user base: biologists, physicians, engineers, physicists, software designers, manufacturers, etc. – who may or may not have an awareness of digital biosecurity risks (or, in fact, traditional biosecurity risks). As the following chapters in this volume attest, there are many hands touching the elephant, and each offers original and valuable perspectives across digital biosecurity interpretations and needs. This introductory chapter is devoted to the identification of important overarching themes that help to better see the whole elephant and is intended to more finely tune existing interpretations of digital biosecurity. More importantly, given the need to protect the tremendous benefits offered by the digital bioeconomy, it provides a balanced consideration of "what are we protecting against" to educate best practices and feasible, sensible governance in the digital biosecurity realm.

2 The Landscape of Digital Biosecurity: From Data to Systems

Different definitions [23] have emerged to describe the digital aspects of biosecurity, depending on the definer and their specific purpose. To get beyond these disparities, envisioning the broadest scope of concern ensures all interpretations, and needs can be included. In attesting to the speed of advances in biotechnology, an attempt to define this broad landscape [5] over a year ago already requires additional refinement here.

The basic currency of biotechnology is genomic data and its associated metadata across all aspects of the life sciences; this genomic data varies across categories and uses, and as such, the relative digital risk associated with each differs. A simple, three-tiered breakdown is still relevant, however, and includes *organism biodata*, *human biodata*, and *operational data, devices*, and *systems*. Organism data serves as a primary driver of the bioeconomy, as it provides "chasses" for bioengineering, for example, the creation of biological circuits, parts, and resultant high-value chemicals, designer organisms, or bioproducts. A large component of organism data resides in shared public resources [16], long supported by *government efforts* genebank [21]. It should be readily acknowledged that datasets have uneven quality, integrity, and annotation – and thus not all biodata is of equal value or in need of strong protections. In terms of the progress and collaborative advancement of biotechnology, it is still not clear which data should be broadly shared to maximize bioeconomic benefits vs that which should be more stringently protected. Open science initiatives have sought to ensure broader public availability, and efforts are expanding to amass evermore increasing data on the totality of Earth's genomic biodiversity [10], including virus pathogens [12]. To ensure better quality, many private entities are creating their own internal proprietary genomic databases that form the basis for their biodesigns (to create either designer organisms or cellular–/cell-free systems for industrial bioproduction platforms or novel bioproducts [13]). These databases and associated biodesigns represent risk targets for espionage as countries compete [3] for economic dominance in the global bioeconomy. Countries are still determining how biodata should be governed and shared [19] given their potential economic value, but already it is clear that disparities exist [25] in openness and sharing, creating asymmetric access to biodata internationally. This pertains to all genomic data but biomedical data in particular. This lack of reciprocity puts the US bioeconomy at a competitive disadvantage if foreign datasets are denied to the USA, while those same foreign entities can freely access US datasets. Similar disadvantage exists if US bioeconomy firms are forced to give datasets to foreign firms in return for doing business abroad or following investments or acquisition by foreign entities. While these issues fall within the development of science and technology business norms and diplomacy, certainly the tools of digital biosecurity have a role to play in their outcomes and in future protections for biodata and biodesign writ large.

A special carve out of organism genomic data is that of infectious pathogens, whose sequence data is vital for global biosurveillance, tracking outbreaks, and designing diagnostics and medical countermeasures. The dual-use implication for this subset of genomic data and related information hazards [25] is that it could be used to design pathogens for purposeful harm [8], bioweapons or even create extinct or novel pathogens from scratch [6]. These traditional risks of biosecurity are indirect risks associated with possession of biodata – that is, there are many more downstream steps to developing a bioweapon, and genomic data represents only the first step.

We have long recognized that human genomic data and associated metadata also represent a special category, and potential benefits of its use only continue to grow in public health and medicine. Most broadly, human biodata provides insights into underlying causes of disease syndromes and reveals targets for precision medicines, as well as revealing familial inheritance and risk factors for disease occurrence. Increasingly, the understanding of how pathogens cause disease is becoming inseparable from the human genomic response during infection, and this type of human biodata reveals individual strengths and susceptibilities to infections, as was demonstrated [22] during the COVID-19 pandemic. In the past, this risk would be relegated to human privacy concerns – but the tools of biotechnology have advanced such that the same biodata can be used to design "precision maladies," to target individuals or groups, or cause physical human harms. Further, human biodata underpins human performance augmentation through biotechnology and human-machine interfaces; as healthy individuals pursue enhanced physical and cognitive capabilities, the bioeconomy is also now serving up designer probiotics and other products intended to go beyond rehabilitative remedies. These also have dual-use implications for militaries [7] and democracies, particularly those which can alter human neurobiology, [4] behavior, or decision-making. The latter is increasingly referred to as a specialized "neuroeconomy" [11] and deserves critical analysis for how neurodata should be best protected [14]. Interestingly, some countries have already enacted governance that specifically pertains to neurodata.

The third category of data may be considered "operational." The convergence of biotechnology and cyber technology extends to cyber-physical systems of several different types in the life sciences. Devices which can monitor, assess, or control biological health (of humans, animals, or crops) can rely on biodata collected from the body, livestock animals, or even a field of wheat so that appropriate medical or environmental treatment can be applied or adjusted as needed. Such devices not only constantly collect and compile biodata that could be hacked or exploited, but they utilize software and hardware that could be vulnerable to attack. Cyber-physical systems also extend to research laboratories, where experimentation is becoming increasingly automated. The same is true for biological manufacturing platforms in the bioeconomy – all these systems feature cyber vulnerabilities.

3 What Is Cybersecurity and What Is Biosecurity?

At the outset of discussions on "digital biosecurity" or "cyberbiosecurity," information technology professionals assured that these concerns were essentially everyday cybersecurity issues no different from that experienced in other sectors. In terms of cyber-physical systems, this is largely true, particularly in terms of the remedies that should be applied, such as regular improvements in good cybersecurity hygiene. However, the downstream outcomes of cyber breaches in the life sciences may be adverse or harmful biological events – outcomes that traditional biosecurity policies were intended to mitigate through physical security ("gates, guns, and guards"). These physical biosecurities are becoming less relevant in the digital age where controlled access to physical pathogens no longer prevents their potential dual-use exploitation. In this context, traditional cybersecurity applies not only to traditional protection of biodata databases but to the protection of the entire bioeconomic enterprise. In addition to biotech datasets, there are softwares, algorithms, and cloud networks which allow the use of datasets, as well as automation that occurs in research laboratories, industry settings, manufacturing platforms, and supply chains. In short, anywhere the bioeconomy intersects with information technology could be a point of cyber risk.

The distinctions between cybersecurity and biosecurity have been examined in a recent study by the National Academies of Sciences, entitled Safeguarding the Bioeconomy [20]. It describes risks stemming from harmful use of biodata and recommends the use of best practices to secure information systems from digital intrusion, exfiltration, or manipulation.

It also recommended the creation of, and participation in, a bioeconomy Information Sharing and Analysis Center (ISAC) for members of the bioeconomy. Since the publication of the report, a new ISAC has been initiated, making a significant stride forward in addressing digital biosecurity risk. More discussion of these distinctions is detailed in this volume.

4 What Is the Adversary's Intent: Dual Use vs. Direct or Kinetic Use?

Given that the life sciences are, for the most part, open and foster an environment of sharing and collaboration, the development of feasible and enforceable biosecurity policies must always first address, "what are we protecting against"? The benefits of public health and medicine, abundant agriculture, and a robust bioeconomy are seen to far outweigh risks of unwanted outcomes by many who work in the life sciences; in keeping with that, many are unaware of potential risks hidden in beneficial life science endeavors. Thus, an open and transparent view of feasible or demonstrated risks is necessary and promotes norms of responsibility in the life sciences and the development of trusted risk/benefit analyses and enables risk mitigation solutions and protections that can be shared. A good to approach risk discussions is to base them on a realistic estimation of an adversary's intent. Many risk discussions are dominated by "science fiction," media hype, or worst-case scenarios, which can contribute to a lack of serious consideration by life science practitioners. However, when presented with realistic potential risks, discussions can become more normalized and generate shared solutions. This approach also affords the provision of "hardening" critical infrastructure – which could be seen as a deterrent to would-be cybercriminals. Risks could vary across a wide range of dual uses or kinetic vs non-kinetic outcomes – from the attempt to develop bioweapons or human "precision maladies" to the targeting of manufacturing facilities for sabotage, espionage, economic gain, or public health infrastructure damage. Figure 1 below represents just a snapshot of various potential adversary goals and how they might be canvassed for just such discussions. Forums for identifying cybersecurity risks should thus include consideration of harmful biological outcomes in addition to traditional cybersecurity concerns.

Critical infrastructure at risk:	Beneficial intent	Dual Use (potential) harms	Cyber intrusions with kinetic harms	Cyber intrusions with harmful biological outcomes
Organisms' biodata	Research, Animal Modeling; public health; Bioeconomy; biodesign	IP data theft; Altered, recreated or novel pathogens; bioweapons;		
Human Genomic Data	Research; Precision Medicine; Human performance augmentation	Precision Maladies (individuals or groups); False forensics; Counterintelligence; Human performance degradation		
Human Devices	Precision Medicine; Human performance augmentation	Precision Maladies based on stolen data from devices	Corrupt wearables; impacts to military operations	Hack wearables to harm or control users; includes military users
Operational Data, devices and platfoms	Automated & accelerated laboratory research; Reproduceability; Standards setting; Biomanufacture; Biodesign; ML/AI	IP Theft, espionage	Industrial Sabotage: Disrupt facilities or hold hostage; Corrupt or disrupt supply chains.	Tamper with biological products; Compromise medical capabilities; Cause environmental spill;

Fig. 1 Depicts the different types of "data as critical infrastructure" in column 1. Each row across then depicts the features associated with that type of data infrastructure to include: benefits, dual use, potential kinetic cyber harms, and harms with potentially harmful biological outcomes

5 Conclusions

This special volume devoted to cyberbiosecurity offers timely discussions of the themes highlighted in this introduction. Its authors cover topics related to all three areas of the digital biosecurity landscape: vulnerabilities across the life cycle of molecular biodata and human biodata and across the most prominent types of robots used in the operation of a biological laboratory.

Other authors tease out descriptions of bio- vs cyber- vs information security and how experience-tested knowledge and applications pulled from these well-established fields could accelerate adoption of digital biosecurity in the bioeconomy. Included is an examination of the role of artificial intelligence – not only the ways in which data science is improving genetics and how that could lead to potential weaponization but also how AI-led cyberthreats could disrupt biodata and systems crucial to bio-medicine and biotechnology, disrupting operations.

Several important tools are featured as well: the introduction of a vulnerability scoring system that quantifies the risk and impact of vulnerabilities in digital systems and a case study demonstration of the need for better quality assurance for biological software. Authors also look at digital biosecurity through the adversarial lens – from criminal actions using synthetic biology to actions which could negatively impact freedom and equity for humans.

It is noted that while many federal departments and agencies play a role in cyberbiosecurity, none have primary responsibility for its coordination, and challenges to integration at the organizational level are discussed in the interest of clarifying ways forward.

Overall, the contributions offered here provide a host of insights and recommendations that can begin to move digital biosecurity from theory and description to reality and solutions.

References

1. N. Bajema, D. DiEuliis, Y. Lim, C. Lutes, The Digitization of Biology: Understanding the New Risks and Implications for Governance. National Defense University report, July, 2018. The Digitization of Biology: Understanding the New Risks and Implications for Governance. Center for the Study of Weapons of Mass Destruction. Publication View (ndu.edu)
2. M. Chui, M. Evers, J. Manyika, A. Sheng, T. Nisbet, The bio revolution: Innovations transforming economies, societies, and our lives. McKinsey Global Institute Report, May, 2020. The Bio Revolution: Innovations transforming economies, societies, and our lives | McKinsey
3. Congressional Research Service Report. The bioeconomy: A primer. Updated September 19, 2022. https://crsreports.congress.gov/product/pdf/R/R46881. Accessed 28 Sept 2022
4. J. DeFranco, D. DiEuliis, J. Giordano, Redefining Neuroweapons: Emerging Capabilities in Neuroscience and Neurotechnology. PRISM 8, No. 3. NDU Press, January 2022. https://ndupress.ndu.edu/Portals/68/Documents/prism/prism_8-3/prism_8-3_DeFranco-DiEuliis-Giordano_48-63.pdf
5. D. DiEuliis, Parsing the digital biosecurity landscape. Georgetown J. Int. Affairs. **21,** 166–172 (2020). Project MUSE, https://doi.org/10.1353/gia.2020.0031. Project MUSE - Parsing the Digital Biosecurity Landscape (jhu.edu)
6. D. DiEuliis, K. Berger, G. Gronvall, Biosecurity implications for the synthesis of horsepox, an Orthopoxvirus. Health Sec. **15**, 629–637. (2017). https://doi.org/10.1089/hs.2017.0081Earth
7. D. DiEuliis, J. Giordano, Balancing act: precision medicine and national security. Mil. Med. **187**(Supplement_1), 32–35 (2022). https://doi.org/10.1093/milmed/usab017
8. D. DiEuliis, J. Giordano, Gene editing using CRIPSR/Cas9: Implications for dual-use and biosecurity. Protein Cell **9**(3), 239–240 (2018). https://doi.org/10.1007/s13238-017-0493-4. https://link.springer.com/content/pdf/10.1007/s13238-017-0493-4.pdf
9. D. DiEuliis, C.D. Lutes, J. Giordano, Biodata Risks and Synthetic Biology: A Critical Juncture. J. Bioterror. Biodef. **9**: 159 (2018). https://doi.org/10.4172/2157-2526.1000159
10. Earth Biogenome Project. https://www.earthbiogenome.org/. Accessed 28 Sept 2023
11. J. Giordano, J. DeFranco, D. DiEuliis, Neurodata & defence: Part I – Realities and risks. Defence IQ. 2020. Accessed 30 Sept 2022. https://www.defenceiq.com/defence-technology/articles/neurodata-defence-part-i-realities-and-risks
12. Global Virome Project. https://www.globalviromeproject.org/. Accessed 28 Sept 2023
13. D. Grushkin, What is biodesign? Issues in Science and Technology, NASEM, Arizona. June 2021. https://issues.org/biodesign-challenge-synthetic-biology-grushkin/

14. D. Hallinan, P. Schultz, M. Friedwald, P. deHert, Neurodata and neuroprivacy: Data protection outdated? Surveill. Soc. **12**(1), 55 (2014). https://doi.org/10.24908/ss.v12i1.4500
15. T.X. Hammes, Technological change and the fourth industrial revolution. Schultz Beyond Disruption. (2018). https://www.hoover.org/sites/default/files/research/docs/beyonddisruption_chapter_2.pdf
16. J. Hutchins, Genome plasticity in health and disease, in *Translational and Applied Genomics*, (2020), pp. 47–62. https://doi.org/10.1016/B978-0-12-817819-5.00004-8
17. R. Murch, D. DiEuliis, Editorial: Mapping the cyberbiosecurity enterprise. Front. Bioeng. Biotechnol. **7**, 235 (2019). https://doi.org/10.3389/fbioe.2019.00235
18. R.S. Murch, W.K. So, W.G. Buchholz, S. Raman, J. Peccoud, Cyberbiosecurity: An emerging new discipline to help safeguard the bioeconomy. Front. Bioeng. Biotechnol. **6**, 39 (2018). https://doi.org/10.3389/fbioe.2018.00039
19. Nagoya protocol on access to genetic resources and the fair and equitable sharing of benefits arising from their utilization to the convention on biological diversity, text and annex. 2011. https://www.cbd.int/abs/doc/protocol/nagoya-protocol-en.pdf
20. National Academies of Sciences, Engineering, and Medicine, *Safeguarding the Bioeconomy* (The National Academies Press, Washington, DC, 2020). https://doi.org/10.17226/25525
21. National Library of Medicine genome database list. Genome List – Genome – NCBI (nih.gov). Accessed 28 Sept 2022
22. B.B. Oude Munnink, N. Worp, D.F. Nieuwenhuijse, R.S. Sikkema, B. Haagmans, R.A.M. Fouchier, M. Koopmans, The next phase of SARS-CoV-2 surveillance: Real-time molecular epidemiology. Nat. Med. **27**(9), 1518–1524 (2021). https://doi.org/10.1038/s41591-021-01472-w. Epub 2021 Sep 9. Erratum in: Nat Med. 2021 Nov;27(11):2048. PMID: 34504335
23. J. Peccoud, J.E. Gallegos, R. Murch, W.G. Buchholz, S. Raman, Cyberbiosecurity: From naive trust to risk awareness. Trends Biotechnol. **36**(1), 4–7 (2018). https://doi.org/10.1016/j.tibtech.2017.10.012
24. L.C. Richardson, N.D. Connell, S.M. Lewis, E. Pauwels, R.S. Murch, Cyberbiosecurity: A call for cooperation in a new threat landscape. Front. Bioeng. Biotechnol. **7**, 99 (2019). https://doi.org/10.3389/fbioe.2019.00099
25. E. Yasinski, China clamps down on foreign use of Chinese genetic material and data. News and opinion, The Scientist. June 17, 2019. https://www.the-scientist.com/news-opinion/china-clamps-down-on-foreign-use-of-chinese-genetic-material-and-data-66016

Security Vulnerabilities and Countermeasures for the Biomedical Data Life Cycle

Eric Ni, Gamze Gürsoy, and Mark Gerstein

Abstract The biomedical data life cycle starts from data collected from human subjects and encompasses its annotation, storage, dissemination, analysis, and transformation into insights for human health. The rapid digitalization of health data and a movement toward personalized medicine have caused this data to grow tremendously in the last decade. Consequently, the security and privacy of biomedical data have become a growing concern throughout its entire life cycle, as health data inherently contains sensitive information from patients.

In this chapter, we provide a broad overview of the types of security vulnerabilities that exist in each stage of the biomedical data life cycle. We cover defenses, both against attacks from bad actors and from accidental leakage of private information. Finally, we conclude with future perspectives and best practices going forward for this quickly evolving field of cyberbiosecurity.

Keywords Bioinformatics · Databases · Privacy · Security

1 Introduction

The last decade has shown explosive growth in the number of biomedical databases, including personal genomes, electronic health records, and wearable device signals. Accordingly, we see numerous clinicians, biologists, and informaticians interpret these data, publish their findings, and build them into tools. These have resulted in many web applications for accessing, uploading, and computing data and improved production pipelines such as DNA or protein synthesis. Many advancements regard-

E. Ni · M. Gerstein (✉)
Program in Computational Biology and Bioinformatics, Yale University, New Haven, CT, USA

Department of Molecular Biophysics and Biochemistry, Yale University, New Haven, CT, USA
e-mail: mark.gerstein@yale.edu

G. Gürsoy
Department of Biomedical Informatics, Columbia University, New York, NY, USA

New York Genome Center, New York, NY, USA

D. Greenbaum (ed.), *Cyberbiosecurity*, https://doi.org/10.1007/978-3-031-26034-6_6

ing this movement of data are focused on storage, transformation, and distribution, but security issues are less well understood.

In this chapter, we give an overview of the life cycle for human biomedical data, from the acquisition of physical samples into a database to data dissemination, to data analysis and product development, and ultimately to how these products are used by consumers. We describe how cyberbiosecurity vulnerabilities exist in each step of this life cycle and how techniques used in bioinformatics and computational biology can help to protect the data from adversaries. Securing the privacy of patient data is especially important, in order to maintain public trust in biomedical research.

Regarding the storage of biomedical data, data integrity is needed, since errors early on can create cascading effects downstream as that data is copied, analyzed, and incorporated into other's research. For example, a mislabeled sequence in a database could lead to incorrect analysis in a pathogen classification algorithm, leading to the synthesis of harmful DNA and proteins [1].

Machine learning methods are vital to the field of bioinformatics for parsing the velocity of data produced, but researchers should consider their potential malicious usage. A well-crafted adversarial attack can directly target machine learning models, leading to wrong interpretations or disguising malicious content. Furthermore, simple or non-robust models may leak training data, which, if containing sensitive information, poses a risk to privacy (Fig. 1).

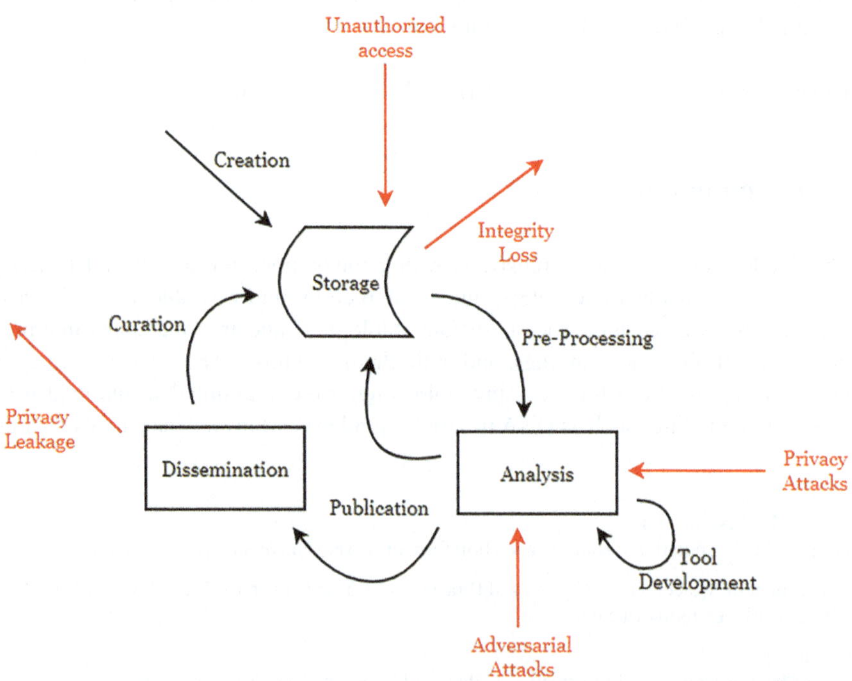

Fig. 1 Scope of this chapter in terms of the biomedical data life cycle and its vulnerabilities

The landscape of cyberbiosecurity challenges is constantly changing as a result of ongoing biomedical research. Genomics, for example, seeks to elucidate relationships between genotype and phenotypes, but these same associations can be reversed to infer potentially private information. The creation of privacy and security vulnerabilities is thus a natural consequence of scientific exploration. While most of these vulnerabilities are only conceptual as of today, these risks will escalate over time as more and more biomedical data becomes available.

Much like cybersecurity, cyberbiosecurity is becoming an arms race between the attackers and the security systems we put up in biomedicine. Even as we have improved guidelines and defenses, data breaches on hospitals are higher than ever, both in numbers and sophistication [2, 3]. There is a need for creating systems with security by design, rather than participating in a purely reactive arms race. We can achieve better security by preventing attacks before they are ever developed or seen through recognizing vulnerabilities, simulating attacks, and developing solutions before publishing our data and models. In most research studies currently, security of the data and models produced tends to be an afterthought, so in this chapter, we hope to highlight vulnerabilities and defensive measures to protect against them.

2 Creation, Curation, and Storage of Datasets

Biomedical data comes in many forms, such as textual data for electronic health records (EHRs), tables to contain various omics data, signals recorded from wearable devices, or Magnetic Resonance Imaging (MRI) and histopathology results. Primarily, these data all originate from a patient and are therefore considered sensitive information when linked to the identity of the patient. In the wrong hands, this information can be used to embarrass, blackmail, or discriminate against a patient. Breaches of healthcare data also lead to losing the trust of patients, which harms the healthcare industry as a whole. Security measures are therefore needed for protecting the privacy of biomedical data.

A basic principle of cybersecurity for protection of data is known as the CIA triad, a model comprising confidentiality, integrity, and availability [4]. Confidentiality protects information such that it is only seen by people who either own the data or have been given permission to access it. Integrity protects information from accidental or unauthorized alteration. Availability ensures that information is available to authorized users by preventing downtime or denial of service attacks. Availability and confidentiality often come at a trade-off. The maximum amount of confidentiality is achieved when nobody has access to data, but this would put a halt to any biomedical research. Therefore, these notions must be balanced.

This section focuses on "data at rest," that is, data that is currently being stored and being shared or processed. While these vulnerabilities are not exclusive to biomedical data, healthcare service providers make up the majority of data breaches among all industries [2, 3]. This is largely due to the high financial value of healthcare data.

3 Data Integrity

Accurate data that can be reliably accessed is necessary for biomedical research. Any errors in stored data get propagated further downstream in the data cycle, leading to erroneous analyses and results. Even small changes in a dataset can have a large effect on models trained on that data [5–7]. Though the problem of data integrity is not exclusive to cyberbiosecurity as to cybersecurity, the consequences can be much greater, since these models could be used as clinical decision support systems for treating and diagnosing patients.

Ensuring consistency with the large data sets common in biomedicine (genomic reads, imaging, biometrics, and many other omics) is difficult. Data can be modified unintentionally, due to disk errors such as random bit flipping, or intentionally and maliciously through spyware, hacking, or other means. Additionally, more and more biomedical datasets are being hosted on cloud platforms to save costs, which create additional confidentiality risks if hosting sensitive data on third-party servers. New requirements from the National Institutes of Health (NIH) state that all cloud computing services should reach compliance with the Federal Risk and Authorization Management Program (FedRAMP) [8], which is notoriously a slow and costly process for data distributors [9].

4 Privacy Concerns: Focusing on Ownership

Another aspect of data storage security are the questions of who gets the right to decide access to data and whose responsibility it is to store and protect that data. Policies like the General Data Protection Regulation (GDPR) [10] and the California Consumer Privacy Act (CCPA) [11] have advocated for people to have control over their own personal data, which puts into discussion the idea of ownership. Though there is some debate in its relation to property law, there currently is no legal definition to data ownership [12]. In most current healthcare systems, the healthcare providers are responsible for storing, securing, and managing data, which includes sharing the data on behalf of the patient [13]. If we are to transition to a fully patient-owned model, security against unauthorized access needs to be a guarantee, no matter who is responsible for that security. Thus it is not advisable, or practical, for patients to store their own healthcare data on personal computers. Moreover, managing consent in such a model could become cumbersome, especially with granting researchers access to de-identified data. Patients are not incentivized to share their data for secondary uses nor are they likely able to accurately assess the privacy risks involved with every possible use.

5 Data Dissemination

The ability to share biomedical datasets has contributed greatly to health science for understanding disease, improving quality of care, and advancing personalized treatments. However, it is a challenge to protect the privacy of participants without impeding scientists' access to the data. The most common way of providing security is by controlled access models. Typically, this requires an agreement that identifies the accessors and their intent [14], but lengthier bureaucratic processes are needed for certain datasets, which could take months before data is accessible [15]. As a result, the majority of published research uses open data portals rather than controlled access databases [14].

Open sharing is possible when the privacy risk is low, but it is not clear what kind of data can be shared safely without imposing privacy risks. The Health Insurance Portability and Accountability Act (HIPAA) only protects identifiable data, containing name, address, social security number, etc. [16]. De-identified data is not protected under HIPAA and can be openly shared. Yet, it is well known that anonymized DNA sequencing data can still be linked back to individuals through their genomic information [17–20]. Nowadays, most human genome projects use controlled access for sharing raw genomic data.

Over the years, research has exposed more types of health data that are vulnerable to re-identification, including summary-level data like gene expression values [21, 22], functional genomics data [23, 24], and clinical records [25]. These privacy attacks on anonymized datasets can be performed using a linking attack, which combines that data with another independent, public dataset containing identities and some quasi-identifiers [26]. These quasi-identifiers could be simple demographic data or a set of values that correlate to a unique identifier [27]. By matching these identifiers in both datasets, one can link identities to the anonymized data.

6 Data Analysis and Product Development

With the advent of electronic health records as well as lower costs to collect genomic, imaging, biosensor, and other healthcare data, the volume of biomedical data is scaling faster than Moore's law. Along with a drive toward developing personalized healthcare, there is higher demand for data analysis than ever. Machine learning has therefore seen huge growth.

Amazon, Google, Microsoft, and IBM provide machine learning as a service (MLaaS), a collection of services that offer machine learning tools on cloud-based platforms. These are made to be highly accessible and can be used for training and tuning models for doing predictive analyses, clustering, classification, natural language processing, etc. Such frameworks have led to a rise in the adoption and usage of machine learning. However, many studies have shown that naively trained models can be exploited by malicious users even if the data used to train the model is protected [7, 28–30]. There are two main categories of vulnerabilities: adversarial attacks that destroy integrity or poison models, that is, to make wrong predictions, and attacks on privacy, which try to infer sensitive information on either the model or its training data.

7 Adversarial Attacks

Adversarial attacks are most well known for studies where minimally perturbed images are used to fool deep neural network image classifiers into producing incorrect predictions with high confidence [5, 6]. Examples of these attacks have been shown in the biomedical domain as well. Histology image classifiers for predicting cancerous tissue are on par with pathologists [31], but the addition of small adversarial noise can flip the prediction [7]. An infectious pathogenic sequence classifier can be exploited by bioterrorists to either stage or hide an outbreak [32]. Other data types such as audio [30], text [7], or wearable device signals [33] can all be affected with minimal changes that would be imperceivable to a person but would change the prediction of machine learning algorithms. Also, adversarial attacks do not just have to be at inference time. If these adversarial examples were somehow inserted into the training set, the model becomes poisoned and may produce weaker or incorrect predictions.

Such attacks are defined as white box attacks, where an adversary has total knowledge of the model architecture, its parameters, and its training data, as is the case with most academically published models. Black box attacks are more difficult to execute, with only access to the model, but no knowledge about it, an attacker would have to explore the model by crafting inputs and inferring the model through outputs. However, black box constraints are not enough to entirely prevent adversarial attacks [34], so there needs to be some robustness built in at training time for machine learning models.

8 Privacy Attacks

We categorize privacy attacks into two types for machine learning models: model inversion attacks that target the sensitivity of the training data and model extraction attacks that target the privacy of the model itself. All of these attacks assume that an attacker cannot access the training data.

The simplest form of a model inversion attack is a membership inference attack, where the goal of an adversary is to determine whether a particular data point was included in the training dataset for a model. This can directly violate the privacy of an individual if their participation in a study or dataset is sensitive information. For example, inferring that a patient's histopathology records were used in a cancer classification model may expose their disease state. Shokri et al. [35] showed that membership inference can be accomplished in a black box setting by training shadow models to mimic the real model. Input/output pairs are evaluated for many shadow models until they match outputs for the real model. Once optimized, membership inference can be performed by measuring the difference between evaluation on datasets with and without the data record. This works most effectively with naively trained models that are overfit to its training data [36], whereas better-generalized models require more advanced techniques to perform the attack [37].

Reconstruction attacks, also known simply as model inversion attacks, aim to either fully or partially reconstruct the training data. This concept was first introduced in a study showing that a dosing model leaks genomic information based on a patient's dosage level of warfarin and basic demographic data [38]. Reconstruction attacks can be done in a probabilistic manner, by calculating an input that maximizes the likelihood of observing a set of outputs [39], or in an iterative manner, by querying how a model's output changes depending on different sets of training data [40]. Recovery of training data is more accurate for simpler models, such as logistic regression, but recognizable faces have been extracted for facial recognition models, given only the model and a name [39].

Model extraction attacks instead aim to expose the private information on the model itself, by attempting to duplicate its parameters or architecture in black box settings. This is notably a security issue when the model itself represents intellectual property. Although model extraction attacks do not directly invade the privacy of the training dataset, this can easily be accomplished as a byproduct by using the extracted model. Methods for performing the attack are similar to those mentioned for model inversion: optimizing a model equation from an output, given a crafted set of inputs. Model parameters can be stolen for a known architecture by querying the model with enough inputs [41], and more recent work has shown that both architectures [42] and hyperparameters [43] can also be inferred if those are unknown.

9 Security Solutions for the Biomedical Data Life Cycle

While most of the dangers discussed so far are only hypothetical, the purpose of highlighting such attacks is to understand the potential for harm before they happen. While certain biomedical systems may seem safe now, it's uncertain what the future will bring as technology continues to advance. Proactive approaches are necessary that build security by design, by simulating conceivable attacks and devising countermeasures while implementing databases, sharing strategies and machine learning models.

In the next section, we review various defenses against the vulnerabilities discussed in this document. Since this text is focused on cyberbiosecurity, we will not go in depth on every existing cybersecurity technique, and we refer the reader to previous reviews covering such material in depth [44–47]. We will instead cover cybersecurity measures relevant for the biomedical sector, particularly ones that focus on securing the privacy of sensitive data as it journeys through the biomedical data life cycle.

10 Blockchain Solutions for Data Integrity and Dissemination

Blockchain technology is most well known in its first implementation in the financial sector as Bitcoin [48] but has recently gained attention in many diverse fields, including healthcare and biomedicine [49, 50]. Blockchain provides security and immutability by using a distributed, decentralized ledger on a peer-to-peer network. Data, which is often in the form of transactions but can be any type of data, is added to the ledger in the form of blocks, which get validated by the entire network. Every block is linked to its previous block by a cryptographic hash that is unique to the block's data and the previous block's hash. Each node in the blockchain network maintains a copy of this ledger, and all nodes are updated simultaneously when a valid block is added. Once data is added to the blockchain, it cannot be altered, creating a strong method of immutability useful for data integrity and efficient auditing.

While blockchain's immutability is an advantage for data integrity, it is not ideal for storing sensitive data that one may want to remove at some point. Additionally, storage does not scale well on the blockchain, making it unsuitable for large files. Accordingly, most applications of blockchain in healthcare do not store sensitive files, such as electronic health records (EHR), directly on the blockchain itself. Instead, these data are put into another secure location, which can be on the cloud or an InterPlanetary File System (IPFS) in some encrypted format, and instead store a hash of the data and its address on the blockchain, which can be used to check if that data has been altered.

One notable use of blockchain in EHR is in Estonia, with technology provided on the HSX platform developed by Guardtime [51]. Citizens are issued with a digital ID linked to many e-services such as digital signatures, online voting, and checking medical records. While EHR is not stored directly on chain, updates to the EHR are hashed and recorded to the blockchain. Access control is also maintained by the platform, with an audit trail that shows who and when any access is made. Researchers have access to any data that patients (the data owners) have consented to be set to "visible," which patients can revoke at any time [51]. This also demonstrates blockchain solving the problem of proof of ownership. On the blockchain, ownership is immutable unless the owner verifies a change using their digital signature. Smart contracts enable patients to manage access to their health records, so that there is no need to trust any central organization.

11 Solutions Against Privacy Attacks

Differential privacy provides a mathematical guarantee for the degree of privacy to individuals contributing to a dataset [52]. This guarantee states that an adversary that can see the output of a differentially private computation is unable to distinguish whether or not an arbitrary individual's data was included. The strength of the privacy guarantee is controllable, to adjust how statistically indistinguishable the results are [53]. Tighter control over the privacy results in some loss of accuracy or utility, thereby explicitly defining a privacy-utility trade-off. Differential privacy is implemented by adding noise either to the dataset or, for example, in a neural network model doing tumor classification; this noise could be implemented by randomly flipping labels, adding a stochastic term to the loss function, or noising the gradients during training [54].

As privacy attacks rely on some degree of overfitting of a model to extract sensitive information about the data, regularization is another useful way to prevent them while also creating a more generalizable model [36]. Techniques such as L1 and L2 regularization have proven effective against membership inference attacks for logistic regression models [35], as well as dropout for neural networks [54].

12 Solutions Against Adversarial Attacks

A number of defenses exist against adversarial attacks. Adversarial training is performed by training a model against the adversarial samples themselves and is one of the most effective ways to defend against adversarial attacks [55]. Randomization borrows ideas from differential privacy by adding random noise within the model that smooths out any adversarial noise from the input [56, 57]. Denoising algorithms also are useful if they can detect adversarial noise in inputs before they are put into the model [58].

13 Data Sanitization

One solution to preventing model inversion attacks is data sanitization to selectively obfuscate or remove parts of the data. Sanitization methods seek to remove sensitive attributes of data that contribute to re-identification, without significantly altering the results of any analyses done on the sanitized dataset [59]. As in differential privacy, the degree of sanitization results in a privacy and utility trade-off, which is important to quantify. Various methods have been demonstrated in sanitizing textual medical records [60–62], but all require high computation cost for an effective privacy threshold. Sanitization has also been achieved with raw reads for functional genomic data to create privacy-preserving Binary Alignment Map (BAM) files, which preserve utility by prioritizing the masking of identifying variants that are unnecessary for most analysis methods [23].

14 Cryptography Solutions for Data Analysis and Dissemination

Encryption is one method of controlling access to data, by storing encrypted data publicly and giving decryption keys to authenticated users. While its main use is to provide confidentiality against unauthorized access, cryptography can also enable confidential computing for machine learning on private data. Several privacy-enhancing techniques take advantage of cryptographic principles, which will be reviewed in the following section. Note that many of these technologies are still in an exploratory stage and not yet at the point of being practical or efficient enough for widespread adoption in biomedical data science.

Recent trends have pushed researchers to perform analyses directly on datasets shared via cloud platforms to bypass expensive data transfer costs, but if these datasets contain private information, they must be trusted with the third-party cloud server. Rather than relying on this trust, we can store encrypted data on the cloud and use homomorphic encryption to compute directly on encrypted data, producing an encrypted output. The output can be decrypted by private key to yield the same result as if computed on plaintext. Examples of homomorphic encryption-based solutions for tasks in biomedical data include performing genome-wide association study (GWAS) [63], genotype imputation [64–66], and tumor classification [67]. Currently, homomorphic encryption is still impractical for use in real-world applications due to a number of issues. Computations involved can only be composed of Boolean or arithmetic operations, limiting the types of tasks that can be done [68]. Fully homomorphic encryption can handle an unlimited number of operations but suffers from low efficiency [69]. Other types of homomorphic encryption, such as partially or somewhat homomorphic encryption, improve on efficiency at the cost of how many operations they can handle, as well as their accuracy [70].

A problem also exists in collaborative data analysis when two or more parties, such as hospitals, want to jointly compute on their combined set of data but cannot directly share that data with each other due to privacy concerns. Federated learning and secure multiparty computation are two solutions that take advantage of cryptography to solve this problem.

In federated learning, different parties train models locally on their own data and share model parameters with a central server that merges them for the full model. Security is required to make sure that the parameters shared do not leak information about each local model's training data [71]. This is most often done by differential privacy techniques, though homomorphic encryption can also be used to protect the underlying data [72].

Secure multiparty computation protocols carry out a joint computation on data distributed among many parties, without any party learning the input of anyone else. This is done by using techniques like garbled circuits [73] or secret sharing [74], which split, scramble, and distribute data among all parties for joint computation. However, these protocols are currently not practical for use in many applications due to a high network overhead, requiring magnitudes greater amounts of data transferred than the original dataset sizes [75]. Also, similar to homomorphic encryption, many types of analyses are not possible using secure multiparty computation, as the joint computation only allows for simple arithmetic operations [76].

So far, all cryptographic techniques mentioned focus on protecting the input and output data but may not protect the model. Trusted execution environments (TEEs) are a hardware-based solution that provide a secure environment for executing private applications [77]. Users can establish a secure channel with a protected unit inside a processor, where confidential data and code cannot be observed by outside applications. Security is guaranteed by "remote attestation," a cryptographic protocol that demonstrates proof to a user that all of their code, data, and the communication channel are secure and tamper-proof inside the isolated environment [78]. Intel Software Guard Extensions (SGX) [79] is one example of a TEE that has found use for privacy-preserving analysis on genomic data [80–82]. Since computation is done on plaintext, there is no significant computational overhead to TEEs like in homomorphic encryption or secure multiparty computation. However, TEEs have severely limited memory size, requiring specialized applications for optimal memory usage. Recently, NVIDIA has introduced the first TEE for a graphical processing unit (GPU), which could lead to even greater performance gains for certain types of computation [83].

15 Conclusion

Although cyberbiosecurity is a relatively new field, its landscape of vulnerabilities is quickly evolving due to the rapid pace of advances in cybersecurity, machine learning, and biomedical research. Each of these respective fields offers much to learn in terms of security solutions as new privacy risks and attacks arise. We predict that issues with privacy and security will grow as biomedicine continues to move toward a personalized healthcare model and patient data becomes more necessarily sensitive.

Researchers and data distributors should remain aware of laws and policy regarding biomedical data security, but also be aware that such policies are slow to adapt. Rather, coordinated efforts and communication among global healthcare communities can help tremendously by establishing a common framework. Initiatives like the GA4GH that set best practice standards for data sharing are important to complement policy surrounding privacy and data protection [84].

This chapter outlines vulnerabilities throughout the biomedical data life cycle in order to promote understanding of the limitations of cyberbiosecurity attacks. We need to keep up with emerging technologies and assess how they may be used to threaten or improve security measures. In particular, it is important to discern how to balance confidentiality, integrity, and availability for solving each specific problem rather than broadly applying one-size-fits-all solutions.

References

1. R. Puzis, D. Farbiash, O. Brodt, Y. Elovici, D. Greenbaum, Increased cyber-biosecurity for DNA synthesis. Nat. Biotechnol. **38**, 1379–1381 (2020)
2. A.H. Seh et al., Healthcare data breaches: Insights and implications. Healthcare (Basel) **8**, 133 (2020)
3. S.I. Khan, A.S.M. Hoque, Digital health data: A comprehensive review of privacy and security risks and some recommendations. Comput. Sci. J. Moldova **24**, 273 (2016)
4. M.S. Olivier, Database privacy: Balancing confidentiality, integrity and availability. SIGKDD Explor. Newsl. **4**, 20–27 (2002)
5. C. Szegedy et al., Intriguing properties of neural networks. arXiv [cs.CV] (2013)
6. A. Nguyen, J. Yosinski, J. Clune, Deep neural networks are easily fooled: High confidence predictions for unrecognizable images, in *Proceedings of the IEEE conference on computer vision and pattern recognition* (2015), pp. 427–436
7. S.G. Finlayson et al., Adversarial attacks on medical machine learning. Science **363**, 1287–1289 (2019)
8. L. Taylor, FedRAMP: History and future direction. IEEE Cloud Comput. **1**, 10–14 (2014)
9. M. McLaughlin, Reforming FedRAMP: A guide to improving the federal procurement and risk management of cloud services. (2020)
10. R.N. Zaeem, K.S. Barber, The effect of the GDPR on privacy policies: Recent progress and future promise. ACM Trans. Manage. Inf. Syst. **12**, 1–20 (2020)
11. P. Jurcys, C. Donewald, M. Fenwick, M. Lampinen, A. Smaliukas, Ownership of user-held data: Why property law is the right approach. Harv. J. Law Technol. Digest (2020). https://doi.org/10.2139/ssrn.3711017
12. P. Hummel, M. Braun, P. Dabrock, Own Data? Ethical reflections on data ownership. Philos. Technol. **34**, 545–572 (2021)
13. B.J. Evans, Much ado about data ownership. Harv. JL Tech. **25**, 69 (2011)
14. Y. Joly, S.O.M. Dyke, B.M. Knoppers, T. Pastinen, Are data sharing and privacy protection mutually exclusive? Cell **167**, 1150–1154 (2016)
15. G. Gürsoy et al., Functional genomics data: Privacy risk assessment and technological mitigation. Nat. Rev. Genet. **23**, 245–258 (2022)
16. S.A. Tovino, HIPAA compliance, in *The Cambridge Handbook of Compliance*, (2021), pp. 895–908
17. E.C. Hayden, Privacy protections: The genome hacker. Nature **497**, 172–174 (2013)
18. M. Gymrek, A.L. McGuire, D. Golan, E. Halperin, Y. Erlich, Identifying personal genomes by surname inference. Science **339**, 321–324 (2013)
19. L. Sweeney, A. Abu, J. Winn, Identifying participants in the personal genome project by name. 2013. Available at SSRN (2013)
20. N. Homer et al., Resolving individuals contributing trace amounts of DNA to highly complex mixtures using high-density SNP genotyping microarrays. PLoS Genet. **4**, e1000167 (2008)

21. A. Harmanci, M. Gerstein, Quantification of private information leakage from phenotype-genotype data: Linking attacks. Nat. Methods **13**, 251–256 (2016)
22. E.E. Schadt, S. Woo, K. Hao, Bayesian method to predict individual SNP genotypes from gene expression data. Nat. Genet. **44**, 603–608 (2012)
23. G. Gürsoy et al., Data sanitization to reduce private information leakage from functional genomics. Cell **183**, 905–917.e16 (2020)
24. A. Harmanci, M. Gerstein, Analysis of sensitive information leakage in functional genomics signal profiles through genomic deletions. Nat. Commun. **9**, 2453 (2018)
25. Y. Nakamura et al., KART: Parameterization of privacy leakage scenarios from pre-trained language models. arXiv [cs.CL] (2020)
26. D.C. Barth-Jones, The 're-identification' of Governor William Weld's medical information: A critical re-examination of health data identification risks and privacy protections, then and now. SSRN Electron. J. https://doi.org/10.2139/ssrn.2076397
27. A. Narayanan, V. Shmatikov, Robust de-anonymization of large sparse datasets, in *2008 IEEE Symposium on Security and Privacy (sp 2008)* (2008), pp. 111–125
28. Robert Philipp Economics and Statistics, University of Vienna, Austria, Andreas Mladenow Economics and Statistics, University of Vienna, Austria, Christine Strauss Economics and Statistics, University of Vienna, Austria & Alexander Völz Economics and Statistics, University of Vienna, Austria. Machine Learning as a Service. ACM Other conferences https://dl.acm.org/doi/abs/10.1145/3428757.3429152
29. Manish Kesarwani IBM Research, India, Bhaskar Mukhoty Indian Institute of Technology, Kanpur, Vijay Arya IBM Research, India & Sameep Mehta IBM Research, India. Model Extraction Warning in MLaaS Paradigm. ACM Other conferences https://dl.acm.org/doi/abs/10.1145/3274694.3274740
30. F.R. Battista Biggioab, Wild patterns: Ten years after the rise of adversarial machine learning. Pattern Recognit. **84**, 317–331 (2018)
31. A. Hamidinekoo, E. Denton, A. Rampun, K. Honnor, R. Zwiggelaar, Deep learning in mammography and breast histology, an overview and future trends. Med. Image Anal. **47**, 45–67 (2018)
32. A. Meiseles, I. Rosenberg, Y. Motro, L. Rokach & J. Moran-Gilad, Adversarial vulnerability of deep learning models in analyzing next generation sequencing data, in *2020 IEEE International Conference on Bioinformatics and Biomedicine (BIBM)* (IEEE, 2021). https://doi.org/10.1109/BIBM49941.2020.9313421
33. A. Aminifar, Minimal adversarial perturbations in mobile health applications: The epileptic brain activity case study, in *ICASSP 2020–2020 IEEE International Conference on Acoustics, Speech and Signal Processing (ICASSP)* (2020), pp. 1205–1209
34. S. Bhambri, S. Muku, A. Tulasi, A.B. Buduru, A survey of black-box adversarial attacks on computer vision models. arXiv [cs.LG] (2019)
35. R. Shokri, M. Stronati, C. Song, V. Shmatikov, Membership inference attacks against machine learning models, in *2017 IEEE Symposium on Security and Privacy (SP)* (2017), pp. 3–18
36. Y. Long et al., Understanding membership inferences on well-generalized learning models. arXiv [cs.CR] (2018)
37. M. Nasr, R. Shokri, A. Houmansadr, Comprehensive privacy analysis of deep learning: Passive and active white-box inference attacks against centralized and federated learning, in *2019 IEEE Symposium on Security and Privacy (SP)* (2019). https://doi.org/10.1109/sp.2019.00065
38. M. Fredrikson et al., Privacy in pharmacogenetics: An end-to-end case study of personalized warfarin dosing. Proc. USENIX Secur. Symp. **2014**, 17–32 (2014)
39. M. Fredrikson, S. Jha, T. Ristenpart, Model inversion attacks that exploit confidence information and basic countermeasures, in *Proceedings of the 22nd ACM SIGSAC Conference on Computer and Communications Security*, pp. 1322–1333 (Association for Computing Machinery, 2015)
40. A. Salem, A. Bhattacharya, M. Backes, M. Fritz, Y. Zhang, ${Updates-Leak}$: Data set inference and reconstruction attacks in online learning, in *29th USENIX Security Symposium (USENIX Security 20)* (2020), pp. 1291–1308

41. F. Tramèr, F. Zhang, A. Juels, M.K. Reiter, T. Ristenpart, Stealing machine learning models via prediction ${APIs}$, in *25th USENIX security symposium (USENIX Security 16)* (2016), pp. 601–618
42. S.J. Oh, B. Schiele, M. Fritz, Towards reverse-engineering black-box neural networks, in *Explainable AI: Interpreting, Explaining and Visualizing Deep Learning*, ed. by W. Samek, G. Montavon, A. Vedaldi, L.K. Hansen, K.-R. Müller, (Springer, 2019), pp. 121–144
43. B. Wang, N.Z. Gong, Stealing hyperparameters in machine learning, in *2018 IEEE Symposium on Security and Privacy (SP)* (2018), pp. 36–52
44. G. Sivathanu, C.P. Wright, E. Zadok, Ensuring data integrity in storage: Techniques and applications. in *Proceedings of the 2005 ACM workshop on Storage security and survivability* (Association for Computing Machinery, 2005), pp. 26–36
45. M. Jegorova et al., Survey: Leakage and privacy at inference time. arXiv [cs.LG] (2021)
46. I.H. Sarker et al., Cybersecurity data science: An overview from machine learning perspective. J. Big Data **7**, 41 (2020)
47. M. Lezzi, M. Lazoi, A. Corallo, Cybersecurity for Industry 4.0 in the current literature: A reference framework. Comput. Ind. **103**, 97–110 (2018)
48. Nakamoto, Bitcoin: A peer-to-peer electronic cash system. Decentralized Bus. Rev. **21260** (2008)
49. M. Mettler, Blockchain technology in healthcare: The revolution starts here, in *2016 IEEE 18th International Conference on e-Health Networking, Applications and Services (Healthcom),* (2016), pp. 1–3
50. R. Jabbar, N. Fetais, M. Krichen, K. Barkaoui, Blockchain technology for healthcare: Enhancing shared electronic health record interoperability and integrity, in *2020 IEEE International Conference on Informatics, IoT, and Enabling Technologies (ICIoT)* (2020), pp. 310–317
51. Guardtime Health. https://m.guardtime.com/files/Guardtime_whitepaper_A4_april_web.pdf
52. C. Dwork, A. Roth, The algorithmic foundations of differential privacy. Found. Trends Theor. Comput. Sci. **9**, 211–407 (2013)
53. C. Dwork, F. McSherry, K. Nissim, A. Smith, Calibrating noise to sensitivity in private data analysis. J. Priv. Confid. **7**, 17–51 (2017)
54. X. Liu et al., Privacy and security issues in deep learning: A survey. IEEE Access **9**, 4566–4593. (undefined 2021)
55. A. Madry, A. Makelov, L. Schmidt, D. Tsipras, A. Vladu, Towards deep learning models resistant to adversarial attacks. arXiv [stat.ML] (2017)
56. X. Liu, M. Cheng, H. Zhang, C.-J. Hsieh, Towards robust neural networks via random self-ensemble, in *Proceedings of the European Conference on Computer Vision (ECCV)* (2018), pp. 369–385
57. M. Lecuyer, V. Atlidakis, R. Geambasu, D. Hsu, S. Jana, Certified robustness to adversarial examples with differential privacy, in *2019 IEEE Symposium on Security and Privacy (SP)* (2019), pp. 656–672
58. K. Ren, T. Zheng, Z. Qin, X. Liu, Adversarial attacks and defenses in deep learning. Proc. Est. Acad. Sci. Eng. **6**, 346–360 (2020)
59. S.R.M. Oliveira, O.R. Zaiane, Protecting sensitive knowledge by data sanitization, in *Third IEEE International Conference on Data Mining* (2003), pp. 613–616
60. C. Iwendi et al., N-Sanitization: A semantic privacy-preserving framework for unstructured medical datasets. Comput. Commun. **161**, 160–171 (2020)
61. I. Neamatullah et al., Automated de-identification of free-text medical records. BMC Med. Inform. Decis. Mak. **8**, 32 (2008)
62. Z. Liu, B. Tang, X. Wang, Q. Chen, De-identification of clinical notes via recurrent neural network and conditional random field. J. Biomed. Inform. **75S**, S34–S42 (2017)
63. T.-T. Kuo et al., iDASH secure genome analysis competition 2018: Blockchain genomic data access logging, homomorphic encryption on GWAS, and DNA segment searching. BMC Med. Genom. **13**, 98 (2020)
64. M. Kim et al., Ultrafast homomorphic encryption models enable secure outsourcing of genotype imputation. Cell Syst. **12**, 1108–1120.e4 (2021)

65. G. Gürsoy, E. Chielle, C.M. Brannon, M. Maniatakos, M. Gerstein, Privacy-preserving genotype imputation with fully homomorphic encryption. Cell Syst. **13**, 173–182.e3 (2022)
66. F.M. Chan et al., Genotype imputation with homomorphic encryption, in *2021 6th International Conference on Biomedical Signal and Image Processing* (Association for Computing Machinery, 2021), pp. 9–13
67. S. Hong, J.H. Park, W. Cho, H. Choe, J.H. Cheon, Secure tumor classification by shallow neural network using homomorphic encryption. BMC Genom. **23**, 284 (2022)
68. A. Acar, H. Aksu, A.S. Uluagac, M. Conti, A survey on homomorphic encryption schemes: Theory and implementation. ACM Comput. Surv. **51**, 1–35 (2018)
69. C. Gentry, *A Fully Homomorphic Encryption Scheme* (Stanford University, 2009)
70. M. Naehrig, K. Lauter, V. Vaikuntanathan, Can homomorphic encryption be practical? in *Proceedings of the 3rd ACM workshop on Cloud computing security workshop*, (Association for Computing Machinery, 2011), pp. 113–124
71. L. Melis, C. Song, E. De Cristofaro, V. Shmatikov, Exploiting unintended feature leakage in collaborative learning, in *2019 IEEE Symposium on Security and Privacy (SP)*, (2019), pp. 691–706
72. B. Pfitzner, N. Steckhan, B. Arnrich, Federated learning in a medical context: A systematic literature review. ACM Trans. Internet Technol. **21**, 1–31 (2021)
73. A.C. Yao, Protocols for secure computations, in *23rd Annual Symposium on Foundations of Computer Science (sfcs 1982)* (1982), pp. 160–164
74. A. Shamir, How to share a secret. Commun. ACM **22**, 612–613 (1979)
75. J.I. Choi, K.R.B. Butler, Secure multiparty computation and trusted hardware: Examining adoption challenges and opportunities. Secur. Commun. Netw. **2019** (2019)
76. R. Cramer, I.B. Damgard, J.B. Nielsen, *Secure Multiparty Computation and Secret Sharing* (Cambridge University Press, 2015)
77. C. Shepherd et al., Secure and trusted execution: Past, present, and future – A critical review in the context of the internet of things and cyber-physical systems, in *2016 IEEE Trustcom/BigDataSE/ISPA* (2016), pp. 168–177
78. A. Vasudevan, J.M. McCune, J. Newsome, *Trustworthy Execution on Mobile Devices* (Springer, New York, 2014)
79. I. Anati, S. Gueron, S. Johnson, V. Scarlata, Innovative technology for CPU based attestation and sealing, in *Proceedings of the 2nd international workshop on hardware and architectural support for security and privacy* vol. 13 (ACM New York, 2013)
80. F. Chen et al., PRESAGE: PRivacy-preserving gEnetic testing via SoftwAre Guard Extension. BMC Med. Genom. **10**, 48 (2017)
81. F. Chen et al., PRINCESS: Privacy-protecting rare disease international network collaboration via encryption through software guard extension S. Bioinformatics **33**, 871–878 (2017)
82. C. Kockan et al., Sketching algorithms for genomic data analysis and querying in a secure enclave. Nat. Methods **17**, 295–301 (2020)
83. NVIDIA, NVIDIA H100 tensor core GPU architecture overview. https://resources.nvidia.com/en-us-tensor-core (2022)
84. H.L. Rehm et al., GA4GH: International policies and standards for data sharing across genomic research and healthcare. Cell Genom. **1**, 100029 (2021)

Cybersecurity Across the DNA-Digital Boundary: DNA Samples to Genomic Data

Peter Ney, Arkaprabha Bhattacharya, Luis Ceze, Karl Koscher, Tadayoshi Kohno, and Jeff Nivala

Abstract Technological advances in biotechnology, especially next-generation DNA sequencing and direct-to-consumer genotyping, have created exponentially more biological data. To reach this scale, biotechnology pipelines have increasingly relied on automation and computation in the molecular data processing workflow: Biological samples are processed at scale using robotic equipment; molecular sensors, like DNA sequencers, have become specialized computers with peripheral sensors designed to read molecules; and extensive data processing and digital storage are required to manage and make use of this data. All of this computation raises security issues that are more typically associated with computer systems. Here, we explore how the entire DNA data processing workflow, from physical sample processing through reading DNA into digital information and eventual data analysis, is plagued by a number of security vulnerabilities, including a lack of data integrity, poor software security practices, and hardware that is insecure by design. In standard DNA sequencing pipelines, DNA samples are presumed to be derived from natural sources without manipulation. In this work, we show how simple synthetic DNA constructs can be used as vectors for computer malware or as commands to backdoored software or firmware, enabling communication across air gaps. DNA sequencing hardware, including flow cells, is also vulnerable by design to data recovery and corruption attacks. Finally, we show how a lack of data integrity checks in genetic databases can lead to catastrophic data breaches and other security concerns. We conclude with some broader themes and lessons from this work that apply to the larger cyber-biosecurity domain.

P. Ney (✉) · A. Bhattacharya · L. Ceze · K. Koscher · T. Kohno
Paul G. Allen School of Computer Science & Engineering, University of Washington, Seattle, WA, USA
e-mail: neyp@cs.washington.edu

J. Nivala
Paul G. Allen School of Computer Science & Engineering, University of Washington, Seattle, WA, USA

Molecular Engineering and Sciences Institute, University of Washington, Seattle, WA, USA

D. Greenbaum (ed.), *Cyberbiosecurity*, https://doi.org/10.1007/978-3-031-26034-6_7

Keywords DNA security · DNA sequencing · Molecular malware · Bypass air gap · Data integrity · Data corruption

1 Introduction

The biotechnology sector along with companion fields like synthetic biology and bioengineering has raised dual-use security concerns due to the possibility that technologies, such as DNA synthesis or gain-of-function research, could be used maliciously to produce dangerous compounds or organisms. Traditionally biosecurity has relied on security through obfuscation, restricting access or expertise, and high costs as barriers to limit risk. In recent years, the biotechnology domain has increasingly relied on automation and computers more generally to accelerate the pace of discovery, to improve scalability, and to lower costs. This has greatly expanded the possible attack surface to now include cybersecurity issues such as remote compromise of robotic equipment and laboratory management systems, data and IP theft, and the bypass of other security controls. The old biosecurity paradigm of relying on obfuscation and expertise is no longer sufficient to handle emerging cybersecurity problems, which are much easier to target and require a more active security mindset. This is especially true as unhardened laboratory equipment becomes connected into computer networks.

Together the biosecurity and cybersecurity risks, which we broadly refer to as cyber-biosecurity (CBS), are more than the sum of their parts. In particular, we have concerns that cyber-vulnerabilities may enhance existing biosecurity concerns—for example, compromised machines being used to bypass security controls to permit the production of dangerous compounds—or that biological material can even work backward to cause cybersecurity problems. Grappling with the scale of these challenges is difficult because existing biotechnology pipelines are integrated workflows containing many phases ranging from molecular synthesis, sample processing, assay automation and orchestration, molecular sensing, data collection and storage, and finally analysis. In some cases, the molecular digital dichotomy is explicitly blurred like in the case of molecular information storage systems that use molecules to store digital information. To have a better sense of how CBS problems can manifest in real biotechnology pipelines, we focus on the protocols and procedures used with end-to-end DNA processing.

DNA has been read using methods like Sanger sequencing for close to 50 years, but the modern era of reading DNA began with the development of high-density genotyping arrays and high-throughput "next-generation" DNA sequencers, which exponentially increased the scale of DNA analysis. Reading DNA at scale involves computers in all phases of the workflow: Biological specimens are prepared for reading using automated assays with liquid handlers, sequencers and microarray instruments are computers with specialized flow cell attachments and cameras, raw data needs to be preprocessed into a usable form (e.g., DNA alignment or assembly), and finally the digital DNA data needs to be analyzed specific to the desired application and stored. Since scalable DNA reading is already used in a wide variety of applications like medicine, genomics, forensics, and consumer testing, a thorough CBS analysis can give insight into risks that may develop in other less polished emerging technologies and provide lessons for the broader industry.

To see how security manifests in existing DNA reading and processing pipelines, we discuss in the following sections how novel and unexpected CBS problems can appear at almost every phase of DNA processing. First, we demonstrate how DNA's role as a reliable information carrier can be leveraged to encode malicious information directly into synthetic DNA. Maliciously designed DNA can be used to target vulnerable computers downstream of sequencing and even be used as a covert communication channel that bypasses air gaps to send information to backdoors or trojaned software. Next, we consider how hardware components in sequencing instruments, such as sequencing flow cells, have important implications for secure data deletion or in multiuser environments. Lastly, we discuss the role that data integrity (or lack thereof) plays in securing genotyping data, especially when that data is used in consumer facing applications like genetic genealogy.

2 DNA as a Malicious Information Carrier

When viewed abstractly, DNA sequencing is the conversion of data encoded physically (in DNA molecules) into a digital form suitable for computer storage and analysis. In the vast majority of cases, the information is derived from biological sources like genomes or readouts of biological processes. However, it is possible to artificially synthesize DNA de novo using chemical processes (e.g., oligonucleotide synthesis). This means that DNA used in sequencing cannot be presumed to have natural origin. It is the potential for wholly artificial data that creates novel CBS threats to the DNA sequencing pipeline.

Since the early days of cybersecurity, software that insecurely processes input has been a major source of cybersecurity problems [1]. Memory vulnerabilities, like buffer overflows, can allow an adversary to execute malicious code on remote machines giving them full control. In traditional computing, many of these vulnerabilities were latent and only became a significant problem once computers became widely networked and Internet accessible. The lesson here is that it is the existence of software vulnerabilities in combination with the ability of adversaries to target and send data to vulnerable systems that lead to problems.

In many ways, DNA sequencing shows similarities to the early days of cybersecurity. As we will show later, widely used bioinformatics utilities show many signs of insecure software design, including the use of memory unsafe languages like C/C++, improperly checked memory buffers, and the use of deprecated function calls [2]. In particular, the parts of the software that process the DNA data itself have not been hardened, and the data they process is presumed trustworthy. Once DNA has been sequenced and the sequencing data is being processed by software, it is represented in some data encoding and analyzed like any other form of data.

This raises the possibility that synthetic DNA could be intentionally designed to encode malicious information including exploitable computer code that could target any vulnerability or be used as a method to send information covertly to software backdoors running on sequencers.

To more concretely understand the risks of synthetic DNA being used as a malicious information carrier, we developed two prototype demonstrations. The first example is a simple bioinformatics utility that reads raw sequencing information, early in the sequencing data processing pipeline, and contains an intentionally inserted buffer overflow vulnerability, similar to what is routinely found in unhardened software. The objective was to design a synthetic DNA construct that after sequencing with a modern Illumina sequencer (NextSeq 500) would be able to compromise the vulnerable software and give an adversary full remote control [2]. The second example is a backdoor program we developed that, if covertly run on a sequencer, would be able to decode and respond to commands received from specially crafted DNA molecules. The backdoor was designed so that synthetic DNA messages could be spiked into typical DNA samples to make them covert. An adversary could use this functionality to communicate across air gaps and even exfiltrate data via the resulting sequencing data.

2.1 Encoding Malware into DNA

The modern DNA sequencing process (often called next-generation DNA sequencing) happens in three phases: First, a DNA sample is prepared for sequencing using standard wet lab assays, then the sample is run through a sequencer which reads the linear sequence of bases in each DNA molecule and stores the raw sequences digitally, and finally, the sequenced data is processed into a usable form with bioinformatics software. Next-generation DNA sequencers cannot read long DNA strands and are limited to reading short strands no more than a few hundred bases in length (each of these small sequences is called a *read*). Typical genomic DNA can be hundreds of thousands of bases in length, so to accommodate the size limitation, during the sample preparation phase, DNA is *fragmented* into smaller pieces via mechanical shearing. Thus, to sequence longer strands, sequencers actually break the strands into small pieces that are read by the sequencer in random order. If necessary for analysis, these random reads produced by the sequencer can be reconstructed into the original, longer sequence. For example, to determine whether a person has a given genetic trait (so-called variant calling), the raw DNA reads will need to be cleaned up for quality control, aligned to a reference human genome sequence (effectively ordering the strands), and individual bases determined at each genomic position. This method of pipelining different small utilities together—each with a distinct purpose—to process the sequencing reads in stages is typical in sequencing analysis, and this design makes securing a bioinformatics workflow difficult because each program may be written by different authors and not well supported.

Our objective was to understand what challenges an adversary would face when trying to synthesize and sequence DNA-encoded malware, so we could better understand the feasibility of DNA as a malicious attack vector. Our goal was not to identify and target actual vulnerabilities in bioinformatics software. (Although, as described later, the DNA processing pipeline does have especially antiquated software security practices.) Therefore, we begin our study assuming we have already identified a buffer overflow vulnerability in a piece of bioinformatics software. In this case, we took an open-source tool written in C designed to compress raw sequencing data files and modified a memory buffer to create a classic buffer overflow vulnerability. Buffer overflow vulnerabilities are a common class of software insecurity that allow adversaries to run their own software on victim computers by sending malformed data. Our goal was to design DNA malware that could be made physically using a low-cost DNA synthesis service and survive the sample preparation and sequencing process as intact malware. Any Internet-connected machine that reads the malicious sequencing data using the compression utility would be compromised and give an adversary remote access and control.

Our first attempts at designing DNA malware ran into a number of roadblocks due to DNA synthesis limitations, randomness inherent in next-generation sequencing, and challenges with the DNA encoding scheme used by the vulnerable software. Buffer overflow malware will imbed machine code (called shellcode) and other computer instructions like memory addresses inside the corrupted data. Since the data being sent into the vulnerable utility is raw sequencing data, the malware must be written into the standard DNA bases (A, C, G, and T). The way the bases are encoded when inside the software determines how the shellcode must be written to become functional computer code; in this software, each base was encoded using two-bit DNA (A, 00; C, 01; G, 10; T, 11). This creates two immediate issues: (1) standard malware when translated into a two-bit DNA encoding scheme results in DNA strands that are difficult to synthesize, and (2) even if the shellcode can be synthesized into DNA, it is unlikely to be read by the sequencer in a way that will result in functional malware.

There were three main synthesis limitations we encountered when we attempted to encode standard shellcode into two-bit DNA. The first was an excessive number or repeated bases which are difficult to synthesize; typical synthesis services limit repeated bases to 10 bases or less. DNA repeats result from shellcode because it is normally repetitive and contains memory pointers with long stretches of 0 s or 1 s. The second issue was skewed GC-content, the ratio of G/C to A/T that has to be relatively balanced for stable, and easy-to-synthesize DNA molecules. Finally, the repetitive property of shellcode would result in secondary structures in the synthesized DNA—single-stranded DNA folding on itself—again due to common repeating patterns.

Even if shellcode could be synthesized, it would not function as intended after sequencing. The shellcode needs to be very small to fit within the length of a single synthesizable DNA (approximately 100–300 bases) or else be stitched back together later down in the data processing pipeline. Another problem was randomness and error tolerance: many reads contain miscalls (incorrectly read bases) that would alter

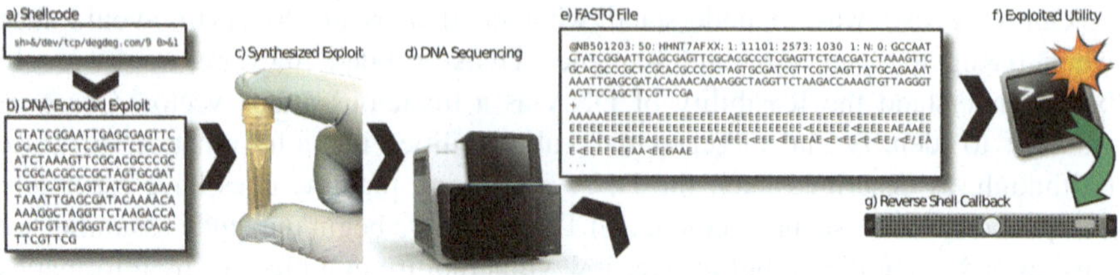

Fig. 1 End-to-end exploit of a computer using DNA. The shellcode was designed, converted into DNA, synthesized, sequenced, and processed by the vulnerable software. After execution, the machine was compromised via a reverse shell

computer instructions, and you cannot predict in advance which direction a strand will be read, which means the shellcode could be reversed.

After repeated design and testing, we were able to construct shellcode that could successfully navigate the synthesis and sequencing limitations; however, this shellcode was much smaller and less robust than shellcode and malware seen in more realistic scenarios. The shellcode strand was ordered using a commercial synthesis service, sequenced using an Illumina NextSeq 500 and run through the vulnerable utility program. The DNA malware compromised the machine which ran the software and gave us full remote code execution via a reverse shell. (See Fig. 1 for a graphical overview of the process).

While we were able to construct an end-to-end exploit in DNA, the difficulty we had in making functional malware for such a rudimentary example means that DNA-encoded malware is not an active concern in most sequencing applications. Yet, as we discuss next, the state of bioinformatics software security is quite poor, and emerging technological trends and use cases in biotechnology make this an important vector to study in the future. Many of the constraints we faced, like short read length, are changing with newer sequencing technologies (e.g., long-read Nanopore sequencers) and may make this type of attack more practical in the future. This demonstration highlights how all information vectors into computer systems, including those from physical molecules themselves, should be considered when analyzing the overall security of a system.

2.2 *Insecurity of Bioinformatics Software*

Here, we shift our attention from future-looking threats, like DNA-encoded malware, to the more immediate issue of software security in bioinformatics software. DNA sequencing is still an emerging technology domain and so much of the popular software, including software used in commercial applications, is written or maintained by small research groups as open-source projects. Much of this software is written in less secure programming languages like C/C++, which are known sources of major security vulnerabilities. Since bioinformatics is not usually considered a cybersecurity risk, compared to web servers or databases, we expect that bioinformatics software developers have less incentive to write high-security software.

To quantify this intuition, we evaluated the software security practices in 13 common sequencing applications used throughout the sequencing workflow, including several present on sequencers. (See [2] for the specific bioinformatics programs and versions we evaluated as of 2017.) All 13 of the programs were open source and written in C/C++. To create a baseline control, we also evaluated other popular open-source software written in C/C++ that we expected to be under adversarial pressure, like Internet-accessible server software. We ran both groups of software through static analysis tools to see if there was a difference in secure software practices and to identify any vulnerabilities. Compared to the controls, the sequencing software had an 11-fold increase in insecure function calls; these function calls are software libraries that have been deprecated because they have known security problems.

More significantly, the static analysis tools were able to identify a number of potential buffer overflow vulnerabilities. After manual inspection, we were able to confirm that three of these vulnerabilities could be used to crash the software, a strong sign that they may be exploitable by malware (see Fig. 2). In some of the code comments, it was clear that the authors recognized that buffer overflows were possible but did not consider them to be an immediate concern. However, we stress that it is important to patch seemingly nonthreatening vulnerabilities like these because they can cause problems in the future as the technology changes. Our security analysis was far from exhaustive, and the ease with which we were able to identify significant vulnerabilities suggests that latent security problems are probably common in bioinformatics and sequencing software.

```
#define MAX_SEQ_LINE_LENGTH (25000)
...
#define MAX_SEQUENCE_LENGTH (2000) //that's pretty arbitrary... should be enough for now
...
struct cycle_data cycles[MAX_SEQUENCE_LENGTH];
...
while ( fastx_read_next_record(&fastx) ) {
    if (strlen(fastx.nucleotides) >= MAX_SEQ_LINE_LENGTH)
        errx(1, "Internal error: sequence too long (on line %llu). Hard-coded max. length is %d",
             fastx.input_line_number, MAX_SEQ_LINE_LENGTH ) ;
    //for each base in the sequence...
    for (index=0; index<strlen(fastx.nucleotides); index++) {
        ....
        cycles[index].nucleotide[ALL].count += reads_count; // total counts
        cycles[index].nucleotide[nuc_index].count += reads_count ; //per-nucleotide counts
        ....
    }
```

```
// header->text is a string with the entire header
char * newtext = header->text;
...
// This is parsed incorrectly if the header
// included multiple LN:<num> in the same line
sprintf(len_buf, "LN:%d", header->target_len[tid]);
strcat(newtext, len_buf);
```

```
int gLineLen = 5000;
...
int lineLen = gLineLen;
char tmpStr[lineLen];
char * str; // = tempStr
...
memcpy ( str, &buf[p + 1], m - p - 1 );
```

Fig. 2 Buffer overflow vulnerabilities in sequencing software. Code fragments with buffer overflow vulnerabilities in three different next-generation programs: fastx-toolkit-v0.0.14 (top), samtools-v1.5 (bottom left), and SOAPdenovo2-v2.04 (bottom right). Text in red highlights buggy code, and text in green denotes comments we included for clarification

2.3 *Using Synthetic DNA to Communicate with a Backdoor and Bypass Air Gaps*

Using synthetic DNA as a vector for executable computer code is surprising, but this is just another example of the growing trend of using synthetic DNA for non-biological purposes. For example, DNA data storage systems are capable of archiving large digital databases into DNA and even computing with molecules [3]. This highlights how scalable DNA reading and writing—through improvements in sequencing and synthesis—make DNA an effective universal information carrier and raise the possibility that other forms of information encoded in DNA could be used maliciously. In particular, we are interested in how DNA molecules or DNA sequencing data can be used as a covert means to communicate to a malicious backdoor on sequencers or computers downstream of sequencing.

Backdoors are covert programs, alterations to software, or even physical hardware modifications that are used to bypass typical security controls and give an adversary unauthorized access and control to computer systems. Backdoors have a long history in cybersecurity and are used by a wide variety of actors for different purposes [4]. They are useful in statecraft to gain intelligence on high-value targets, as a tool for corporate espionage to steal intellectual property and as a means to destroy or incapacitate computers and equipment. Hardware backdoors are often placed into systems during manufacturing or upstream in the supply chain, while software backdoors may be inserted directly into compromised computers or even via small modifications to open-source software or data. Given the challenge of preventing and detecting covert backdoors, one approach to ensure security in high-risk environments is to air gap any critical systems; air gapping is a method of isolating machines on a secure network without direct connectivity to wider area networks like the Internet or avoiding networks all together. Air gapping reduces the cyberattack surface by physically blocking malicious messages from reaching potentially vulnerable machines.

In the context of DNA sequencing, there are a number of reasons to believe that DNA sequencers would be useful targets for backdoors. DNA sequencers process lots of sensitive medical or genetic information and are often used in research facilities to sequence valuable intellectual property, like engineered organisms or therapeutics. Sequencers may also be located in secure networks within wet labs which may give adversaries a foothold to compromise other important systems. Given the cost of high-end sequencers, they may be valuable targets for ransomware or denial-of-service attacks.

Sequencers are essentially regular computers with specialized hardware and software for sequencing—current-day Illumina sequencers run Windows 10 and have network capability, mouse and keyboard peripherals, and monitors like conventional desktop computers. Therefore, when sequencing in sensitive circumstances, it is recommended that sequencers are only connected to secure networks or disconnected altogether (i.e., air gapped). However, for a sequencer to meet its basic functions, it must be able to take in DNA samples and produce sequencing data as output, even if it is otherwise completely isolated. It is this sequencing interface—DNA samples as input and files as output—that creates a possible communication channel between adversaries and backdoors running on (supposedly) air gapped sequencers.

Consider the case where an adversary successfully inserts a backdoor into a DNA sequencer that communicated only through the sequencing interface (i.e., sequencer is air gapped). For this limited backdoor to be useful, the adversary would need to be able to direct DNA to a compromised sequencer, and if return communication is necessary (e.g., to exfiltrate data), the adversary would need the resulting sequencing data. Even with these constraints, there are a number of situations today where adversaries can direct DNA to sequencers. Outsourced sequencing facilities allow third parties to submit samples to be sequenced, and direct-to-consumer genetics enables anyone the ability to submit DNA samples through the mail to be sequenced or genotyped. (See the later section on data integrity in DNA processing for other cybersecurity issues in the consumer genetics industry.) Other possibilities include insiders with direct access to sequencers, situations where an adversary could anticipate that DNA will be sequenced (like with forensics), and even emerging technologies like DNA data storage systems where users may be able to directly specify "DNA files" to be read by sequencers. As sequencing, and DNA processing more generally, becomes more ubiquitous, we expect these possibilities to grow.

2.4 Developing a Backdoor

We developed a prototype backdoor to run inside the Illumina iSeq 100 to better understand the feasibility and challenges of constructing a backdoor that only communicates to an adversary via the sequencing interface. The backdoor was designed to receive commands from covert instructions encoded in DNA strands and write output into the DNA sequencing data. Backdoor messages written into DNA could be stealthily mixed into normal DNA samples (e.g., genomic DNA) and still be parsed and executed; sequencing regular DNA would be unaffected by the backdoor. Once decoded, the messages sent to the backdoor are treated as arbitrary commands that can be executed. Possible commands could include network mapping if the sequencer is on a private network, copying or exfiltrating data present on the sequencer, or even wiping the sequencer to render it unusable. Any output would be covertly encoded back into DNA bases and written into the sequencing output file along with the legitimate sequencing data.

When a DNA sample is processed, all of the separate DNA strands are sequenced together on a piece of hardware called a flow cell. (The flow cell has interesting security properties that will be discussed later.) The sequence of bases (i.e., A, C, G, and T) in each strand is read one at a time with a camera that detects fluorescent light emitted during the sequencing process. This data is eventually written into intermediate binary files called BCL (binary base call) files. These BCL files are later converted to plain text files that are more suitable for bioinformatics analysis;

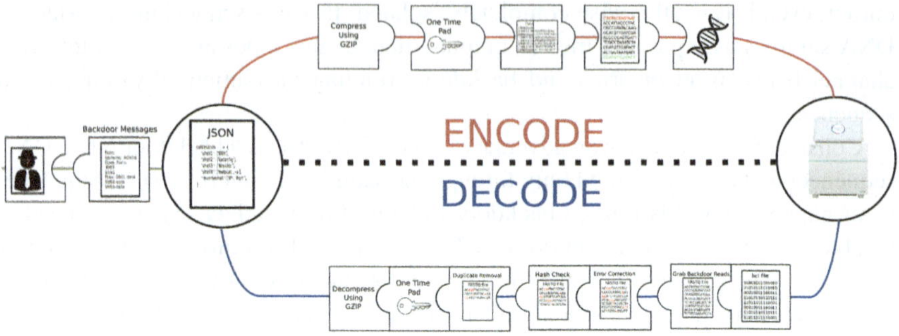

Fig. 3 *Communication life cycle for a backdoor on a DNA sequencer.* Backdoor messages and commands are sent and received from the backdoor as JSON messages. *Encoding*: (1) JSON commands are compressed; (2) data is XORed with a fixed string for "randomization"; (3) data are converted into DNA bases and broken into small chunks; (4) strands are appended with a tag and hash; and (5) reads are synthesized into physical DNA. *Decoding*: (1) BCL files are parsed and message reads removed; (2) reads are checked for proper hash and errors corrected; and (3) duplicates are removed, the "randomization" is reversed, and the data is uncompressed into the JSON commands.

however, since these files are the direct output of sequencing and are in an obtuse binary format, they make a good location for the backdoor to inject itself into the sequencing pipeline.

The backdoor functions as follows: After being inserted onto the sequencer, it reads the batches of BCL files generated during sequencing. (These are typically placed in a fixed location in the sequencer's file systems so they are easy to locate.) From these BCL files, the backdoor extracts any DNA reads meant for the backdoor, decodes those DNA reads back into the original message, executes any commands that are required, and appends any return messages or data back into the same BCL files. Viewed abstractly, this is just a specific way of encoding/decoding arbitrary digital data into and out of DNA, which is what existing DNA data storage pipelines already accomplish. So to make a backdoor, we repurposed existing DNA data storage software to function in a new context. (See Fig. 3 for an overview of this process.) While a detailed treatment of the backdoor architecture is beyond the scope of this chapter, we describe the high-level principles used by backdoor encoder, decoder, and exfiltration module below.

2.5 *Encoding Module*

The encoding module takes data messages as input—text-based JSON messages—and outputs a set of DNA sequences representing the input message. To be an effective encoder for this backdoor, it needs to meet three requirements: (1) the output DNA sequences must be possible to synthesize (see the previous section for challenges that can occur with DNA synthesis), (2) each output strand must be small enough to fit within a single read (approximately 100–200 bases), and (3) there needs to be some tag included so that the backdoor can distinguish message reads from other DNA strands, such as genomic DNA, that might be mixed in. Since sequencers read DNA strands in no particular order, this means that output strands must be designed in a way so that larger messages can be broken into pieces and later reconstructed.

To encode a message, it is sent through the following pipeline: First, desired commands are written into a text-based JSON object and compressed using gzip to reduce the message size. Next, the compressed message is XORed with a fixed string to fully "randomize" the message. This is necessary because random-like DNA bases with this encoding are more easily synthesized because it makes repeats and GC-content issues less likely. The randomized data is then broken into fixed length, 18-byte chunks, so it can fit within a single short strand, with each chunk assigned with a 3-byte index used to signify the strand ordering. The binary-to-DNA encoding scheme we used comes from an existing DNA data storage encoding scheme [5]. After encoding, each chunk+index (denoted as the *payload)* fits in 96 DNA bases. Finally, we include two last pieces of information in each strand: a fixed tag to signify the strand as designated for the backdoor and a hash of the payload that can be used to filter out reads with errors. At this point, the adversary has a set of strands, each 152 bases in length, that encode the intended message and are ready for synthesis.

2.6 Decoder Module

Decoding is roughly the reverse of the encoding process with some important differences. The decoder takes BCL files produced by a sequencer and returns the decoded message or instructions to be run by the backdoor. The decoder is filtering for messages with the correct tag prefix in the BCL files, filtering reads with errors (determined via the hash), converting the DNA bases into binary data, and placing them in the correct order to produce the final JSON message. Each specific strand will be present in multiple copies, which is helpful for redundancy; these duplicates also need to be filtered out.

2.7 Exfiltration Module

Some communications to the backdoor do not require a response. For example, if the goal is denial-of-service (DoS), the destruction of data, or a means to send malware, the command may not require a response. However, we were also interested in cases when the adversary wants to receive messages back from the backdoored sequencer. In some cases, the sequencer may be networked and so it can provide a response

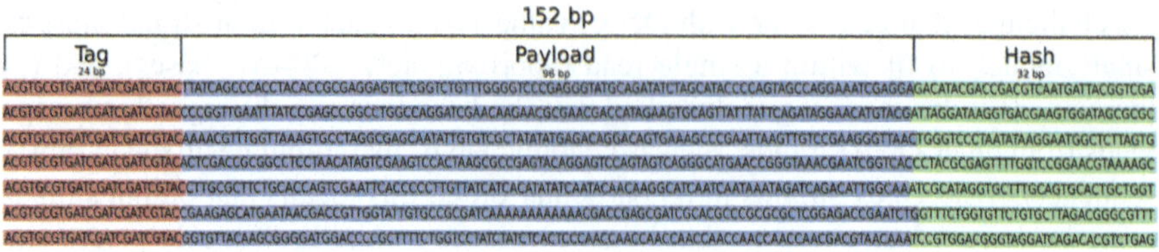

Fig. 4 Backdoor commands encoded into DNA. A seven-strand sequence encoding a JSON message instructing the backdoor software to scan the network, open a reverse shell to the adversary, run OS scans, and exfiltrate data on the sequencer into sequencing data files

back over the Internet, but to return a message via the sequencing interface, the response would have to be returned to the adversary in the sequencing data files (e.g., at an outsourced sequencing facility). To accomplish this, we encoded the response into DNA using the encoder module and appended the data to the end of the BCL files produced by the sequencing run. After receiving the BCL files, the adversary would just use the decoder as usual to parse the response.

2.8 *Prototype*

We developed a prototype backdoor designed for the Illumina iSeq in 800 lines of python code implementing the encoder, decoder, and exfiltration module described above. To keep this simple, we built it as a Windows 10 application to be run on the sequencer that would be pointed to the BCL files; however, in realistic conditions, the backdoor could be designed more covertly, for example, by including it in the OS or embedded in the sequencing software.

To demonstrate end-to-end functionality of the backdoor, we encoded five commands into a single JSON message that included network scanning, reverse shell, two OS scanning commands, and an instruction to exfiltrate data off the sequencer. When encoded, these commands could fit within DNA strands (see Fig. 4) that were ordered using a commercial synthesis service. After sequencing, the strands were properly decoded and executed, and the exfiltrated data was appended to the BCL files.

3 Cyber-Physical Security and the Molecular-to-Digital Hardware Interface

In this section, we consider the cybersecurity aspects of a different component of the sequencing workflow: the flow cell hardware used to convert physical DNA into digital data [6]. Flow cells are small, disposable fluidic devices contained within sequencers to hold the samples and reagents during sequencing. The enzymatic process of sequencing—called sequencing by synthesis in Illumina sequencers—happens on the surface of the flow cell. DNA strands stick to the surface of the flow cell and enzymes, like DNA polymerase, are used to read the DNA in the strands one base at a time; each time a base is added, it emits a fluorescent signal that can be read by high-resolution cameras. The flow cell, therefore, is the primary piece of hardware responsible for the molecular-to-digital conversion.

An important theme in cybersecurity is that interfaces and boundaries between different systems are common sources of security issues. The reason is that engineers may have unspoken assumptions or different mental models about what is possible on either side of the boundary. Regarding flow cells, we suspected that the molecular nature of DNA could lead to unanticipated information security risks because of properties inherent to physical DNA molecules. For example, DNA molecules are stable for long periods at room temperature and can be enzymatically amplified (i.e., copied). This may have implications for unauthorized data recovery because usable information may be recoverable from improperly disposed flow cells. Another security concern has to do with how different DNA samples are mixed together to improve sequencing efficiency. So-called multiplex sequencing is a technique whereby DNA samples are combined into one solution, sequenced together, and data from the individual samples are separated out later in software. However, we show how small errors in this process, due to sequencing chemistry and flow cell design, can be leveraged by adversaries to maliciously alter the genetic interpretation of other samples.

3.1 Data Remanence

Flow cells used by the most popular Illumina sequencers are meant to be single use and discarded after sequencing. However, there is little guidance on how to properly dispose of flow cells, and so oftentimes flow cells are just thrown into the trash. We hypothesized that discarded flow cells contain enough residual DNA to recover sensitive information from a previous sequencing run. This is related to data remanence attacks against traditional magnetic hard drives; residual representations of data still remain on disks even after file deletion, and improperly wiped and discarded hard drives create an information security risk.

We developed a simple protocol to collect the residual DNA stuck to the flow cell by flushing laboratory-grade water multiple times through the flow cell's fluidic channel and collecting the wastewater. This waste contains a portion of the residual DNA on the flow cell in solution where it can be amplified (i.e., exponentially copied) using polymerase chain reaction (PCR). The amplified product is resequenced to read out the "improperly deleted" data. We tested this protocol out using iSeq flow cells on two different DNA inputs: a high redundancy DNA data storage file used to evaluate error rates and file recovery and a human genome sample sequenced at low coverage (low redundancy) to let us explore the limits of data recovery.

DNA data storage files are highly redundant and tolerant to errors, so the file can still be fully recovered when there are missing strands. In this case, 96.5% of the unique strands representing the file were present in the residual sample, which was sufficient to fully reconstruct the file without error. In the human genome sample, which had much lower redundancy, we were able to recover approximately 1.8 million unique DNA reads compared to 4.4 million unique reads in the original sequencing run. This makes for a 40.1% residual recovery rate for low redundancy samples. For most genomic sequencing applications, including medical diagnosis, the 40% yield we recovered from the used flow cell would be sufficient to predict the bases in a person's genome.

These experiments show that, similar to hard drive data remanence problems [7], residual data recovery is possible on discarded flow cells. The security risk of this disclosure will depend on the specific sequencing application, but it is substantial for typical genomics and DNA data storage pipelines. Laboratories should consider flow cell remanence in their sequencing pipelines, and simple solutions like the physical destruction of flow cells may prevent unintended information leakage.

3.2 Data Leakage Between Samples

High-throughput DNA sequencers improve throughput and reduce per-sample sequencing cost by sequencing multiple samples concurrently. This is accomplished using short DNA barcodes (6–8 base strands) that are appended to all DNA strands and made unique for each sample. After sequencing, the barcodes can be used to separate each DNA read into the corresponding file for each sample (called demultiplexing). Barcoding and demultiplexing cause DNA reads to be assigned to the correct sample over 99.9% of the time. In sequencing runs producing over 100 million reads, less than 1000 reads will be assigned to the incorrect sample. This low level of improper assignment is due to particular flow cell architectures (non-patterned) and unintended enzymatic side effects during sample preparation. While this level of error is negligible in most routine sequencing applications, it can be utilized by adversaries as a way to alter other samples in a reproducible and specific manner.

To show this, we conducted the following experiment. We began with two DNA samples: one an actual human genome sample and the other a synthetic sample designed to look like the genetic variant responsible for sickle cell disease (a single base substitution of A to T). The synthetic sample was a short fragment of DNA identical to the wild-type human β-*globin* gene, except that it included the T base reflecting sickle cell trait. This fragment was inexpensively synthesized using a commercial DNA synthesis service. These two samples were sequenced in a multiplex fashion according to the usual protocols. As anticipated, enough DNA encoding the sickle cell trait leaked from the synthetic sample into the human genome index to cause the human sample to appear like a sickle cell carrier. This is possible because at any given genomic position, there is only a small amount of coverage (<200 read depth on average). Therefore, even a small amount of incorrect assignment during demultiplexing (<0.01%) can be sufficient to alter genetic interpretation.

What this simple demonstration shows is that seemingly independent samples, when sequenced together, can lead to undesirable side effects. When this side effect is not intentionally misused, its effects are negligible in routine sequencing applications; however, when directed with intelligence, side effects like this can be used adversarially. While not detailed here (see [6] for details), the synthetic DNA used in the sickle cell sample can even be spiked into tissue samples like saliva and have a similar effect after sequencing. This type of theoretical attack highlights how sequencing operators should be wary when untrusted samples are sequenced concurrently, especially if those samples can be submitted by consumers.

4 Data Integrity in DNA Processing

Previously, we have explored CBS aspects of the earlier stages of the DNA processing pipeline. Here, we focus on the security of the last component of DNA processing: data generation, analysis, and storage.

The most mature and widely accessible DNA analysis services come from the direct-to-consumer (DTC) genetic testing industry. DTC companies like 23andMe and AncestryDNA have processed genetic samples from 10s of millions of customers collected through the mail [8]. The vast majority of DTC testing is done using high-density genotyping arrays that measure individual genetic markers (known as SNPs) at approximately 500,000 locations in the genome. These tests are low cost (<$100) and give customers insights into their health, ethnicity, and other traits. The ability of DTC tests to identify close genetic relatives has been one of the most exciting applications of genetic testing and spawned the field of genetic genealogy, which combines genetic data with existing datasets, like family trees, to identify unknown relatives. Genetic genealogy has proved so successful that it has been co-opted for other uses including as an aid to forensics to identify the source of DNA samples from crime scenes, known as investigative genetic genealogy (IGG).

Data security in genetic genealogy is made especially hard because there are lots of data sharing. Some of the most popular genetic genealogy tools are third-party applications that accept genetic uploads from users directly; users are tested with a DTC service, download their raw genetic data, and then upload it to a third-party service for analysis. This approach to data sharing raises many security concerns, but we focus on the security risks derived from one problem: the lack of data authentication. We show how this fundamental problem can lead to catastrophic security risks to genetic genealogy services. The lack of authentication gives adversaries the opportunity to steal private user genotypes from genetic databases and upload corrupted results to appear like fake relatives.

In 2019, we studied GEDmatch, the most popular third-party genetic genealogy service [9]. GEDmatch runs as a web service that lets users upload files to a central

Fig. 5 DNA comparison of chromosome 3 between two users on GEDmatch. Colors indicate the degree of DNA sharing

database to run genetic genealogy analysis. GEDmatch is a favorite of genetic genealogists and law enforcement because it is an open platform that gives users fine-grained control over queries and returns extensive results and visualizations. Most significant for us is that there is no control over what kind of information can be uploaded to the service. (In fact, when law enforcement use GEDmatch for IGG, they upload artificially generated files constructed from crime scene samples.) We hypothesized that this design could lead to serious vulnerabilities for a few reasons. First, the fact that users can upload any data so long as it is formatted like a typical DTC file means that an adversary has significant flexibility over what can be uploaded to GEDmatch, including pathologically designed data. This can be combined with significant user control over what queries can be run to give an adversary a lot of leeway to target any identified vulnerabilities. The results returned by queries also include high-resolution chromosome images of DNA comparisons, a basic technique in genetic genealogy that can reveal a lot of potential sensitive information (see Fig. 5 for an example of a chromosome comparison).

To explore how the GEDmatch service architecture could lead to cybersecurity problems, we created two users on the service, one representing an adversary and the other a victim. Under the victim user, we uploaded five genetic profiles constructed from open-source data (GEDmatch allows users to upload more than one genetic data file), and to the adversary user, we uploaded artificial data designed to attack the victim profiles. We configured the privacy settings of the uploaded files to not interfere with or view any real user data. All vulnerabilities we discovered were disclosed to GEDmatch and patched prior to the publication of our work.

We were first interested how DNA comparisons used to predict ancestry can be used to exfiltrate sensitive genetic data from other users in the database. To predict a relationship, the genetic profiles of two users are compared to find long stretches of chromosomal DNA that are nearly identical (so-called matching segments). The closer the relationship, the more matching segments there will be between the two files, and the degree of the relationship can be predicted by the distribution of matching segments. To find unknown relatives, GEDmatch lets users run matching queries between files they owned and any other files in the database. However, the chromosome visualizations returned in these comparisons (Fig. 5) leak too much information about how the two files differ from one another.

GEDmatch takes significant steps to obfuscate the underlying genetic data, so to actually take advantage of this vulnerability is an involved process (see [9] for details). However, the high-level attack is straightforward. The adversary uploads specially designed artificial data files that return deterministic results depending

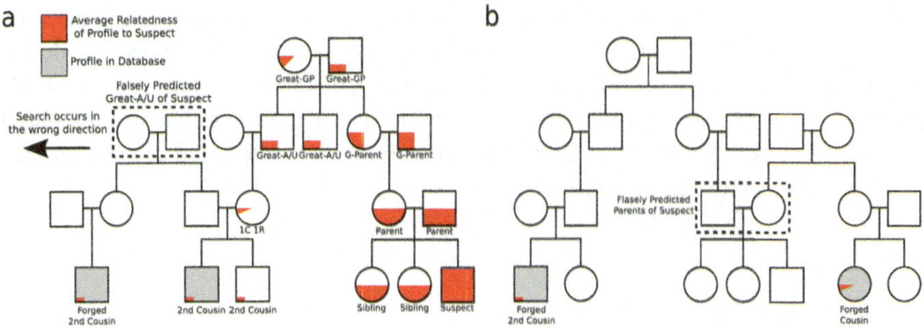

Fig. 6 Example attacks using forged relatives to imply false ancestry. (**a**) An adversary wants to avoid identification when their second cousin is already in a third-party database. The adversary uploads a falsified second cousin under the identity of a second individual that is related to the second cousin but not the adversary. This falsely implies that the adversary is on a different branch of the family tree. (**b**) The adversary uploads two falsified relatives on different branches to falsely imply that a couple was the adversary's parents

on which specific genetic markers are present in the other file—the color of the visualization can be analyzed at the pixel level to get SNP-level resolution. These "malicious" artificial files can then be compared to any file in the database to reconstruct its private genetic markers, and any missing gaps can be filled in using publicly available genetic information (via imputation). In total, we were able to predict 92.6% of the markers in each of the experimental profiles we uploaded with 98.4% accuracy using this technique. This attack could then be automated to extract data from every file in the database.

The second question was whether the relationships predicted by GEDmatch could be manipulated or forged by an adversary. This could be useful to someone trying to impersonate a lost relative or even as a tool to evade detection in an IGG investigation. Since genetic relationships are found by looking for users with shared DNA segments, any artificial data that matches segments with another user will appear like a new relative. Using an approach similar to the data extraction attack and by taking advantage of a form of compression used by GEDmatch, we were able to generate wholly artificial genetic files that could appear like arbitrary relatives for any user on GEDmatch. In Fig. 6, we show two simplified examples where forged relatives can be submitted to a genetic genealogy database to imply incorrect genealogical inferences.

The fundamental vulnerability in both of these examples comes from the lack of data authentication. There is no confirmation that uploaded genetic data actually originated from a legitimate DTC service. As long as the data is properly formatted like DTC data, any kind of file can be uploaded. When artificial uploads are allowed on a feature-rich service with complex, user-driven analysis, major security risks are almost inevitable. GEDmatch relied on techniques like obfuscation and compression, which they believed were sufficient to deter attackers. This type of faulty reasoning is a common theme in cybersecurity; obscurity is not a substitute for rigorous security design principles. For example, digital signatures or direct file transfers from the DTC companies to third-party services would have been sufficient to avoid these problems. Moving forward, we believe the lessons for the DTC industry, and any field with significant genetic data sharing is that it is essential that there are some assurances about the authenticity or provenance of any data that is shared and analyzed.

5 Conclusion

In this chapter, we performed an extensive security analysis on many stages of the DNA processing pipeline from raw DNA samples to sequencing and eventual storage and analysis. While this is not an exhaustive list of potential issues, these results show how challenging CBS can be when applied to mature technologies. Security in the biotech domain is much more complex than simply layering cybersecurity into a wet lab context because there are unique security dimensions and risks that do not exist in other domains. Like traditional cybersecurity, each technology and system will have its own unique security problems. However, there will be common security paradigms and lessons that apply across domains. We conclude this chapter with a few broad CBS takeaways that we hope are useful for biotech engineers to consider as they design and implement new technologies.

Biotechnology has a complex threat surface: Many biotechnology pipelines, like DNA sequencing, are long and complex. They involve different hardware components and take data and commands from many sources. This type of design makes security particularly hard because design choices made in one stage of the process can have effects on seemingly independent stages later on. We saw this many times in the DNA processing pipeline. For example, flow cell design affected data integrity, digital information could be encoded in physical molecules and affect downstream analysis, and the lack of authentication by DTC providers affected user-driven data sharing. Any security analysis of a biotech pipeline will need to view the entire process holistically and consider how the process and data may be used unexpectedly.

Pay attention to interfaces and data boundaries: One issue we saw repeatedly was problems at the boundary between different phases and stages of a biotechnological process, for example, places where physical data was converted into a digital form. Issues at interfaces are common in cybersecurity; they occur because there are mismatched assumptions between the designers on either side of a boundary. As with other cyber-physical systems, such as automobiles [10], biotech engineers should pay special attention to any interface or conversion point and make sure the expected behavior of the interface is understood in advance.

Confirming authenticity and integrity are critical for security: It is much harder to build security into a process when engineers do not consider the trustworthiness or validity of any input data or commands. Any part of a biotech process that is downstream of data input, whether physical or digital, should consider: (1) Is the source of this data trustworthy, and (2) is the data formed as expected? This can be accomplished by regulating access only to trusted parties when possible via authentication and authorization, but when that cannot be assured, data should be sanitized before being processed by software.

Current biotech and bioinformatics software do not follow cybersecurity best practices: Our analysis showed that bioinformatics software does not meet security standards. However, we suspect this problem is much broader across the industry because there are few incentives for secure software design. For example, laboratory management systems, control software, and robotics equipment, like liquid handlers, all have potential security problems. If true, latent software vulnerabilities could exist throughout many current and future biotechnology processes. This is reminiscent of the early stages of computers where latent vulnerabilities were not exploited until computers were later connected. The industry should get ahead of this problem and begin hardening and patching software before problems manifest.

Current trends in biotechnology make security important in the future: Biotech is becoming more automated, integrated with computers, connected, and accessible. This means that CBS problems will only become more feasible over time. While it might be tempting to address security risks through restricting information about and access to biotech systems, such an approach is insufficient. In the biological sector, there is extensive data sharing, the need for connectivity and remote access, and demand by end customers for services. Further, as we demonstrated, even the core input into many bio-systems (like DNA) is arbitrarily writable and creates additional attack surfaces not present in other security domains. Engineers need to take a more active security approach to get ahead of security problems, especially in domains with dual-use biosecurity risks.

Acknowledgments This research was supported in part by a grant from the DARPA Molecular Informatics Program, NSF Grant CNS-1565252, the University of Washington Tech Policy Lab (which receives support from the William and Flora Hewlett Foundation, the John D. and Catherine T. MacArthur Foundation, Microsoft, the Pierre and Pamela Omidyar Fund at the Silicon Valley Community Foundation), the Short-Dooley Professorship, and the Torode Family Professorship.

References

1. H. Orman, The Morris worm: A fifteen-year perspective. IEEE Security & Privacy **1**(5), 35–43 (2003)
2. P. Ney , K. Koscher, L. Organick, L. Ceze, T. Kohno, Computer security, privacy, and DNA sequencing: compromising computers with synthesized DNA, privacy leaks, and more. In *26th USENIX Security Symposium (USENIX Security 17)* (2017), pp. 765–779
3. L. Ceze, J. Nivala, K. Strauss, Molecular digital data storage using DNA. Nat. Rev. Genet. **20**(8), 456–466 (2019)
4. S. Adee, The hunt for the kill switch. IEEE Spectr. **45**(5), 34–39 (2008)
5. C.N. Takahashi, B.H. Nguyen, K. Strauss, L. Ceze, Demonstration of end-to-end automation of DNA data storage. Sci. Rep. **9**(1), 1–5 (2019)

6. P. Ney, L. Organick, J. Nivala, L. Ceze, T. Kohno, DNA sequencing flow cells and the security of the molecular-digital Interface. Proceedings on Privacy Enhancing Technologies **2021**(3), 413–432 (2021)
7. P. Gutmann, Secure deletion of data from magnetic and solid-state memory. In *Proceedings of the 6th USENIX security symposium*, vol. 14, (San Jose, 1996), pp. 77–89
8. A. Regalado, "More than 26 million people have taken an at-home ancestry test," MIT Technology Review, (2019)
9. P. Ney, L. Ceze, T. Kohno, Genotype Extraction and False Relative Attacks: Security Risks to Third-Party Genetic Genealogy Services Beyond Identity Inference. In *Network and Distributed System Symposium (NDSS 2020)*, (2020)
10. S. Checkoway, D. McCoy, B. Kantor, D. Anderson, H. Shacham, S. Savage, ..., T. Kohno, Comprehensive experimental analyses of automotive attack surfaces. In the 20th USENIX Security Symposium (USENIX Security 11), (2011)

Applying CVSS to Vulnerability Scoring in Cyber-Biological Systems

Rami Puzis and Isana Veksler-Lublinsky

Abstract With the advent of synthetic biology, security concerns are rapidly emerging spanning both the biological and the digital realms. These concerns materialize into concrete weaknesses and vulnerabilities in biological and biomedical systems and in their supply chains. Cybersecurity risks and their biological impact on biosafety and health must be considered when developing new protocols, biological systems, and supporting machinery. It is very important to assess the risk and impact of exploiting cyberbiosecurity vulnerabilities in a systematic and methodological way. The common vulnerability scoring system (CVSS) quantifies the risk and impact of vulnerabilities in digital (software and hardware) systems. Although vulnerabilities in the machinery supporting synthetic biology can be reported in a standard way, their severity scoring does not encompass the biosafety and health impacts. Furthermore, no current scoring systems exist for vulnerability assessment in the biological systems themselves (i.e., synthetic genes, biosensors, DNA chips, etc.). In this chapter, we challenge the ability of CVSS to address biosecurity and cyberbiosecurity concerns in synthetic biology by showcasing three different cyberbiosecurity attacks. We conclude that CVSS v3.1 scale is general enough to accommodate biological systems after minor adjustments of its specification. Specifically, we generalize the environmental metrics of CVSS to consider the security requirements of biological processes the same way they are considered for digital software or hardware. We further discuss a potential issue with the scope change metric of CVSS and the definition of security authority when it comes to living organisms.

Keywords Vulnerability scoring · CVSS · Cyberbiosecurity · Rubric

---

R. Puzis (✉)
Software and Information Systems Engineering, Ben-Gurion University of the Negev, Beer-Sheva, Israel

Cyber@BGU, Ben-Gurion University of the Negev, Beer-Sheva, Israel

I. Veksler-Lublinsky
Software and Information Systems Engineering, Ben-Gurion University of the Negev, Beer-Sheva, Israel

D. Greenbaum (ed.), *Cyberbiosecurity*, https://doi.org/10.1007/978-3-031-26034-6_8

1 Introduction

Assessment of vulnerabilities and their risk factors is a critical task in security management. Vulnerabilities are weaknesses in the design or the implementation of a system that can be exploited by an attacker. Vulnerabilities may be found in software [1], cyber physical systems [2], medical systems [3], and even civil construction [4]. Biologists constantly search for vulnerabilities of cancer cells to design medications, for example [5].

The advancement of synthetic biology tightens the interconnection between digital and biological processes. It includes a variety of tools used by engineers throughout the development cycle of synthetic biological systems, for example, integrated development environments for compiling genes [6], libraries of engineered genes [7], desktop DNA synthesizers and sequencers [8], bio chips [9], and more. A vulnerability in an electronic component of a system may adversely affect processes in a biological component [10] and vice versa [11].

While exploitable vulnerabilities exist in various domains, these concepts are the most mature in the domain of cybersecurity. Common Vulnerability Scoring System (CVSS) [12, 13] is a scale for prioritizing vulnerabilities according to the severity of potential attacks. Multiple criticisms exist arguing against the ambiguity of CVSS and its unjustified formulae [14, 15]. Nevertheless, it is standardized and the most widely used vulnerability scoring system available today. CVSS defines the exploitability and impact metrics, modifiers that account for the exploit code maturity, protections, report confidence, and security requirements (see Sect. 2 for details).

CVSS focuses on the CIA triad (confidentiality, integrity, and availability) as the basic security impacts but provides an extension mechanism allowing the incorporation of additional impacts. Carreon et al. [16] have recently proposed extending CVSS with health impact and data sensitivity impact to produce a Medical Vulnerability Scoring System (MVSS). In contrast, the rubric for applying CVSS to medical devices provided by the MITRE corporation refers to adverse therapeutic/medical effects as well as to potential exposure of personal health information as a part of the CIA impacts. There are also multiple additional attempts to provide vulnerability frameworks for medical devices which are not based on CVSS [17–19]. We provide additional details on these efforts in Sect. 3.

CVSS is not yet adapted to biological systems mainly because the notions of confidentiality, integrity, and availability are not defined for biological processes. Schabacker et al. [20] sketch the roadmap of cyberbiosecurity assessment in the increasingly digitized environment of synthetic biology. They also provide example interpretations of CIA in cyber-biological systems. In this chapter, we provide a biological perspective on CIA (see Sect. 4). This perspective facilitates the assessment of CVSS for biological vulnerabilities as well as vulnerabilities in digital systems that affect biological process. In Sect. 5, we demonstrate three case studies of cyber-biological vulnerabilities while motivating the choices of the CVSS metric values:

1. A weakness in synthetic DNA screening guidance allowing unauthorized production of select agents and toxins.
2. A weakness in the synthetic biology supply chain allowing replacement of DNA sequences within synthetic DNA orders.
3. A vulnerability in a cardiac monitor allowing privileged access to a cardiac implant.

In Sect. 6, we summarize the main highlights and propose the next steps in the development of a holistic vulnerability scoring system.

2 CVSS Background

The CVSS is an open framework for communicating the characteristics of a vulnerability and producing a numerical score reflecting its severity. It is used in many applications, mostly related to software and computer systems. Scores are calculated based on a formula that takes into account several metrics that approximate the ease and the impact of an exploit. Scores range from 0 to 10, where 10 represents the most severe vulnerability. The numerical score can then be translated into a qualitative representation (such as low, medium, high, and critical) to help organizations prioritize responses and vulnerability management resources according to the threat.

The CVSS is composed of three metric groups; each group is related to an area of concern: *base*, *temporal*, and *environmental*. The *base* metrics group includes *exploitability* metrics and *impact* metrics. It represents the intrinsic characteristics of a vulnerability that are constant over time and across user environments and always assume the worst-case impact on the system. The *temporal* metrics group reflects the characteristics of a vulnerability that may change over time but not across user environments. The *environmental* metrics group represents the characteristics of a vulnerability that depend on a particular implementation or environment. Next, we briefly describe the metrics. Possible values for each metric are enclosed in square brackets [.].

The exploitability metrics represent the properties of the vulnerability that lead to a successful attack (relatively to the vulnerable component) and include the following:

Attack vector (AV) reflects the context by which the attacker is able to carry out the attack [network (N), adjacent network (A), local (L), or physical (P)].

Attack complexity (AC) depicts the conditions beyond the attacker's control that are required to exploit the vulnerability [low (L) or high (H)].

Privileges required (PR) describes the level of privileges the attacker needs to carry out the attack successfully [none (N), low (L), or high (H)].

User interaction (UI) determines whether the vulnerability can be exploited by the attacker alone or whether a separate user must participate [none (N) or required (R)].

Scope (S) defines whether the vulnerable component is the same as the impacted component or if the impact goes beyond the vulnerable component [unchanged (U) or changed (C)].

The impact metrics reflect the direct consequence of a successful attack to the component that suffers the worst outcome, formally referred to as impacted component:

Confidentiality impact (C) measures the impact on the confidentiality (i.e., disclosure of sensitive information to authorized and unauthorized users) of data stored by the system [none (N), low (L), or high (H)].

Integrity impact (I) measures the impact on the integrity (i.e., assurance and consistency) of the stored data, reflecting whether the protected information has been tampered with or modified in some way [none (N), low (L), or high (H)].

Availability impact (A) measures the impact on the availability (i.e., the ability to access the data when necessary) of the stored data [none (N), low (L), or high (H)].

The temporal metrics represent metrics that change over the lifetime of a vulnerability and measure the current state of exploit techniques or code availability, the existence of any patches or workarounds, and the confidence in the description of a vulnerability. In the following list, we include the descriptions of some metric values.

Exploit code maturity (E) indicates the likelihood of the vulnerability being attacked and depends on the existing state of exploit techniques and code availability [not defined (X) and high (H), a reliable, easy-to-use, functional exploit code is available; functional (F), code is available and works in most situations where the vulnerability exists; proof-of-concept (P), code exists but might require modifications to use such code by a professional attacker; and unproven (U), no code is available].

Remediation level (RL) refers to the availability and maturity of a fix or patch for the vulnerability [not defined (X) and unavailable (U), there is no mitigation or patch available for the vulnerability; work-around (W), an unofficial solution is available or users of the affected technology can create a patch of their own; temporary fix (T), there is an official but temporary fix available; official fix (O), a complete vendor solution is available].

Report confidence (RC) measures the confidence level in the existence of the vulnerability as well as the credibility of the known technical details [not defined (X) and confirmed (C), either the vendor has confirmed that the vulnerability exists, reproduction of the vulnerability has been proven, or source code is available to confirm the issue; reasonable (R), important details are published but the vulnerability has not been independently verified; and unknown (U)]. There are reports that indicate the existence of the vulnerability, but the validity of those reports is questionable or the vulnerability is not consistently reproducible. We skip the report confidence metric (assuming **RC:X**) in all case studies because this metric will be removed in future versions of CVSS.

Not defined (X) in all three metrics above indicates there is insufficient information to choose one of the other values and has no impact on the overall temporal score.

The environmental metrics are designed to account for the aspects of an organization that might affect the severity of a vulnerability. Environmental metrics consist of modified base metrics and security requirements.

Modified base metrics override individual base metrics based on specific characteristics of a user's environment and include modified attack vector (MAV), modified attack complexity (MAC), modified privileges required (MPR), modified user interaction (MUI), modified scope (MS), modified confidentiality (MC), modified integrity (MI), and modified availability (MA). These metrics receive the same values as the corresponding base metric described above, as well as the not defined (X) value.

The impact subscore modifiers – security requirements (CR, IR, AR) help in customization of CVSS score based on the affected IT asset to a user's organization: confidentiality requirement (CR), integrity requirement (IR), and availability requirement (AR). Security requirements are assigned one of four values: [not defined (X) – no impact on the overall environmental score; or high (H)/medium (M)/low (L) corresponding to catastrophic/serious/limited adverse impact on the enterprise due to loss of confidentiality, integrity, or availability].

3 Related Work

CVSS is widely used in all branches of cybersecurity. It is continuously adapted to new cybersecurity ecosystems and new flavors of attacks [16, 21–23]. The introduction of networking capabilities into the medical domain, in particular into medical devices, greatly increased security- and privacy-related attacks. Based on Food and Drug Administration (FDA) guidance, policy, and regulation, medical device manufacturers need to assess the severity of vulnerabilities as part of their risk assessment process, both during product development and as part of post-market surveillance after the product has been cleared or approved [24].

MITRE corporation provided guidance for utilizing CVSS for assessing the risk of attacks on medical devices [22], highlighting its values in providing a consistent and standardized way to communicate the severity of a vulnerability between multiple parties, including the medical device manufacturer, hospitals, clinicians, patients, National Cybersecurity and Communications Integration Center (NCCIC), and vulnerability researchers.

A variety of security risk assessment frameworks for medical devices were proposed both by academia and industry, which either adapt and extend the traditional CVSS framework or suggest an alternative.

The Risk Scoring System for Medical Devices (RSS-MD) was developed [17] to account for the potential impact of a software vulnerability in a medical device on patient safety. This medical device variant of CVSS incorporates two categories, functional impact and vulnerability characterization, to discern risk scoring. The functional impact category considers the impact on patient therapy and the scope of impact. The vulnerability characterization category considers the attributes of the identified vulnerability, similar to the CVSS base metrics, and includes new factors such as the duration of the attack or the chain of exploitation. The rubric also explicitly refers to personal health information, diagnosis, clinical workflows, etc. when considering the CIA impacts.

Stine et al. presented a cyber risk scoring system for medical devices (CRSSMD) [18] which relies on a security questionnaire (based on the spoofing, tampering, repudiation, information disclosure, denial of service, and elevation of privilege (STRIDE) model). Their scoring system has two components: (i) a worst-case assessment of the outcome if the medical device were to be compromised and (ii) an assessment on the security features of the target device. The scoring system is intended to aid healthcare organizations in identifying medical devices with the potential to endanger patient health or disrupt patient care.

Mahler et al. [19] present the threat identification, ontology-based likelihood, severity decomposition, and risk integration (TLDR) methodology for risk assessment of attacks on medical devices. To estimate the likelihood of the attacks, they integrate multiple data sources including the Common Attack Pattern Enumeration and Classification (CAPEC). The impact of the attack is left out of the TLDR scope. Later, Medical Vulnerability Scoring System (MVSS) [16] was proposed to extend the CVSS with health- and privacy-related impacts of attacks on medical devices. MVSS includes two new parameters, health impact and sensitivity impact, as part of the CVSS *base* metrics. In their framework, sensitivity indicates the importance of the type of data that can be stolen from the device which can range from simple device data to patients' personal information. Health impact measures the potential impact on the safety of the patient if the vulnerability is exploited (e.g., potential harm, life-threatening).

However, there are no vulnerability scoring systems that take into account the specifics of synthetic biology and bio-manufacturing ecosystem.

4 Biological Perspective on CVSS

4.1 Impact (Confidentiality, Integrity, and Availability)

In the security of digital systems, *confidentiality* refers to the ability of the system to prevent unauthorized access to information. *Integrity* refers to the ability of the system to prevent unauthorized and unintended alteration of data and computational processes. *Availability* refers to the ability of the system to provide functionality to authorized users continuously and without interruptions.

We would like to examine the concepts of confidentiality, integrity, and availability (CIA) from the perspective of biological systems and hybrid systems involving

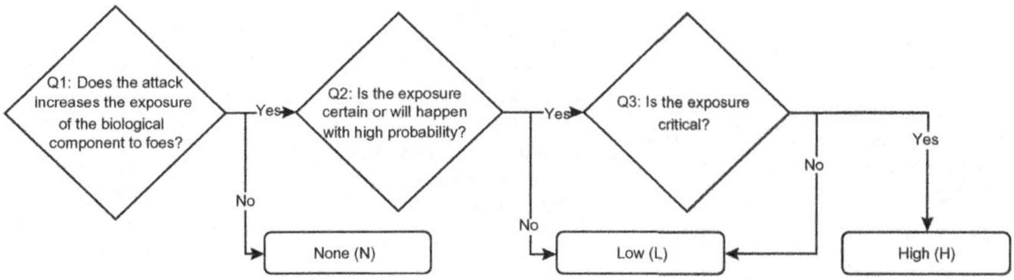

Fig. 1 Proposed confidentiality rubric for cyber-biological attacks

digital and biological components. It is hard to define the concept of data within the scope of a biological system. Genes encoded in the DNA are commonly referred to as data. These genes are transcribed to RNA. Some RNA molecules are translated to proteins and others are involved in the control mechanisms inside the cells. In either case, the information inside the DNA comprises the biological processes in the broadest sense, including intra- and extracellular processes.

In the context of synthetic biological systems, *confidentiality* can be compromised by reverse-engineering the system. For example, a company producing genetically engineered corn would like to reduce the risk of leaking their intellectual property through DNA sequencing of their product [25]. Camouflage and mimicry are examples of techniques used by organisms to conceal their presence. Hindering such abilities at the molecular level or the level of the whole organism can be considered as an impact on biological confidentiality. Figure 1 presents a rubric for assessing the confidentiality impact on a biological component.

The biological concept most closely related to *integrity* of digital systems is homeostasis – an ability of a biological system to maintain structural and functional stability. We, therefore, refer to integrity not only as genome integrity [26] but as the integrity of the entire biological processes. In contrast to confidentiality, integrity is enforced in living organisms through a variety of controls, for example, preserving genome integrity in human cells via DNA double-strand break repair [27]. DNA mutations that lead to cancer are the most prominent example of integrity breach. Cancer cells continue to function but their functioning violates homeostasis. Figure 2 presents a rubric for assessing the integrity impact on a biological component.

Finally, similar to its digital counterpart, *availability* of a biological system refers to its continuous uninterrupted functioning. The most intuitive example of an availability breach is the toxins. For example, cyanide intervenes with the ATP (adenosine triphosphate) production process within mitochondria. It prevents cells from producing energy, causing rapid death [28]. In nature, this toxin is used by plants as a defense against herbivores [29]. Figure 3 presents a rubric for assessing the integrity impact on a biological component.

The rubrics we suggest here for assessing the CVSS impact metrics for biological components are defined along the lines of CVSS 3.1 specification and the corresponding rubrics in CVSS user guide [30]. In general, whenever a compromised

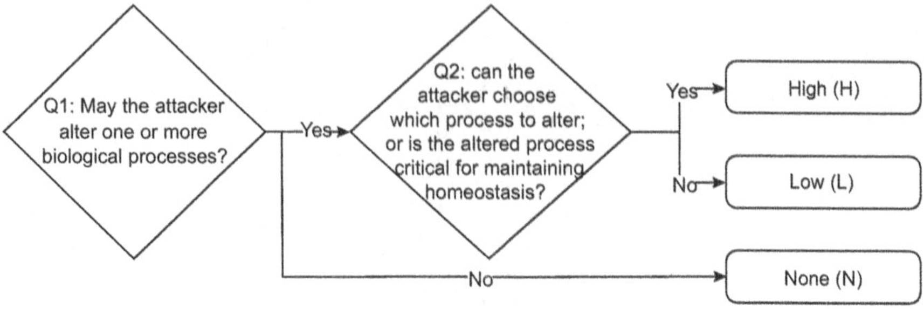

Fig. 2 Proposed integrity rubric for cyber-biological attacks

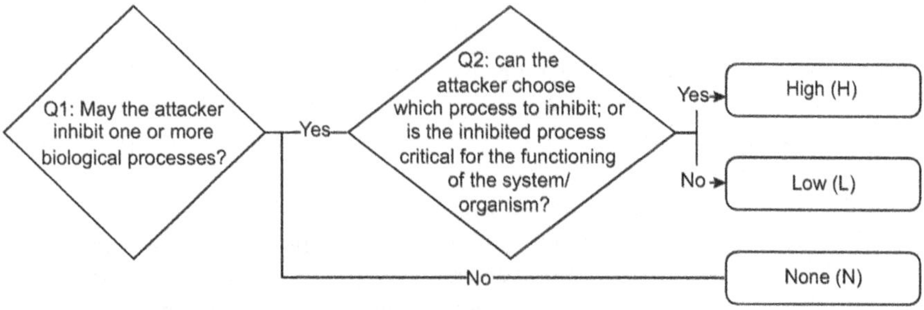

Fig. 3 Proposed availability rubric for cyber-biological attacks

integrity of biological processes or their inhibition may cause disease, injury, or death of the impacted organism, the impact should be considered high.

According to the CVSS specification, the impact subscore modifiers represent the importance of the affected IT asset to a user's organization, measured in terms of confidentiality, integrity, and availability. The guideline should be extended to include any digital, physical, or biological asset. The integrity and availability requirements of biological processes in humans should be considered as high. The CIA requirements in other biological systems should be judged according to the system critically similar to the information technology (IT) systems and industrial control systems (ICS). It is very important to use the modified CVSS impact subscores for all attacks affecting biological components.

4.2 Scope Change in Digital-Biological Environments

According to the specification of CVSS v3.1 [13], scope change occurs when "an exploited vulnerability can affect resources beyond the security scope managed by the security authority of the vulnerable component." This definition remains unchanged in the rubric for applying CVSS for medical devices [22]. Related work that we identified so far, including the original CVSS specification and the rubric for medical devices, refers only to the digital components as vulnerable or impacted. A holistic perspective on cyberbiosecurity requires that we consider biological components as vulnerable or impacted components alongside their digital counterparts.

Similar to the biological perspective on the attack impacts, we argue that scope change specification needs to be interpreted from a biological perspective as well. What is a security authority in the case of a biological process? Can some metabolic regulation mechanisms be regarded as security controls? While today these questions are mostly philosophical, future research and development of engineered biological systems will have to tackle them. At this stage, we can safely make only the following claim: unless a vulnerable digital component directly participates in a biological process, for example, thrombolysis blood nanobots [31], a biological process is outside the scope of the vulnerable digital component. We propose the following rubric, depicted in Fig. 4, to determine the scope change in case of a cyber-biological vulnerability assessment.

According to CVSS specification, when both the vulnerable component and impacted component are affected, the analyst should assess the CIA impact that is most severe [13]. CVSS v3.1 specification does not refer to the effect a scope change may have on CIA requirements. Consider the following example: Exploitability metrics are set to their highest values. The scope is changed. The impact of an attack on the availability of a vulnerable component X is low, but its availability requirement is high (CVSS=6.7). The impact of an attack on the availability of the impacted component Y is high, but its availability requirement is low (CVSS=6.3). CVSS does not specify whether the requirement score needs to be changed alongside the impact and how to choose it. Setting only the availability impact to the highest value among the vulnerable and impacted components will result in an unreasonable value of CVSS = 10 rather than the highest among the

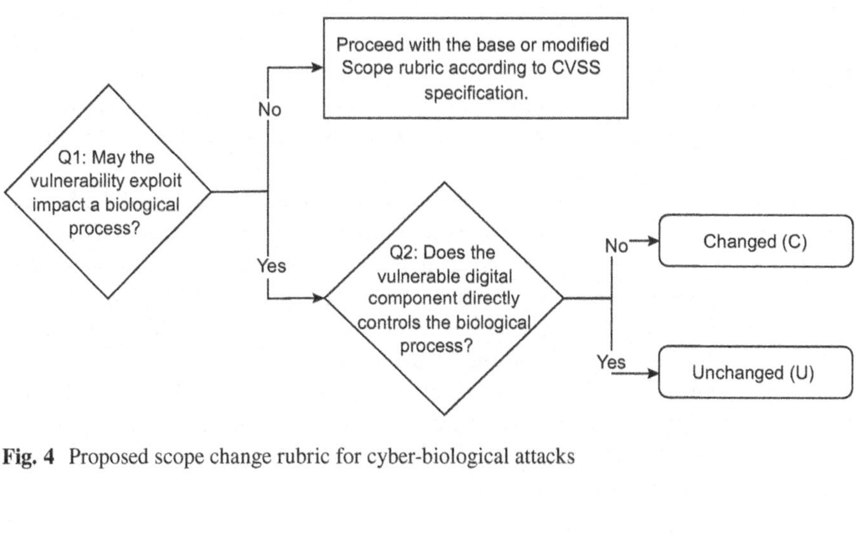

Fig. 4 Proposed scope change rubric for cyber-biological attacks

overall scores. This issue is stressed in cyber-biological and medical domains where the security requirements of the biological components are expected to be high.

In this chapter, we choose the highest overall score for either the vulnerable (digital) or the impacted (biological) component. We do so separately for each impact-requirement metrics pair.

5 Case Studies of Cyberbiosecurity Threats

In this section, we present three case studies assessing the severity of vulnerabilities in synthetic DNA order screening, synthetic DNA order integrity, and privileged access to a cardiac monitor (see Table 1). Each one of the cases is discussed in depth providing motivated severity score metrics. Special attention is paid to the change of security scope from the digital realm to the biological realm, either in a wet lab environment or directly affecting a human organism.

5.1 Case 1: DNA Screening: Best-Match Weakness in the HHS Guidance

DNA synthesis companies, which produce and ship the DNA sequences, are an important element of the growing synthetic biology market. Synthetic DNA is available in multiple ready-to-use forms, such as a plasmid or a retrovirus. A synthesized plasmid can be inserted into an organism by following a simple biological protocol, after which it can start producing proteins [32].

Some DNA sequences may encode extremely dangerous products, such as toxic peptides, viruses, pathogens, etc., collectively called sequences of concern. Multiple synthetic DNA providers have joined forces to limit the availability of sequences of concern for permitted use only. While there are still companies that provide any synthetic DNA, without screening, to all their customers, the legislation has already started to recognize the need for rigorous inspection of the synthetic DNA orders [33].

Some companies that do screen synthetic DNA orders have adopted the Screening Framework Guidance for Providers of Synthetic Double-Stranded DNA (HHS guidelines), published by the US Department of Health and Human Services. HHS guidelines suggest methods to minimize the risk of unauthorized distribution of select agents [34]. The HHS guidelines recommend screening to detect sequences of concern in the most specific manner to avoid false-positive alerts.

It is generally advised to use a sequence alignment tool, such as BLAST (the Basic Local Alignment Search Tool) [35], to compare synthetic gene orders with known sequences in the GenBank database [36]. HHS guidelines recommend the best-match approach to determine the legitimacy of an order based on the

Table 1 Summary of the CVSS scores for three cases in Sect. 5

		Metrics/case study	Case 1	Case 2	Case 3	Used values
1	Environmental score metrics					
1.1	Exploitability metrics	Attack vector (MAV)	P	L	P	P, physical, and L, local
		Attack complexity (MAC)	H	L	H	L, low, and H, high
		Privileges required (MPR)	N	L	N	N, none, and L, low
		User interaction (MUI)	N	R	N	N, none, and R, required
		Scope (MS)	C	C	C	C, changed
1.2	Impact metrics	Confidentiality impact (MC)	N	L	L	N, none, and L, low
		Integrity impact (MI)	H	H	H	H, high
		Availability impact (MA)	H	H	H	H, high
1.3	Impact subscore modifiers	Confidentiality requirement (CR)	L	L	L	L, low
		Integrity requirement (IR)	H	L	M	L, low; M, medium; and H, high
		Availability requirement (AR)	H	L	H	L, low, and H, high
2	Temporal score metrics	Exploit code maturity (E)	U	F	U	U, unproven that exploit exists, andF, functional exploit exists
		Remediation level (RL)	W	W	W	W, work-around
		Report confidence (RC)	X	X	X	X, not defined

classification of the most similar sequence in the database. Specifically, every fragment of 200 bp in the ordered sequence is searched within the database using the sequence alignment tool. If the best match of any fragment is a sequence of concern, the order is deemed a hit, and it is forwarded for further investigation.

Relevant vulnerabilities: A generic weakness in DNA screening guidelines advised by HHS permits an adversary to avoid detection by obfuscating the malicious DNA [37]. The best-match approach coupled with a screening window size of 200 bp introduces a weakness in the design of the HHS guidelines that allows hiding small fragments of the sequences of concern within larger sequences of benign genes. The scores of the alignments with the benign gene and the sequence of concern (SOC) are affected by the size of the sequence of concern within the 200 bp window. Reducing the size of a SOC fragment reduces the likelihood of detecting the sequence of concern.

Threat model: Here we assume a common bio-security scenario where an attacker, for example, a bioterrorist, attempts to produce dangerous substances at his own facility. The goal of the attack phase we focus on here is obtaining a DNA sample sufficient to produce toxins for a small-scale bioweapon. The attacker's sophistication is medium as of an experienced do-it-yourself (DIY) biologist, but his resources are very low. He has no desktop DNA synthesizer and no facilities to produce synthetic DNA from oligos.

Possible attacks: Under the assumption that the attacker must order synthetic DNA from a company adhering to the HHS guidelines, he may exploit the weakness in the guidelines to obtain a toxin-encoding DNA [37].

Exploitability metrics: *Attack vector* is physical (**MAV:P**) since the attacker has physical access to the dangerous substance. The *attack complexity* may be considered as high (**MAC:H**) as the obfuscation method described by Puzis et al. [37] requires experience with biological methods (such as CRISPR gene editing and cell transformation) and requires some effort in preparing the attack DNA. No prior *privileges* are required to exploit the weaknesses in the DNA screening procedures (**MPR:N**) since any new customer may order synthetic DNA. An exploit includes only the *interaction* between the attacker and the online DNA ordering system in most cases (**MUI:N**).

Scope change: The *scope* of the attack may change since the impacted components are the victims against whom the synthesized DNA molecules may be used (**MS:C**) as opposed to the vulnerable component which is the DNA screening machinery. Due to the scope change, at later stages, the victim of the terrorist's attack may be impacted as well.

Impact metrics: There is no impact on *confidentiality* (**MC:N**). Without a change of the attack scope, the vulnerable component (the DNA screening system) does not suffer from availability breach but does suffer from a low integrity breach. The latter is due to the attacker's ability to order toxin-producing DNA. With the change of the attack scope, the impacted components are victims of the potential terrorist attack executed using the ordered toxin DNA. This change of scope requires the attacker to perform multiple additional steps before the actual impact (see attack complexity). But depending on the specific type of the ordered toxin, the victims of

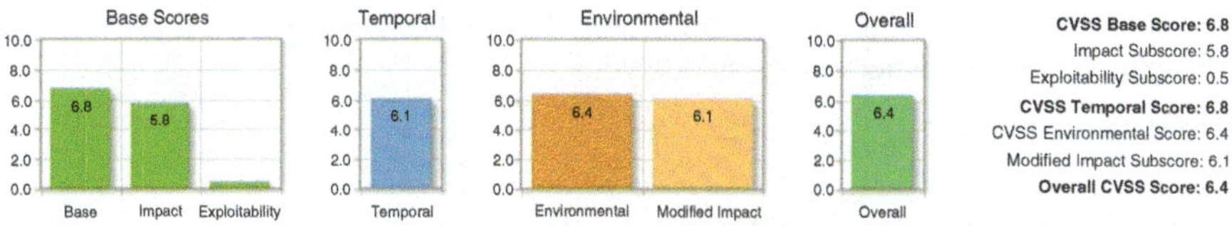

Fig. 5 CVSS score for the DNA screening best-match weakness in the HHS guidance

the attack may suffer from high loss of *integrity* or *availability* of their biological processes leading to serious health impacts. Following the CVSS specification, the most severe outcome should be reflected (**MI:H,MA:H**).

Impact subscore modifiers – (CIA) requirements: The confidentiality of the synthetic DNA orders and the customers should be preserved. Leakage of this information through the DNA screening software to the Internet or competitors may have a limited adverse effect on the customers. Therefore, we set the *confidentiality requirement* to low (**CR:L**). Due to the changed scope of the attack, we consider the potential victim of a terrorist's attack as the subject of the integrity and availability requirements. Toxins may have catastrophic adverse effects on the victim; thus, we set both requirements to high (**IR:H,AR:H**).

Temporal metrics: The *maturity* of the discussed exploits is twofold. First, the digital part of the DNA obfuscation is at the level of a *proof of concept* (E:P)[1]. The DNA was proven to avoid detection using multiple open-source screening tools such as SeqScreen, InterProScan, etc. [37]. The biological feasibility of decoding the obfuscated DNA is not yet published. Thus, we can temporarily set the biological *exploit code maturity* to "unproven that exploit exists" (E:U). Since both parts of the exploit are required for the attack to make impact, we can safely assign the minimal *exploit code maturity* level to the overall attack (**E:U**).

Attacks exploiting the weakness in the HHS screening guidelines can be implemented using a screening algorithm that does not follow the best-match principle of the HHS guidelines. Possible robust solutions may include machine learning algorithms, deep learning, or even adversarial artificial intelligence targeted specifically to withstand obfuscation. Since this is an active area of research, some companies already employ screening tools that may be invulnerable to attacks exploiting the best-match principle of the HHS guidelines. Thus, for specific synthetic DNA providers, the *remediation level* score may be an official fix (RL:O). Yet, their solutions remain mostly unpublished. There are also published workarounds such as the gene edit distance algorithm [37], leading to at least a work-around level of remediation for most synthetic DNA providers (**RL:W**).

Figure 5 summarizes the CVSS score for Case 1. Despite the high impact, the base score is at the medium level (4.0–6.9) according to the CVSS scale due to the physical attack vector and high attack complexity. The overall score is further

[1] Metric values with regular face are used for the discussion. Final metric values are stated with bold face.

lowered due to temporal metrics indicating that no working biological protocol of the exploit was published for the attack yet.

5.2 Case 2: Synthetic DNA Order Integrity

Most communication with gene synthesis companies, including gene orders, takes place through a company's website or email. All synthetic gene orders are validated prior to purchase and during production. Unfortunately, most validation reports are delivered through the same channel which, in the case of an attack, is presumably controlled by the attacker. Standard end-to-end encryption provided by HTTPS does not help when the data is corrupted, for example, by a malicious browser plug-in. Some projects, such as InterProScan [38], provide MD5 checksum for large downloads.

Relevant vulnerabilities: Most synthetic DNA providers do not request electronic signatures for data upload potentially allowing a man-in-the-browser attack [39] that replaces a DNA order with an altered sequence.

Threat model: Here we assume that a victim might be a do-it-yourself (DIY) biology enthusiast or a small bioengineering company that develops its own DNA sequences or combines existing genes to produce fuel, medical components, or resilient plants. We assume that the victim does not use their own facilities to produce the DNA but prefers ordering synthetic DNA strands from synthetic gene providers.

An attacker need only possess the resources of an average individual and an intermediate level of sophistication [40]. For example, they must be able to write a Trojan plug-in for a browser and successfully execute the man-in-the-browser attack technique. The biological sophistication of the attacker is very low, at the level of a script kiddie with high school knowledge in genetics. The goal of the attack is to sabotage the experiments or the production process of the victim.

Possible attacks: Assume an attacker that targets the victims via synthetic biology forums where the attacker promotes a Trojan browser plug-in. In addition to malicious functionality, the Trojan plug-in implements some useful functionality to convince the victims to install it. The useful functionality can be, for example, an automatic annotation of the DNA with common features and visual presentation of the DNA sequence on the web page instead of the incomprehensible text string. The malicious functionality of the Trojan plug-in can be a sabotage of the ordered DNA sequence. Possible sabotages in the DNA order can range from minor changes to the DNA sequence that may disrupt its functionality, for example, removing all ATG subsequences, required for the translation of a DNA sequence into proteins, up to replacing the whole DNA sequence with a predesigned sequence containing toxin-producing DNA. During the production of the DNA and its delivery, the attacker inspects the provider's website pages related to the order including progress and quality reports, for all information concerning the specific DNA sequence ordered.

Exploitability metrics: *Attack vector* is local (**MAV:L**) since the attacker has digital access to the synthetic DNA order by residing on the same local machine. The attack complexity may be considered as low (**MAC:L**). Although the attack requires some preparation to study the vulnerable website of the synthetic DNA provider, the attacker can expect repeatable success in replacing the synthetic DNA orders of many different synthetic DNA customers. Low *privileges* are required on the victim's computer (**MPR:L**) by the Trojan browser plug-in. The plug-in can perform the same HTML manipulations that the regular website can. *User interaction* is required (**MUI:R**) for successful exploitation of the synthetic DNA order. The victim should at least submit the synthetic DNA order.

Scope change: The *scope* of the attack may change (**MS:C**) from the synthetic DNA order to the recipients of the DNA product, depending on the nature of the malicious DNA insert and the biological protocol executed by the victim biologist. For example, the synthetic DNA may be inserted into cells as a part of some biological experiment or used as primers in a PCR test. The scope change depends on the victim's actions stressing that user interaction is required.

Impact metrics: There is a low impact on *confidentiality* (**MC:L**) since the attacker can steal the DNA sequences of the synthetic DNA orders but does not control the DNA sequences submitted by the victim. Without the scope change, there is no impact on the availability of synthetic DNA order (**MA:N**), but there is a total loss of *integrity* (**MI:H**) since the synthetic DNA order can be replaced by the attacker. The scope may change to impact a biological process where the ordered synthetic DNA is used. The scope of the attack may change, for example, through transfection into cells, the DNA amplification process in PCR tests, and other protocols. The new impacted components, for example, the cells or the PCR test, may suffer from increased integrity and availability impact. For instance, an attack that replaces all ATG triplets within the synthetic DNA order with ATC will render useless as most biological experiments involve protein translation.[2] This could be regarded as a classical case of a denial-of-service (DoS) attack in the biological domain. The scope change, although not controlled by the attacker, may result in high loss of *integrity* and *availability* (**MI:H,MA:H**). A factor also contributing to the high scores of the impact metric is the ability of the attacker to sustain the effect during repeated orders of the synthetic DNA from the same provider.

Impact subscore modifiers – (CIA) requirements: Similar to the attack on DNA screening in Sect. 5.1, the *confidentiality requirement* of synthetic DNA orders is low (**CR:L**). Due to the changed scope of the attack, we consider the impacted component to be the biological substance affected by the ordered DNA. Although this biological substance can be destroyed (availability) or altered (integrity), the adverse impact of this event on the organization (the biological lab that ordered the DNA) or on the individuals involved is limited (**IR:L,AR:L**). We disregard in this discussion the hypothetical case of a toxic DNA sequence inserted within

[2] Why ATC? Because C looks like G and the replacement may be regarded as a human mistake rather than a malicious attack.

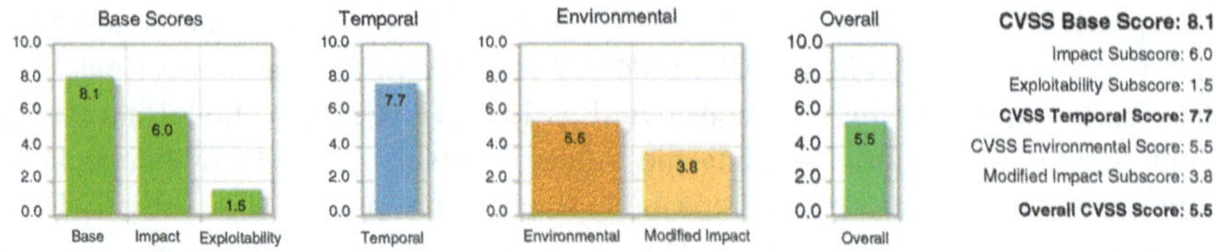

Fig. 6 CVSS score for the synthetic DNA order integrity

the synthetic DNA order because it assumes a second change of scope from the biological substance to the lab personnel.

Temporal metrics: The discussed Trojan plug-in was successfully developed and tested [37]. It works for all vulnerable synthetic DNA providers. However, the code is not widely available and may require adjustments. Thus, we set the *exploit code maturity* metric to "functional code exists" (**E:F**).

Sequencing the delivered DNA order will prevent the scope change of the attack and reduce its impact. In addition, delivering a paper copy of the ordered DNA sequence alongside the order will prevent the scope change since the substitution is likely to be identified at an early stage. We consider such *remediation* as a workaround (**RL:W**). To the best of our knowledge at the time of writing this chapter, no official fix, such as an option to add an electronic signature to the synthetic DNA orders, is provided by companies.

Figure 6 summarizes the CVSS score for Case 2. In this case, the environmental score (5.5) is significantly lower than the base score (8.1) due to the low CIA requirements we set on the integrity and availability of the synthetic DNA orders. In the case of a critical production environment, the CIA requirements may increase leading to an overall score of up to 7.9. Even in this case, the overall score is lower than the base score due to the sequencing mitigation that prevents most of the adverse impacts.

5.3 Case 3: Privileged Access to a Cardiac Monitor

Vulnerabilities and weaknesses in engineered systems may also directly affect an organism leading to health impacts. Here we refer to medical device vulnerabilities reviewed by Carre'on et al. [16]. Consider the privileged access vulnerability in the Medtronic MyCare Patient Monitor detailed in the ICS advisory ICSMA18–179-01.[3] The device monitors and controls the state and function of cardiac implants. It communicates directly with the patient's clinician allowing to collect implant's readings, such as the heart rate, and control its configuration.

[3] https://www.cisa.gov/uscert/ics/advisories/ICSMA-18-179-01

Relevant vulnerabilities: Two MyCareLink monitors (24,950 and 24,952) include hardcoded credentials (CWE-259[4]). If an attacker can remove the case of the monitor and access the debug port, then the attacker can exploit the hardcoded credentials to gain privileged access to the monitor's operating system. Furthermore, such access allows the adversary to use a dangerous function exposed by the monitor (CWE-749[5]), namely, changing the configuration of the cardiac implant provided that the implant is in close proximity to the tampered monitor.

Threat model: The attacker should be a knowledgeable individual well familiar with the internals of the specific MyCareLink monitor. During the attack, the adversary must physically tamper with the monitor and be in close proximity to the victim patient. The attacker should also have some medical knowledge concerning the operation of the implant and possible adverse effects. However, such knowledge can be assumed if the attacker is well familiar with the MyCareLink monitor.

Possible attacks: An adversary tries to physically harm a victim who has a cardiac implant and a vulnerable monitor.

Exploitability metrics: The official Industrial Control Systems (ICS) Advisory exploitability metrics assigned to this vulnerability are physical access (**MAV:P**) and high complexity (**MAC:H**), without required privileges or user (victim) interaction (**MPR:N**, **MUI:N**). Some variability in the metrics is possible due to subjective judgment. For example, the National Institute of Standards and Technology (NIST) assigns this vulnerability a low attack complexity.

Scope change: According to the official ICS Advisory, the CIA impact metrics are all high. Scope change is not indicated. The advisory considers only the impact on the confidentiality, integrity, and availability of the equipment itself, assigning the CVSS base score metric of 6.4. Carre'on et al. [16] highlight the medical impact of the possible attack assigning a higher score (MVSS = 8.89). We consider the environmental CVSS score with a change of scope to the victim patient (**MS:C**).

Impact metrics: While in control of the cardiac implant and under the assumption that the victim is not hospitalized with an external pacemaker, the attack can cause death – full loss of availability of the victim's organism (**MA:H**). We also consider high impact on the integrity of the cardiac implant's main function (**MI:H**) and low impact on confidentiality (**MC:L**) because the adversary may only have access to readings related to the victim's heart and has no control over the available information.

Impact subscore modifiers – (CIA) requirements: The availability requirement for the changed context is the highest because heart is a critical organ and human life is involved (**AR:H**). Breach of integrity, for example, a wrong pace, is likely to have a serious adverse effect but is not as catastrophic as the loss of availability (**IR:M**). Finally, we consider the effect of information about the victim leaked from the cardiac implant itself quite limited (**CR:L**).

[4] https://cwe.mitre.org/data/definitions/259.html

[5] https://cwe.mitre.org/data/definitions/749.html

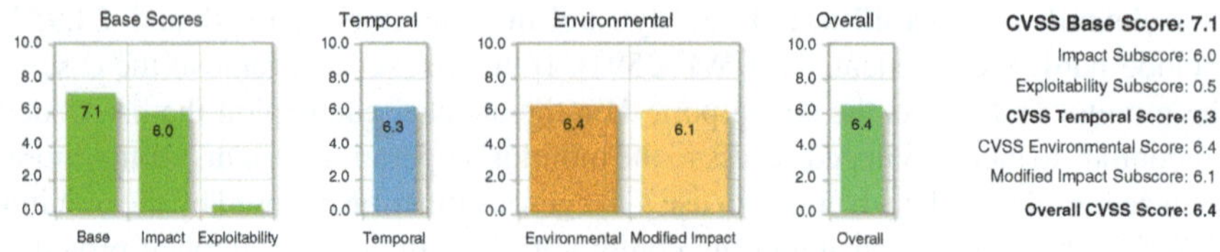

Fig. 7 CVSS score for privileged access to a cardiac monitor

Temporal metrics: The ICS Advisory specifies that no public exploits are known leading to *exploit code maturity* (**E:U**). Medtronic has released general mitigation guidelines to the users and is issuing automatic software updates. Medtronic security bulletins[6] do not contain a security update notification for the vulnerable devices. Thus, we consider the *remediation* level as a workaround (**RL:W**).

Figure 7 summarizes the CVSS score for Case 3. The overall score is significantly lower than the base score due to the temporal metrics. The fact that a workaround mitigation is available and that no proof-of-concept code was published, showing the feasibility of disrupting the implanted cardiac device, reduces the severity of breaching the monitor. In absence of the temporal metrics, the environmental score would be higher than the base score due to the high availability requirement of the implanted cardiac device.

6 Summary and Conclusions

In this chapter, we explored the applicability of CVSS to attacks affecting biological processes. The relevant adjustments to the interpretation of security metrics in the context of biological processes were proposed and applied on three case studies: DNA screening, synthetic DNA order integrity, and privileged access to a cardiac monitor. In contrast to related work, we found that the current CVSS framework is reasonably suited to accommodate future challenges associated with cyber-biological vulnerabilities. While the CVSS specification is not yet adapted for the terminology and concepts of synthetic biology, such adaptation can be performed by a collaborating team of cybersecurity and biosecurity experts. The rubrics proposed in this chapter are the first attempt in this direction.

In addition to adapting the CVSS specification and user guides to synthetic biology, the ambiguity of security requirements in case of scope change needs to be resolved. Finally, it is important to start thinking about security controls and security requirements in the context of synthetic biological systems and living organisms.

[6] https://global.medtronic.com/xg-en/product-security/security-bulletins.html

Acknowledgments This study was partially supported by the Cyber Security Research Center at the Ben-Gurion University of the Negev. "This study was supported by the Israeli Ministry of Defense."

References

1. Ivan Victor Krsul, *Software Vulnerability Analysis* (Purdue University, 1998)
2. O. Andreeva, S. Gordeychik, G. Gritsai, O. Kochetova, E. Potseluevskaya, Sergey I Sidorov, and Alexander a Timorin. Industrial control systems vulnerabilities statistics. in *Kaspersky Lab*, Report, 2016
3. A.H. Patricia, Williams and Andrew J Woodward. Cybersecurity vulnerabilities in medical devices: a complex environment and multifaceted problem. *Medical Devices (Auckland, NZ)***8**, 305 (2015)
4. M.A. Mousa, L. Dong, N. Uddin. Risk-consistent design approach for designing innovative hazard-resistant structures. in *Vulnerability, Uncertainty, and Risk: Quantification, Mitigation, and Management* (2014), 60–73.
5. F. Petrocca, G. Altschuler, S.M. Tan, M.L. Mendillo, D. Haoheng Yan, J. Jerry, A.L. Kung, W. Hide, T.A. Ince, J. Lieberman, A genome-wide siRNA screen identifies proteasome addiction as a vulnerability of basal-like triple-negative breast cancer cells. Cancer Cell **24**(2), 182–196 (2013)
6. P. Boeing, T. Ozdemir, C.P. Barnes. Design tools for synthetic biology. in Synthetic Biology Handbook, pages (CRC Press, 2016), 278–299.
7. M. Herscovitch, E. Perkins, A. Baltus, M. Fan, Addgene provides an open forum for plasmid sharing. Nat. Biotechnol. **30**(4), 316–317 (2012)
8. T. Ybert, New tools are democratizing the life sciences and enabling entrepreneurial biology: DNA script believes that the life sciences is emulating the computer sciences by deploying technology that allows small, nimble operations to engage in iterative development. Genetic Engineering & Biotechnology News **41**(4), 21–22 (2021)
9. N. Azizipour, R. Avazpour, D.H. Rosenzweig, M. Sawan, A. Ajji, Evolution of biochip technology: A review from lab-on-a-chip to organ-on-a-chip. Micromachines **11**(6), 599 (2020)
10. R. Puzis, D. Farbiash, O. Brodt, Y. Elovici, D. Greenbaum, Increased cyber-biosecurity for DNA synthesis. Nat. Biotechnol. **38**(12), 1379–1381 (2020)
11. Peter Ney, Karl Koscher, Lee Organick, Luis Ceze, and Tadayoshi Kohno. Computer security, privacy, and {DNA} sequencing: Compromising computers with synthesized {DNA}, privacy leaks, and more. in *26th {USENIX} Security Symposium ({USENIX} Security 17)* (2017), 765–779.
12. P. Mell, K. Scarfone, S. Romanosky, Common vulnerability scoring system. IEEE Security & Privacy **4**(6), 85–89 (2006)
13. Inc. FIRST.Org and the CVSS Special Interest Group (SIG), Common vulnerability scoring system v3.1: Specification document. https://www.first.org/cvss/v3.1/specification-document. Accessed 25 March 2022.
14. J. Spring, E. Hatleback, A. Householder, A. Manion, D. Shick, Time to change the CVSS? IEEE Security & Privacy **19**(2), 74–78 (2021)
15. H. Howland, CVSS: Ubiquitous and broken. Digital Threats: Research and Practice, (in press).
16. N.A. Carre'on, C. Sonderer, A. Rao, R. Lysecky, A medical vulnerability scoring system incorporating health and data sensitivity metrics. *International journal of computer and information*. Engineering **15**(8), 458–466 (2021)
17. QED Secure Solutions. Risk scoring system for medical devices (rss-md)technical specification guide

18. I. Stine, M. Rice, S. Dunlap, J. Pecarina, A cyber risk scoring system for medical devices. Int. J. Crit. Infrastruct. Prot. **19**, 32–46 (2017)
19. T. Mahler, Y. Elovici, Y. Shahar, A new methodology for information security risk assessment for medical devices and its evaluation. *arXiv preprint arXiv:2002.06938*, (2020)
20. D.S. Schabacker, L.-A. Levy, N.J. Evans, J.M. Fowler, E.A. Dickey, Assessing cyberbiosecurity vulnerabilities and infrastructure resilience. Front. Bioeng. Biotechnol. **7**, 61 (2019)
21. V.M. Vilches, E. Gil-Uriarte, I.Z. Ugarte, G.O. Mendia, R.I. Pis'on, L.A. Kirschgens, A.B. Calvo, A. Hern'andez Cordero, L. Apa, C. Cerrudo, Towards an open standard for assessing the severity of robot security vulnerabilities, the robot vulnerability scoring system (RVSS). *arXiv preprint arXiv:1807.10357*, (2018).
22. M.P. Chase and S.M. Cristey Coley. Rubric for Applying CVSS to Medical Devices. MITRE Corp., *Tech. Rep., Jan*, (McLean, VA, USA, 2019)
23. Art Manion, *Modifying CVSS for ICS and Other Meaningful Uses* (Technical report, Carnegie Mellon University Software Engineering Institute, 2019)
24. Food, Drug Administration, et al. Postmarket Management of Cybersecurity in Medical Devices. Guidance for Industry and Food and Drug Administration Staff. (Silver Spring: Food and Drug Administration, 2016).
25. E. Welch, M.A. Bagley, T. Kuiken, S. Louafi, Potential implications of new synthetic biology and genomic research trajectories on the international treaty for plant genetic resources for food and agriculture. Emory Legal Studies Research Paper (2017).
26. S. Nik-Zainal, From Genome Integrity to Cancer, (2019).
27. R.B. Jensen, E. Rothenberg, Preserving genome integrity in human cells via DNA double-strand break repair. Mol. Biol. Cell **31**(9), 859–865 (2020)
28. R. Gracia, G. Shepherd, Cyanide poisoning and its treatment. *Pharmacotherapy: The journal of human pharmacology and drug*. Therapy **24**(10), 1358–1365 (2004)
29. M.Van Ohlen, A.M. Herfurth, U. Wittstock. Herbivore adaptations to plant cyanide defenses. *Herbivores; Shields, VDC, Ed.; InTech: Rijeka, Croatia*, (2017), 29–57.
30. Inc. FIRST.Org and the CVSS Special Interest Group (SIG). Common vulnerability scoring system version 3.1: User guide. https://www.first.org/cvss/v3.1/user-guide. Accessed: March 25, 2022.
31. P. Malhotra, N. Shahdadpuri. Nano-robotic based thrombolysis: Dissolving blood clots using nanobots. in *2020 IEEE 17th India Council International Conference (INDICON)*, (IEEE, 2020), 1–4.
32. W. Mandecki, M.A. Hayden, M.A. Shallcross, E. Stotland, A totally synthetic plasmid for general cloning, gene expression and mutagenesis in *Escherichia coli*. Gene **94**(1), 103–107 (1990)
33. Rachel West and Gigi Kwik Gronvall, California shows the way for biosecurity in commercial gene synthesis. Nat. Biotechnol. **38**, 1–1 (2020)
34. Department of Health and Human Services. Screening framework guidance for providers of synthetic double-stranded DNA https://www.phe.gov/Preparedness/legal/guidance/syndna/Documents/syndna-guidance.pdf, 2010
35. S.F. Altschul, W. Gish, W. Miller, E.W. Myers, D.J. Lipman, Basic local alignment search tool. J. Mol. Biol. **215**(3), 403–410 (1990)
36. D.A. Benson, M. Cavanaugh, K. Clark, I. Karsch-Mizrachi, D.J. Lipman, J. Ostell, E.W. Sayers, Genbank. Nucleic Acids Res. **41**(D1), D36–D42 (2012)
37. Dor Farbiash and Rami Puzis. Cyberbiosecurity: DNA injection attack in synthetic biology. *arXiv preprint arXiv:2011.14224*, (2020).
38. P. Jones, D. Binns, H.-Y. Chang, M. Fraser, W. Li, C. McAnulla, H. McWilliam, J. Maslen, A. Mitchell, G. Nuka, et al., Interproscan 5: genome-scale protein function classification. Bioinformatics **30**(9), 1236–1240 (2014)
39. ICEBRG. Justin Warner. Man in the browser. In *MITRE ATT&CK*, number T1185. The MITRE Corporation.
40. B.J.R. Piazza, J. Wunder. *STIX™ Version 2.0. Part 1: STIX Core Concepts*. OASIS Committee Specification 01, 2017. http://docs.oasis-open.org/cti/stix/v2.0/stix-v2.0-part1-stix-core.html

Biocrime, the Internet-of-Ingestible-Things and Cyber-Biosecurity

Mariam Elgabry

Abstract As biotechnology continues to develop and the way that science is practised evolves, so too does the nature of crime. This chapter discusses how a crime science lens can be used to identify new forms of offending that might be facilitated by synthetic biology and related biotechnology with the aim of developing informed strategies to prevent them through an active design process. It uses an example of a future biotechnology crime – bio-malware – as identified through multiple methodologies employed, including a systematic review and a Delphi study, to demonstrate the hyBrid hAcKathon dElphi (BAKE) framework and its prospect toward a cyber-biosecurity by design policy.

Keywords Crime science · Security by design · Framework · Crime prevention · Responsible innovation

1 Introduction

The way science is practised is changing [80]. Life science is ever more integrated within the cyber-domain as laboratories become "connected" and scientific research is dependent on Internet-connected systems, tools and devices [64]. Broader community groups work in the biotechnology space due to declining costs of synthetic biology and more accessible community facilities [40, 70].

At the same time, we live in an increasingly health-centred global economy [77] but one in which security is often overlooked (e.g. Gittleman et al. [33]). The emerging field of cyber-biosecurity becomes important in the safeguarding of valuable biological information and material [58, 63].

M. Elgabry (✉)
Bronic Ltd, London, UK

DAWES Centre for Future Crime at UCL, Jill Dando Institute for Security and Crime Science, London, UK

UCL Biochemical Engineering, London, UK
e-mail: Mariam@bronic.co

D. Greenbaum (ed.), *Cyberbiosecurity*, https://doi.org/10.1007/978-3-031-26034-6_9

However, forecasting crime trends remains a challenge [3, 26]. Current evidence is fragmented and often distorted, which causes a difficulty in understanding what proportion of the known synthetic biology offending problems require prioritisation [54, 67]. The discussion of security implications is limited to siloed expertise from traditional professions, and there has been no engagement with diverse communities (e.g. Dyson and Harris [11], Minshull and Wagner [56], Feldman [30] and Lewis et al. [50]). Moreover, manufacturers continue to place security as a second priority to a fast product market launch [47], and there continues to be uncertainty in conformity for start-ups and small-medium-enterprises (SME) [45]. Security risk assessments are treated as a compliance check, and security design is not considered or taught early enough where innovation originates such as at universities [59].

The scale of identified and predicted biocrime is unknown and is expected to be more sophisticated in the future [19, 26, 58]. In the literature, authors have referred to "biocrime" as the use of a biological agent to cause harm for personal reasons [37] and "biowarfare" as the international misuse of biological agents as weapons [42]. In this chapter, I suggest that a broader definition is used to cover crime types at the intersection and consequently use biocrime synonymously with "biotechnology crime" to mean the exploitation of susceptibilities in biological tools, data/databases, devices or techniques for criminal purposes. These can be either categorically new or a combination of current crime types, enabled by both the increase in biological data created and the decreasing costs of the technology used [35]. The identification of biocrime, of course, depends on the definition of "crime." What is considered a crime can differ from border to border, highlighting the challenge of universal laws that could provide global protection, as crime is a social construct and local jurisdiction reflects the integration of ethics, culture and societal perspectives. And this is continually informed by public opinion and adoption; what may be "illegal" today may be legal tomorrow and vice versa (e.g., see Evans [29]). Therefore, in this chapter, the term "crime" takes into account these complexities and is used interchangeably with the term "misuse" to include both currently illegitimate activities that are punishable by law [4] but also emerging issues in biotechnology activities. Moreover, this broader scope of the term "crime" extends the otherwise dominant discussion in the biosecurity literature about weaponised misuse, which focuses primarily on biowarfare and bioterrorism [76]. Other aspects of biocrime are relatively neglected but important, and hence it is necessary to increase attention to this [25, 78].

It is thus critical to systematically synthesise empirical evidence from peer-reviewed primary data from a range of disciplines, including life science, computer science and criminology, in doing so, assessing the maturity of the field, to identify gaps and synthesise what is known to inform the design of crime prevention strategies. Equally so, it is imperative to elicit potential threats from a diverse panel of experts to include "biohackers", who represent a different population that experiments with these types of technologies in unexpected ways [20, 80]. By eliciting the opinions on emerging crime and security trends that may be facilitated by synthetic biology, a more informed and multi-sector policy framework can be designed and put in place to address these. Biotechnology products are often designed and manufactured to be portable (e.g. Oxford Nanopore's MinION DNA/RNA sequencing device), movable (e.g. polymerase chain reaction (PCR) machine), mobile (e.g. container laboratories), part of another object (e.g. software) or installed in place (e.g. biosafety cabinet) [13, 15]. Risk factors can therefore be considered in the design of the product to safeguard against potential crime.

The rest of the chapter is organised as follows. In the next section, the crime science lens is introduced including the main theoretical frameworks. Then, insights on the criminogenic potential of biotechnology are summarised from a systematic review and Delphi study conducted. The crime science concepts of designing against crime are presented. To conclude the chapter, a framework developed, the hyBrid hAcKathon dElphi framework (BAKE), is discussed that was used to identify "security by design" principles for a selected emerging technology test case, ingestible devices. The aim of this proof-of-concept framework was for government stakeholders such as the UK Parliament Joint Committee on National Security to consider this – or a similar model – as a "red teaming" approach that can be introduced into national security decision-making and towards cyber-biosecurity by design policy [10].

2 Looking Through the Crime Science Lens

As biotechnology continues to develop and the way that science is practised evolves, so too does the nature of crime. Consequently, following the traditional dogma of the criminal justice system is necessary but is unlikely to be sufficient. Traditionally, studying crime has been centred around the offender focusing on the aetiology of why they offend and distal influences such as genetics, parenting, early childhood experiences and neighbourhoods [1]. Crime science, on the other hand, embarks on the perspective of crime as an event and attempts to understand it through the immediate environment and less on how the offender developed [7]. The field of crime science acknowledges the offender as one element of the criminal event and aims to prevent crime by addressing its dynamics and through empirical enquiry [79]. It investigates the proximal circumstances, situational contingencies and crime opportunities to reduce crime through prevention and/or early detection [48]. In an age of great technology dependence, current and traditional criminal justice systems (e.g. police, courts, prisons, community service) are necessary but not sufficient to control crime. It, therefore, represents a shift from offender-oriented responses to event-focused ones [71].

Forms of biotechnology crime are considered here from a crime opportunity perspective, in particular the Conjunction of Criminal Opportunity (CCO) [17]. I take this approach as it provides a theoretical framework for thinking about what and who might influence the likelihood of a crime event, while also drawing together the major types of relevant interventions to block, weaken or divert these influences to prevent the crime event from occurring. The CCO was assembled from crime causation theories, including the rational choice perspective [8] and routine activities theory [31].

2.1 The Rational Choice Perspective

According to Cornish and Clarke [8], when deciding whether to commit crimes, offenders weigh the perceived costs and benefits of so doing, generally deciding to offend when the latter exceed the former. This rationality is bounded in that the true costs and benefits will be unknown – offenders will only have a perception of what they are. Situational crime prevention (SCP) therefore aims to reduce crime opportunity by reducing (perceived) rewards and increasing associated risks [5].

2.2 The Routine Activities Theory

According to Felson and Cohen [31], for a crime to occur, a motivated offender and suitable target need to converge in space (physical and/or virtual) and time in an unguarded place [57]. Absent this convergence, crime is unlikely or even impossible. For instance, when car security was introduced, there was a significant decline in vehicle theft [66]. Each element (motivated offender, suitable target and unguarded place) has a "controller" that can influence these interactions locally (e.g. a place manager and the policies adopted), which are in turn influenced by "super-controllers" (e.g. governments and internationally agencies) who have an influence on, for example, place managers and hence influence crime more indirectly [69]. Considering the role of each of these actors is thus useful in the context of preventing new or emerging crimes (hereafter "crime harvests") [61, 62] since each can influence the likelihood of crime in different ways.

One way that crime is influenced is by design and hence by designers and, in turn, design decision-makers in management and marketing who commission the designs. All of these are super-controllers in the chain of influence with many new technologies, a recurrent pattern is that of the retrospective fitting of security solutions to address the new and overlooked crime opportunities that they introduce [60]. This cycle is typically necessary because manufacturers of new products do not – and are not required to – consider security implications at the design phase. A crime harvest (emerging crime opportunity) may also occur as a consequence of the early adoption of the technology by criminals [12, 60]. These crime harvests should, in theory, be anticipated, up to a point, during the prototyping of the product. Where manufacturers consider security a priority (albeit secondary), situational crime prevention (SCP) [5] can be built in from the outset. SCP represents an opportunity-reducing approach to specific forms of crime through the manipulation of the immediate environment in a systematic way to make crime more risky, difficult and less rewarding.

Exploring how these theories from crime science inform the rapid developments of biotechnology and the commercialisation of synthetic biology is one purpose of this chapter. Accessibility to the tools of biotechnology widens participation in biology, which allows experimentation outside of the regulated institutional premise and encourages widespread use through available kits, for example, that can be found online (e.g. genetic engineering kits, The Odin) [34]. While this may have clear benefits in education, there is no registration in place to keep track of product use post-marketisation, for example, and, as will become evident in later sections, products and systems are not designed with security in mind to begin with. Historically, tools of synthetic biology were used in physical isolation and were air-gapped from the Internet [49]. Today, the design, build and test cycle of synthetic biology is computer-dependent and in some case automated and Internet connected (e.g. Linshiz et al. [51]). Unfortunately, there is an absence of cyber hygiene required to keep these systems (and the other services they use such as the cloud) secure. To illustrate the point, it is well known that connected medical devices (essentially Internet-of-Things (IoT) devices integrated for use on/within the body) have suffered from overlooked information technology security such as coding defects that have led to "hacked" insulin pumps [43]. These risks, and others, will be discussed in the context of biotechnology and synthetic biology.

3 Insights on the Criminogenic Potential of Biotechnology

The nature of crime is constantly evolving as emerging technologies [55] – such as biotechnology – may generate new crime opportunities. However, to date, there exists no synthesis of the varied malicious opportunities enabled or generated by biotechnology, either currently occurring or forecasted. To perform a systematic review across disciplines that could capture evidence of emerging crime trends, I designed a protocol[1] that adhered to the Preferred Reporting Items for Systematic Review and Meta-analysis Protocols (PRISMA-P) guidelines (PROSPERO CRD42019131685) and can be found in Elgabry et al. [18]. This systematic review approach was used to map out what new forms of offending might be facilitated by developments in synthetic biology [19]. Systematic reviews formulate research questions and identify and synthesise studies that directly relate to the systematic review question [36]. They are designed to provide a complete, exhaustive summary of current evidence relevant to a research question (e.g. Curtis and Cairncross [9]). Systematic reviews frequently inform government delivery of health care,

[1] Systematic review study overview: a three-step article identification procedure was implemented across five databases. Only 15 articles were considered for the thematic synthesis from the initial 794 hits as they did not meet the inclusion criteria (articles that explicitly mention synthetic biology/biotechnology can be a threat to person(s) in a community, have negative security implications, are/can be involved in crime or criminal exploitation or are/can be hacked). For more details, see Elgabry et al. [19].

public health and public policy (e.g. Cockbain et al. [6]). To complement the systematic review, and where the literature was lacking, a Delphi study[2] was also conducted with experts to generate further insight into future potential trends and ways of addressing them. The Delphi method, also known as Estimate-Talk-Estimate, is a structured forecasting tool which relies on a panel of experts [68, 74]. I elicited opinions from traditional and nontraditional experts [23, 26]. Together the systematic review and Delphi study were submitted as evidence to the UK Parliament Joint Committee on National Security and Biosecurity and informed the First Report [21, 38].

For the purpose of this chapter, a single crime type is selected as an example. The remaining seven distinct crime types identified from the systematic review, the rest of the findings and resulting policy brief can be found in Elgabry et al. [19, 26], and Elgabry [22], respectively.

3.1 Bio-malware: A Future Biotechnology Crime

Bio-malware or *bio*logical *mal*icious soft*ware* was identified as a future biotechnology crime, and the possibility of compromising a target software system using malware stored in physical DNA was demonstrated [81]. The practice of synthetic biology consists of integrated cyber- and bio workflows for the synthesis of complex systems with functions non-existent in nature or that modify natural systems for useful purposes. Hence, there exists the risk of bio-malware in the form of "Trojans" or malicious code used to obtain unauthorised access to or otherwise compromise systems [46].

In a laboratory setting, Ney et al. (2017) converted a known computer exploit into the four nucleotides of DNA (A, C, T, G) to make "DNA-encoded malware". The authors then artificially introduced a vulnerability into the DNA analysis software such that it would be triggered by the DNA-encoded malware.

As per typical workflows, the sample was then synthesised using Illumina sequencing to generate the reconstructed sequences in digital form (FASTQ files). Once read, the files were executed, and the DNA-encoded malware enabled remote access to the system. While this was an orchestrated attack in that the authors introduced the vulnerability for the deployed DNA-encoded malware, this approach,

[2] Delphi study overview: a parallel study was conducted to elicit opinions on emerging crime trends that may be facilitated by biotechnology from two groups, traditional and nontraditional experts. Traditional experts were identified stakeholders within academia, industry and government, which were recruited by stakeholder mapping, industry conference and security crime science network database. Nontraditional experts were biohackers who are individuals who perform scientific experimentation outside institutional premises, who may or may not have traditional (academic) qualifications [80]. Biohackers were recruited by fieldwork conducted by the first author of Elgabry and Camilleri [23]. For more details, see Elgabry et al. [26].

of anticipating or simulating scenarios of adversarial behaviour, remains rare in both bioinformatics and synthetic biology.

4 Designing Against Crime and the Internet-of-Ingestible-Things

To properly design a product against crime, according to Ekblom [13, 14], requires incorporating stakeholder's interests early in the design process. In so doing, potentially contradictory design requirements can be resolved through creative leaps that enable both security and aesthetics, for example. For this, designers of products need to adapt a different perspective and extend the "for function" thinking to "think thief", making the product not only "user friendly" but also "*ab*user *un*friendly" [13]. Moreover, this process is not static, in that offenders can adapt to existing preventative measures, "For any given preventive measure, therefore, eventual obsolescence is not a possibility but a certainty" [12].

To implement these concepts in practice, I developed a framework that comprises of the inclusion of nontraditional experts and a red teaming approach (or adopting adversarial methodology, [52, 75]) to national security through an active design process. Arguably, it is a crucial piece for achieving cyber-biosecurity [32].

4.1 BAKE: A hyBrid hAcKathon dElphi Framework

Initially published at the UK's national security machinery First Report [24, 39], the hyBrid hAcKathon dElphi (BAKE) framework [27, 28] was developed and deployed to couple the scenario building of the Delphi process with the prototyping of the hackathon exercise. The aim of BAKE is to capture insight from experts regarding the threats posed by the tested technology earlier in the product development life cycle and to systematically consider security design in the tested technology ahead of its widespread use.

Briefly, the BAKE framework involved a three-month format with three stages of prototyping (hackathon), scenario building (Delphi) and assessing technology implications (policy briefing). "Cross-pollinated" and diverse teams according to their skills (e.g. technical, theoretical) were assembled that are exposed to state-of-the-art presentations from leading stakeholders in cyber biosecurity, future crime and consumer market research. Participants were then provided with training intended to help them prototype the design of their ideated technology, after which they participated in a Delphi study with four survey rounds regarding the intended misuse of the technology in focus.

To demonstrate, an emerging technology of ingestible devices was selected as a use case. In parallel to its functional purpose, insights from the systematic review and Delphi study discussed above were used to inform the design of the ingestible device. The aim of the design process employed was to minimise the threats already identified in previous research [19, 26], and hence what follows is intended to also illustrate *how* a design process can be structured to proactively address such risks. Here, I first provide an overview of what the Internet-of-Ingestible-Things are and what it means for a product to be "secure by design". I then discuss an example design criteria applied and how this was informed by the work presented in the previous sections. This section thus serves as a test or proof of concept as to how the ideas discussed in previous sections might be applied.

4.2 The Internet-of-Ingestible-Things

Ingestible devices are the size of normal vitamin pills but are Internet connected to provide insight into the state of the gut – a relatively inaccessible [53] and highly unexplored location of the human body [72]. Part of the wider set of Internet-of-*Ingestible*-Things, ingestible devices promise the future of personalised health by measuring the impact of food, medicine and supplements as well as environmental and lifestyle changes through "in-body" biotelemetry [2, 41, 44]. However, widespread access to the Internet allows for data that can be readily seen and reviewed online (both by patient and physician), which may increase the wider environment of crime opportunity, if security is overlooked – rendering secure by design vital to the product development life cycle. A product is "secure by design", when its engineers have taken appropriate steps to ensure that the overall design of the product is free of vulnerabilities and is impervious to attack as possible from the outset, through such measures as continuous testing, authentication safeguards and adherence to best programming practices [65].

4.3 Designing Out Risk of Bio-malware Using BAKE

In the systematic review, 53% of the articles identified criminally exploitable biotechnologies associated strongly with synthetic biology to include those concerning the modification of organisms (33%). The risks associated with genetically modified living cells/microorganisms include the possibility of them being "hijacked" and "re-programmed" to cause harm. Considering this and using BAKE, the election of a biological sensing system that was "cell-free" was designed and deployed such that it mitigates this risk, as the biological sensing unit is not alive and therefore cannot be receptive to new external (malicious) "commands", for example, in the systematic review, "bio-malware" or biological malicious software, which was identified as a crime type.

This identified route for criminogenic activity was actively designed out during the early product development phase of the ingestible device and is used as an example to the application of the BAKE framework. Experimental work can be first informed by theoretical scenario-building exercises, to bi-laterally combine the complimentary benefits of each to informing understanding of the criminogenic potential of emerging technology.

5 Towards Cyber-Biosecurity by Design

The security task of crime science or of designers in arranging the situation or design of a product to favour preventers over offenders is perturbed with disruptive trends such as automation, remote monitoring, mass customisation and miniaturisation [16]. Crime accessibility, productivity and diversification are increased, as more individuals can offend, in more than one way, and, in new forms, using developing technology [73]. Other accelerants to technology include co-evolving factors such as increased population size ("more people to invent things"), pre-existing technology that can be combined, dissemination of inventors and techniques on the Internet and capitalistic competition and incentives [15, 16].

In this chapter, we focused on how changing technology can both create and prevent crime opportunity and understanding how can help anticipate, detect and respond to the many changes in the crime and security in today's integrated world. If we are to anticipate future crime opportunities and address them, the engagement with nontraditional experts to fight against modern biotechnological threats and the adoption of a red teaming and crime opportunity approach to cyber-biosecurity needs consideration. As discussed throughout this chapter, a crime science and red teaming approach, as embodied by the hyBrid hAcKathon dElphi (BAKE) framework, builds on systematically forecasting biotechnology crime forms, from both a theoretical and practical understanding, while also empirically shaping adaptive policy and regulatory governance design for disruptive technologies.

References

1. R.L. Akers, *Criminological Theories: Introduction and Evaluation* (Roxbury, Los Angeles, 1997)
2. F.N. Alsunaydih, M.R. Yuce, Next-generation ingestible devices: Sensing, locomotion and navigation. Physiol. Meas. **42**(4), 04TR01 (2021)
3. A. Blumstein, Crime modeling1. Oper. Res. **50**(1), 16–24 (2002)
4. W.A. Bonger, *An Introduction to Criminology* (Routledge, 2015)
5. R.V. Clarke, Situational crime prevention. Crime Justice **19**, 91–150 (1995)
6. E. Cockbain, K. Bowers, G. Dimitrova, Human trafficking for labour exploitation: The results of a two-phase systematic review mapping the European evidence base and synthesising key scientific research evidence. J. Exp. Criminol. **14**, 319–360 (2018). https://doi.org/10.1007/s11292-017-9321-3
7. D.B. Cornish, R.V. Clarke, Understanding crime displacement: An application of rational choice theory, in *Crime Opportunity Theories*, (Routledge, 2017), pp. 197–211
8. D.B. Cornish, R.V. Clarke, *The Reasoning Criminal* (Springer, New York, 1986)

9. V. Curtis, S. Cairncross, Effect of washing hands with soap on diarrhoea risk in the community: A systematic review. Lancet Infect. Dis. **3**, 275–281 (2003). https://doi.org/10.1016/S1473-3099(03)00606-6
10. D. DiEuliis, *Perspective: The Rapidly Expanding Need for Biosecurity by Design* (BioDesign Research, 2022)
11. A. Dyson, J. Harris, *Ethics & Biotechnology* (Routledge, 2002)
12. P. Ekblom, Gearing up against crime: A dynamic framework to help designers keep up with the adaptive criminal in a changing world. Int. J. Risk Secur. Crime Prevent. **2**, 249–266 (1997)
13. P. Ekblom, Designing products against crime, in *Handbook of Crime Prevention and Community Safety*, ed. by N. Tilley, (Willan, Cullompton, 2005), pp. 203–244
14. P. Ekblom, The private sector and designing products against crime, in *The Oxford Handbook of Crime Prevention*, (Springer, New York, 2012), pp. 384–403
15. P. Ekblom, Designing products against crime, in *Encyclopedia of Criminology and Criminal Justice*, ed. by G. Bruinsma, D. Weisburd, (Springer, New York, 2014). https://doi.org/10.1007/978-1-4614-5690-2_551
16. P. Ekblom, Technology, opportunity, crime and crime prevention: Current and evolutionary perspectives, in *Crime Prevention in the 21st Century*, (Springer, Cham, 2017), pp. 319–343
17. P. Ekblom, The conjunction of criminal opportunity theory, in *Encyclopedia of Victimology and Crime Prevention*, (Sage, London, 2010), pp. 140–146. ISBN 9781412960472
18. M. Elgabry, D. Nesbeth, S.D. Johnson, A systematic review protocol for crime trends facilitated by synthetic biology. Syst. Rev. **9**, 22 (2020a). https://doi.org/10.1186/s13643-020-1284-1
19. M. Elgabry, D. Nesbeth, S.D. Johnson, A systematic review of the criminogenic potential of synthetic biology and routes to future crime prevention. Front. Bioeng. Biotechnol. **1119** (2020b)
20. M. Elgabry, *Bio-Crime and COVID-19*, Special Series on COVID-19: No. 14 ISSN 2635–1625 (UCL Jill Dando Institute of Security and Crime Science, 2020c)
21. M. Elgabry, *National Biosecurity: Cyber-Biosecurity Written Evidence* (UK Parliament Joint Committee on National Security and Biosecurity, 2020d)
22. M. Elgabry, *Policy Brief: Synthetic Biology and Future Crime* (Dawes Centre for Future Crime, UCL, 2021a)
23. M. Elgabry, J. Camilleri, Conducting hidden populations research: A reflective case study on researching the biohacking community. Futures **132**, 102769 (2021)
24. M. Elgabry, *National Machinery: Red-Teaming Approach Written Evidence* (UK Parliament Joint Committee on National Security and Machinery, 2021c)
25. M. Elgabry, *Individual NGO Statements to Biological Weapons Convention, United Nations, Meetings of State Parties, Geneva, 22–25 November 2021* (UN Web TV, 2021d)
26. M. Elgabry, D. Nesbeth, S.D. Johnson, The future of biotechnology crime: A parallel Delphi process with NonTraditional experts. Futures **141**, 102970 (2022)
27. M. Elgabry, Towards cyber-biosecurity by design: An experimental approach to internet-of-medical-things design and development. Crime Sci. (2022b)
28. M. Elgabry, *BAKE: A Novel hyBrid hAcKathon dElphi Framework for Mapping the Security Landscape of Biotechnology* (Bronic, 2022c). https://www.bronic.co/white-paper
29. E.P. Evans, *The Criminal Prosecution and Capital Punishment of Animals* (London William Heinemann, 1906). https://www.gutenberg.org/files/43286/43286-h/43286-h.htm
30. R. Feldman, The open source biotechnology movement: Is it patent misuse. Minn. JL Sci. Technol. **6**, 117 (2004)
31. M. Felson, L.E. Cohen, Human ecology and crime: A routine activity approach. Hum. Ecol. **8**(4), 389–406 (1980). https://doi.org/10.1007/BF01561001
32. A.M. George, The national security implications of cyberbiosecurity. Front. Bioeng. Biotechnol. **7**, 51 (2019)
33. J. Gittleman, K. Hasty, S. Schacter, D. Payne, Health data security in clinical R&D: An international security blindspot? Glob. Biosecur. **4**(1) (2022). https://doi.org/10.31646/gbio.152
34. C.J. Guerrini, G.E. Spencer, P.J. Zettler, DIY CRISPR. NCL Rev. **97**, 1399 (2018)

35. G. Gürsoy, A. Harmanci, H. Tang, E. Ayday, S.E. Brenner, When biology gets personal: Hidden challenges of privacy and ethics in biological big data, in *BIOCOMPUTING 2019: Proceedings of the Pacific Symposium*, (2018), pp. 386–390
36. J.P. Higgins, J. Thomas, J. Chandler, M. Cumpston, T. Li, M.J. Page, V.A. Welch (eds.), *Cochrane Handbook for Systematic Reviews of Interventions* (Wiley, 2019)
37. H.-J. Jansen, F.J. Breeveld, C. Stijnis, M.P. Grobusch, Biological warfare, bioterrorism, and biocrime. Clin. Microbiol. Infect. **20**, 488–496 (2014)
38. JCNSS, *The UK's Biosecurity First Report* (Joint Committee on the National Security Strategy Biosecurity and National Security, 2020). Available at https://publications.parliament.uk/pa/jt5801/jtselect/jtnatsec/611/61102.htm
39. JCNSS, *The UK's National Security Machinery First Report* (UK House of Commons and House of Lords, 2021). Available at https://publications.parliament.uk/pa/jt5802/jtselect/jtnatsec/231/23102.htm
40. L.J. Kahl, D. Endy, A survey of enabling technologies in synthetic biology. J. Biol. Eng. **7**(1), 1–19 (2013)
41. K. Kalantar-Zadeh, N. Ha, J.Z. Ou, K.J. Berean, Ingestible sensors. ACS Sensors **2**(4), 468–483 (2017)
42. N. Khardori, T. Kanchanapoom, Overview of biological terrorism: Potential agents and preparedness. Clin. Microbiol. Newsl. **27**, 1–8 (2005)
43. M. Khera, Think like a hacker: Insights on the latest attack vectors (and security controls) for medical device applications. J. Diabetes Sci. Technol. **11**(2), 207–212 (2017)
44. A. Kiourti, K.S. Nikita, A review of in-body biotelemetry devices: Implantables, ingestibles, and injectables. IEEE Trans. Biomed. Eng. **64**(7), 1422–1430 (2017)
45. R. Kitchin, M. Dodge, The (in) security of smart cities: Vulnerabilities, risks, mitigation, and prevention. J. Urban Technol. **26**(2), 47–65 (2019)
46. S. Kramer, J.C. Bradfield, A general definition of malware. J. Comput. Virol. **6**(2), 105–114 (2010)
47. C.D.E. LaGreca, C. Boonthum-Denecke, Survey on the insecurity of the internet of things, in *Symposium on Computing at Minority Institutions (ADMI)*, (2017)
48. G. Laycock, Defining crime science, in *Crime science*, (Willan, 2013), pp. 3–24
49. L.E.J. Lee, P. Chin, D.D. Mosser, Biotechnology and the internet. Biotechnol. Adv. **16**(5–6), 949–960 (1998)
50. G. Lewis, P. Millett, A. Sandberg, A. Snyder-Beattie, G. Gronvall, Information hazards in biotechnology. Risk Anal. **39**(5), 975–981 (2019)
51. G. Linshiz, E. Jensen, N. Stawski, C. Bi, N. Elsbree, H. Jiao, J. Kim, R. Mathies, J.D. Keasling, N.J. Hillson, End-to-end automated microfluidic platform for synthetic biology: From design to functional analysis. J. Biol. Eng. **10**(1), 1–15 (2016)
52. D.F. Longbine, *Red Teaming: Past and Present* (School of Advanced Military Studies, Army Command and General Staff College, 2008). www.dtic.mil/cgi-bin/GetTRDoc?AD=ADA485514
53. J.F. MacKenzie, Push enteroscopy. Gastrointest. Endosc. Clin. N. Am. **9**(1), 29–36 (1999)
54. M. Marien, Futures studies in the 21st century: A reality-based view. Futures **34**(3–4), 261–281 (2002)
55. O. Martinu, G. McEwen, Crime in the age of technology. Eur. Law Enforc. Res. Bull. **4 SCE**, 23–28 (2019)
56. J. Minshull, R. Wagner, Preventing the misuse of gene synthesis. Nat. Biotechnol. **27**(9), 800–801 (2009)
57. F. Miró Llinares, S.D. Johnson, Cybercrime and place: Applying environmental criminology to crimes in cyberspace, in *The Palgrave Handbook of International Cybercrime and Cyberdeviance*, ed. by T. Holt, A. Bossler, (Palgrave Macmillan, Cham, 2018)
58. S. Mueller, Facing the 2020 pandemic: What does cyberbiosecurity want us to know to safeguard the future? Biosaf. Health **3**(01), 11–21 (2021)

59. R. Owen, J. Stilgoe, P. Macnaghten, M. Gorman, E. Fisher, D. Guston, A framework for responsible innovation, in *Responsible Innovation: Managing the Responsible Emergence of Science and Innovation in Society*, vol. 31, (Wiley, 2013), pp. 27–50
60. K. Pease, Predicting the future: The roles of routine activity and rational choice theory, in *Rational Choice and Situational Crime Prevention: Theoretical Foundations*, ed. by G. Newman, R.V. Clarke, S.G. Shoham, (Dartmouth, Aldershot, 1997), p. 233
61. K. Pease, Crime reduction, in *The Oxford Handbook of Criminology*, (Oxford University Press, Oxford, 2002a), pp. 947–979
62. K. Pease, *Cracking Crime Through Design*, Working Paper (The Design Council, 2002b)
63. J. Peccoud, J.E. Gallegos, R. Murch, W.G. Buchholz, S. Raman, Cyberbiosecurity: From naive trust to risk awareness. Trends Biotechnol. **36**(1), 4–7 (2018)
64. J.C. Reed, N. Dunaway, Cyberbiosecurity implications for the laboratory of the future. Front. Bioeng. Biotechnol. **7**, 182 (2019)
65. F. Restuccia, S. D'Oro, T. Melodia, Securing the internet of things in the age of machine learning and software-defined networking. IEEE Internet Things J. **5**(6), 4829–4842 (2018)
66. M. Rogerson, P. Ekblom, K. Pease, *Crime Reduction and the Benefit of Foresight* (S. Ballintyne et al., 2000)
67. R. Rosenfeld, Studying crime trends: Normal science and exogenous shocks. Criminology **56**(1), 5–26 (2018)
68. G. Rowe, G. Wright, Expert opinions in forecasting: The role of the Delphi technique, in *Principles of Forecasting*, (Springer, Boston, 2001), pp. 125–144
69. R. Sampson, J.E. Eck, J. Dunham, Super controllers and crime prevention: A routine activity explanation of crime prevention success and failure. Secur. J. **23**(1), 37–51 (2010). https://doi.org/10.1057/sj.2009.17
70. D. Sarpong, G. Ofosu, D. Botchie, F. Clear, Do-it-yourself (DiY) science: The proliferation, relevance and concerns. Technol. Forecast. Soc. Chang. **158**, 120127 (2020)
71. M.J. Smith, N. Tilley, *Crime Science* (Taylor & Francis, 2013)
72. P. Swain, A. Fritscher-Ravens, Role of video endoscopy in managing small bowel disease. Gut **53**(12), 1866–1875 (2004)
73. V. Topalli, M. Nikolovska, The future of crime: How crime exponentiation will change our field. Criminologist **45**(3), 1–8 (2020)
74. M. Turoff, The design of a policy Delphi. Technol. Forecast. Soc. Chang. **2**(2), 149–171 (1970)
75. U.S. Department of the Navy. HQ, United States Marine Corps. MCWP 5–1, Marine Corps Planning Process. Government Printing Office, Washington, DC, 24 September 2001, 2–6. The Marine Corps Planning Process describes the function of the red cell as assisting the commander in assessing courses of action against a thinking enemy. The Marine Corps red cell primary responsibility is to role play the enemy during the wargame
76. K.M. Vogel, Framing biosecurity: An alternative to the biotech revolution model? Sci. Public Policy **35**(1), 45–54 (2008)
77. Y. Wang, Evaluating digital health care startups: Forecasts and market insights, in *2022 2nd International Conference on Enterprise Management and Economic Development (ICEMED 2022)*, (Atlantis Press, 2022, July), pp. 1298–1304
78. S. Whitby, M. Dando, L. Shang, Nature **609**, 895 (2022). https://doi.org/10.1038/d41586-022-03011-0
79. R. Wortley, L. Mazerolle, *Environmental Criminology and Crime Analysis* (Willan Publishing, Cullompton, 2008)
80. A.K. Yetisen, Biohacking. Trends Biotechnol. **36**(8), 744–747 (2018)
81. P. Ney, K. Koscher, L. Organick, L. Ceze, T. Kohno, Computer security, privacy, and DNA sequencing: Compromising computers with synthesized DNA, privacy leaks, and more. In USENIX security symposium **26**, 765–779 (2017, August)

Potentials of Pathogen Research Through the Lens of Cyberbiosecurity, or What Threat Actors Can Learn from the Covid-19 Pandemic

Siguna Mueller

Abstract After 3^+ years of investigation, the origin of SARS-CoV-2 remains unclear, and both a spillover event from nature and a lab accident are heavily debated. As the Covid pandemic has caused unprecedented loss and damage affecting everyone worldwide, there is no sound rationale that the virus had been deliberately released. While the catastrophic effects of the pandemic remain to be resolved, it is of paramount importance to safeguard the future so that similarly tragic events could be prevented. Many of the discussions surrounding the origin of the virus have centered on dangerous ("gain-of-function") research, noting that the technology to generate pathogens with pandemic potentials is readily in place. Yet, an additional factor that causes concern has not received much attention—the intended weaponization of pathogens via a computer interface.

This work analyzes possible gaps fostered by the computerization and automation of related biotechnologies, taking the pandemic as an example to scrutinize any possible vulnerabilities that could be exploited by malicious actors. In addition to questions about the origin of SARS-CoV-2, actual challenges during the pandemic will be discussed from an adversarial perspective, revealing a disturbing gap between an actual biological/clinical entity in question and their digital information, which is difficult to close even in a well-intended context.

The cyberbiosecurity gaps identified both during the emergence of the pandemic as well as during its unfolding are not meant as judgments of past events, but to demonstrate the reality of these dangers. The potential for intended exploitation in a related situation can be enormous, leading to vulnerabilities related to attribution as well as the integrity, confidentiality, and availability of genetic information, molecular assays, devices, and the interpretation of computerized processes and tools used throughout the public health sector. It is argued here that the inherent dangers stemming from ("dual-use research") get highly exacerbated by cyber interfaces, allowing bad actors to exploit "dual-appearance" gaps that effectively can camouflage the most dangerous research projects and enable the generation and release of more dangerous bioweapons than ever seen before.

S. Mueller (✉)
Independent Research Scientist, Kaernten, Austria

D. Greenbaum (ed.), *Cyberbiosecurity*, https://doi.org/10.1007/978-3-031-26034-6_10

Keywords Covid-19 pandemic · Surrogates · Cyberbiosecurity · Dual-use · Dual-appearance · Bioweapons

1 Motivation

The Covid pandemic has triggered fear about the design of even more deadly pathogens than SARS-CoV-2 resulting from dangerous ("gain-of-function[1]") experiments and their accidental or even deliberate release from a lab [52, 53]. While there is no logically sound reason to believe that SARS-CoV-2 was intentionally released, the devastative impact of the pandemic urges us to rigorously analyze related safety and security gaps from an adversarial perspective.

The potentiality of bioterrorist attacks involving pandemic pathogens is well demonstrated by the fact that it has been possible to re-construct SARS-CoV-2 entirely from digital information alone [50]. As [50] is motivated by the intent to study characteristics of the new virus, it does not include any security considerations, and instead, details of how to manipulate this, and other, pathogens are explicitly described.

The specific vulnerabilities raised in this chapter are not meant to evoke any accusations about specific Covid-19-related events. Rather, they are meant to demonstrate the complexity of the issue and the lack of awareness in this regard. As will be seen, there are major unresolved problems. Rather than serving politicization and speculation of past events, a drastic potential of gaps in knowledge, policy, and oversight is that they could potentially be exploited by malicious actors too.

Thus, the critical question is: what is it that those intending to cause harm could learn from the associated dangers and risks? Bad actors do not play by any rules. They do not care if something is a political offense or if it could violate ethical or humanitarian rules. Their freedom to sidestep bureaucracy and common morals can give them an incredible advantage to realize the very thing that some are trying to accuse the Wuhan Institute of Virology of—the design and release of biological weapons. And thanks to the computerization of synthetic biology, this leads to the concern that even the most well-intended pathogen research could be diverted into a covert biological weapons program.

Attack potentials fostered by the reliance on automation, digitization, and cyber-overlaps throughout the bioscience fields have spurred the development of cyberbiosecurity as a new discipline [34, 38]. Nonetheless, because of the convergence of the fields and the enormous breadth of the risk landscape, the related concerns as played out during the pandemic have not been adequately appreciated.

[1] What exactly constitutes gain-of-function has been heavily debated. This chapter applies the definition of the U.S. government which understands gain-of-function as studies or research that 'improves the ability of a pathogen to cause disease.' Such work inherently entails biosafety and biosecurity risks, as e.g. related to the 'enhancement of a pathogen's transmissibility or virulence in humans' - see https://www.phe.gov/s3/dualuse/Pages/GainOfFunction.aspx

To help fill this gap, this chapter aims to open a discussion of what could have happened during the outbreak and unfolding of the pandemic, or what analogously could happen, when seen through the lens of deliberate attack potentials. While it is not meant to attribute specific historic events to any of these postulated adversarial potentials, it points to the need to fully understand and mitigate these types of situations in order to inform appropriate governance measures and help prevent future disasters.

It will be argued that in addition to inherent dual-use[2] characteristics of pathogen research, technical challenges that are fostered by the computerization of synthetic biology may lead to dire consequences and lend themselves to hard-to-detect forms of bio-crime. Some of the vulnerabilities specifically unfolded during the pandemic. While they arose in the context of necessary responses, utilizing latest advances in modern biotechnologies, the emergence of a new pathogen engenders major risk potentials which have not received adequate attention.

For the common good of humanity, it is indispensable to put politicization aside, leave what happened in the past, and learn our lessons. Just as decades ago, it was an absolute must to expose flaws in computer systems—even if it meant that these flaws were made public in a detailed way—it is an absolute imperative to openly acknowledge the risk potentials of computerized pathogen research. As such, the identification of flaws and misconceptions is a necessity, becoming a part of the solution, to be able to secure what previously had been unidentified or undisclosed.

2 The Gap Between the Digital and Biological/Physical Could Be Exploited in Numerous Ways

The examples below provide specific situations of how cyber- and cloud-based applications in synthetic biology—which were initially geared towards improving practicality and efficacy—provide unrecognized opportunities for both human error and deliberate manipulation. Key developments during the pandemic will be described to depict how they could become subject to intended exploitations.

2.1 Biomarkers and Surrogate Endpoints

Over the years, the identification of pathogens has shifted from the isolation of the purported culprit—requiring the handling of potentially risky biological material—to approaches based on synthetic biology. Among others, the identification of human infectious diseases has more and more been automated.

[2] Dual-use here is understood as research or policy that could be used for good or bad purposes.

More generally, clinical tests have moved away from randomized, placebo-controlled, clinical trials. Aimed to streamline ethical and safety precautions and cut cost and time, specific "endpoints" are used, i.e., certain outcomes to evaluate the safety and efficacy of a specific intervention. Traditionally, these were clinical in nature and thereby provided an understanding of actual clinical outcomes. Albeit, in recent years, leading regulators such as the U.S. Food and Drug Administration (FDA) have more and more relied on surrogate endpoints. In clinical trials, these surrogate endpoints have been used instead of clinical endpoints in situations such as "when the clinical outcomes might take a very long time to study, or in cases where the clinical benefit of improving the surrogate endpoint ... is well understood" [22].

While easier to implement, there are dangers and downsides to using surrogates. They are only a proxy of an actual biological/clinical measure. Recognizing the potential pitfalls, the FDA requires clinical trials that clearly demonstrate that surrogate endpoints can be relied upon "to predict, or correlate with, clinical benefit." Once such surrogate endpoints have undergone such clinical verification testing, they are deemed validated surrogate endpoints.

2.1.1 A Computer Interface Could Intensify the Knowledge Gap

Of note, so far, the focus has only been on showing that specific surrogates could replace actual clinical parameters in terms of speed, cost, etc. But this approach has not been rooted in a security mindset, i.e., the concern that something could be willfully exploited.

The FDA acknowledges that even with validated surrogate endpoints these can give misleading information about the overall risks and benefits, e.g., of a medical product. This is because different settings and contexts—which cannot be captured by a surrogate—can lead to markedly different outcomes. This knowledge gap, and the difficulty to verify the validity of surrogates, may establish unrecognized opportunities for misuse and likely be exacerbated in a digitized, computerized, and automated context. While it is true that no surrogate fully resembles actual/clinical outcomes, reliance on a computerized interface introduces yet another layer of abstraction, intensifying the gap between the digital and the real and thereby increasing attack potentials.

The applicability of surrogates is very broad. For instance, a validated surrogate endpoint is Human Immunodeficiency Virus (HIV) *viral load* as a proxy and predictor of the clinical outcome *of developing AIDS* (with a higher load suggestive of a more severe disease). In this case, this proxy has been extensively validated, as many decades of research have contributed to the isolation and characterization of HIV, and the disease has been intensively studied. However, the entire situation is entirely different with a new pathogen resulting in a new disease.

2.1.2 Unanticipated Attack Potentials

Even with well-studied viruses (such as HIV), the largely computerized processes and devices that measure the viral load of an infectious agent do not seem to have seen much adversarial pressure, and some may question the motivations for malicious exploitation to begin with. However, attacks via a cyber-interface to manipulate the measurement or reporting of this information could be done for a number of reasons, such as for insurance benefit purposes or to gain sensitive patient data.

Attacks involving novel surrogate endpoints could have a disruptive impact on larger groups of individuals, especially when mounted before or during the very process of obtaining regulatory approval of these endpoints and could be aided by computer modeling, in silico clinical trials, and machine learning. The very fact that surrogate endpoints lack real-world clinical data lends itself to a high degree of manipulation. For instance, bad actors intending to fake some new diagnostic endpoints could rely on some bogus computerized output and convincingly looking data. In a context where clinical verification is only meant to be established later, malicious actors could fabricate fictitious relationships to, say promote a totally fake medical product, or even one endowed with harmful features.

2.1.3 The PCR Test to Diagnose Covid-19

Inter alia, surrogates are important epidemiological players for the diagnosis of a disease. With Covid, the most famous example is the PCR test which has been globally employed to measure the viral load—and thereby, the presence or absence of infection with SARS-CoV-2.

Very shortly after the new virus SARS-CoV-2 triggered public attention, the article "Detection of 2019 novel coronavirus (2019-nCoV) by real-time RT-PCR" [15] proposed the first protocol for detection and diagnostics of the new virus. This test was quickly applied widely as the de-facto standard for Covid-19 diagnosis. As the pandemic progressed, however, concerns emerged about the logic and design of this "Corman-Drosten" test [27]. It was not until the beginning of 2021 that these and related problems were acknowledged by the WHO [57]. Then, by the end of 2021, the Centers for Disease Control and Prevention (CDC) also withdrew their (related) process as a valid test for detecting and identifying SARS-CoV-2, urging laboratories to work towards tests that would be able to "facilitate detection and differentiation of SARS-CoV-2 and influenza viruses" [11].

This latter point seems to confirm one of the main controversies that had surrounded the Corman-Drosten test since its first publication: it is merely based on surrogates.

The test was constructed during a time when neither control material of infectious ("live") nor inactivated SARS-CoV-2 nor isolated genomic RNA of the virus was available. Despite the lack of actual isolates of the new virus, the authors report "on the establishment and validation of a diagnostic workflow for 2019—nCoV screening and specific confirmation, designed in absence of available virus isolates or original patient specimens."

The entire design and presentation of the automated Corman-Drosten test were based on a logical model tested via synthetic RNA. More specifically, the justification of the test relied on the assertion that some theoretical/in silico genes could serve as appropriate identifiers of the unknown SARS-CoV-2—which to this point only existed in digital form.

In their previous work, Drosten et al. had established a "standard" to identify an infectious disease agent [18]. Although previously this had been complemented by both extensive validation of the putative pathogen and complementary diagnosis of clinical symptoms, it obviously led to the belief that the same steps could be replaced by computerized models and simulations—even in the complete absence of any clinical verification, and without being able to validate the choice of the primers, cycle-threshold values, GC content, melting curve characteristics, and other sensitive parameters [27, 31].

Moreover, in [15], SARS-CoV-1 is taken as a "positive control," and positive samples are then further distinguished from SARS-CoV-2 via specific genes believed to be unique to the latter. However, the only form of validation related to SARS-CoV-2 is rooted in the assumption that synthetic RNA snippets can be taken as a surrogate for the unknown virus. Thereby, the "verification" is merely a theoretical validation that a synthetically generated RNA string behaves according to specifically modeled limiting dilution experiments.

As also admitted by Dr. Anthony Fauci [33], former Chief Medical Advisor to the President of the United States, PCR tests do not measure if you have live replicating viruses in you, and they may be picking up dead viral debris for months after infection. According to the original invention of the PCR test, this method can only detect proteins that are believed, in some cases wrongly, to be associated with a certain disease, but they cannot detect viruses themselves [5]. This is why traditionally it has been important to infer potential pathogenic candidates from clinical manifestations and then have the actual culprit be confirmed by a specific PCR assay. But with a new disease, an unknown virus, and even more so relying on digital surrogates of that virus, the gap between an actual entity and its purported digital sequence information could easily be exploited. The effect could be either that such a test returns too many false negatives—allowing a new pathogen to spread widely and rapidly without giving those infected the required medical attention—or false positives—creating unsubstantiated public terror and fear, and impairing appropriate response measures.

PCR assays also play a critical role to estimate the epidemiological impact of a new virus, e.g., from wastewater samples. Inherent challenges of PCR testing related to a new virus, even with the best of intentions in mind, can be seen from the observation that SARS-CoV-2 has thereby been identified in human sewage both in Brazil and Spain, albeit much earlier than the first reported cases in these regions [21, 42].

From an adversarial perspective, the gap between the digital and biological/physical inherent to all such computerized molecular assays leads to the concern that attacks could disturb the integrity of the machine or system relied upon, leading to manufactured models to diagnose and treat a new virus. Computerized models or predictions, entirely fabricated, would be very difficult to be recognized as such. This would be next to impossible in wake of a new disease outbreak, and even more so under the pressure of a pandemic, when the need for drugs and medical countermeasures, including those with limited clinical validation, may be extensive.

2.2 *Genetic Information Storage and Pathogen Databases*

Genetic databases may be one of the most intensively studied areas of cyberbiosecurity (see, e.g., [10, 55]). The importance of securing sensitive genetic data of pathogens is well demonstrated by the observation that minor changes can have far-reaching consequences on their pathogenicity or transmissibility. In particular, already in 2008, Ren et al. demonstrated in lab experiments involving bat SARS-like and human SARS viruses that a minimal insert region (amino acids 310 to 518) was found to be sufficient to convert the spike from one that is not binding to angiotensin converting enzyme 2 (ACE2) to one that is [43].

Likewise, as mentioned, the technology is in place [50] to design and manipulate pandemic pathogens from digital sequence information alone. This prompts urgent questions to what extent "gain-of-function" research should be conducted (initially raised by the Cambridge Working Group [46]), as well as potential security and accountability issues that might engender.

2.2.1 The Challenge of Attribution

Concerning SARS-CoV-2, several problems related to genetic (and medical) databases have become a source of intrigue. Notably:

- A Wuhan Institute of Virology (WIV) viral database, the most extensive globally in terms of coronavirus research, went dark in 2019 [29, 54].
- A memorandum of understanding between the University of Texas Medical Branch (UTMB) and three high-level biosecurity labs in China, including one with the WIV, states that each lab can destroy any so-called secret files—which in context seems to imply any documents, communications, or data in general resulting from their collaboration [29, 48].
- The ongoing debate related to the extent to which China has made all the material relevant to the origin of SARS-CoV-2 public, or not, has been stirring the suspicion that researchers in the West currently do not have the data related to the most recent progenitor of SARS-CoV-2 [54]. Meanwhile, China has been asking whether the virus came from a U.S. Army lab at Fort Detrick in Maryland [14].

- Medical reports suggest that early Covid patients were not all connected to the market; on the other hand, some people were originally mistakenly identified as having the disease who did not [14].

This work cannot do justice to all these complex issues and developments. However, the unresolved questions and controversies also point to a major security problem. Amidst growing geopolitical tensions, if it seems impossible to validate the completeness and correctness of relevant data about the origin of a pandemic, then this confirms a double-bind type situation that also could be intentionally exploited.

2.2.2 Deleted and Re-Emerged Coronavirus Genome Sequences

In June 2021, the discovery and recovery of deleted deep sequencing data implied with early SARS-CoV-2 triggered further questions about the origin of SARS-CoV-2 [7, 9]. Specifically, the study [7] describes how partial SARS-CoV-2 sequences from early outbreaks in Wuhan were removed from a U.S. government database by the scientists who deposited them. Knowing the genetic diversity of early SARS-CoV-2 strains is highly relevant to its origin. The fact that partial SARS-CoV-2 genome sequences from the beginning of the pandemic were deposited to an official database, later removed, and then excavated, has important implications.

The recovered sequences led to a conundrum regarding the view that the new virus came from Wuhan's Huanan Seafood Market late in 2019. Ironically, as highlighted in [7], the sequences linked to the seafood market are known to be more distantly related to SARS-CoV-2's closest relatives in bats than later sequences. This observation does not align with evolutionary theory. One would expect the opposite trend: if it is indeed the case that the virus came out of the market, then the viral strains from the early stages of Wuhan's epidemic should be most closely related to SARS-CoV-2's believed natural relatives that infect bats, and not the reverse.

The above is not only relevant in relation to the core controversy of whether the virus came out of, or went into, the Huanan Seafood Market [32, 54, 58]. Without wanting to imply any intent, it is worthwhile to analyze the entire situation from an adversarial perspective.

The incident confirms previous fears [10, 55] about pathogen genomic databases, notably the absence of rigorous security mechanisms against human error (erroneous entries, deletions, etc.) as well as intended manipulations.

As described in [7], the deleted and then recovered sequence information in question was initially deposited to the Sequence Read Archive (SRA), a repository for raw sequencing data maintained by the National Center for Biotechnology Information (NCBI), part of the U.S. National Institutes of Health (NIH). The same information was initially also published by the same group of researchers in a May 2020 preprint. At some point in 2020, this information disappeared. The NIH later confirmed it removed the data at the request of the researchers. This is in itself troubling, as it demonstrates potential gaps in policies that could be exploited by those intending to manipulate such information. Especially with novel pathogens, this points to a critical security issue. Despite their importance, there is no guarantee that only independently validated and correct sequences get uploaded, or that such sensitive information could not get alternated.

In addition to their relevance to the viral origin, manipulated sequence information can easily be misused to construct biased diagnostics or impair the development of prophylactic or therapeutic modalities. This is even more concerning as the SRA, for instance, keeps sequences stored on the Google and Amazon clouds [7], making these a largely unrecognized target of cyber-crime.

2.3 *DNA Sequencing and Sequence Verification*

As summarized in a 2020 article [44], the entire DNA sequencing infrastructure is susceptible to numerous risks and vulnerabilities, virtually at all stages, from material collection, the pre-analytical, analytical, to the post-analysis phase of data storage and dissemination. Yet, the insecurity of genetic information systems remains greatly underappreciated.

2.3.1 The Computer Interface Largely Increases the Insecurity of Genetic Information Systems

Schumacher and collaborators [44] raise alarm that modern biotechnologies such as DNA sequencing instruments have built-in computers and widely rely on connected computers and servers for data storage, networking, and analytics. Notably: (1) Hardware vulnerabilities in the life science fields could be introduced and exploited through various means, and once present, are often unpatchable and remain with those devices until these get replaced by other devices. (2) Like hardware, in-field software upgrades are difficult to do. Moreover, because of the ubiquitous implementation throughout the bioengineering field, software issues are particularly concerning [35]. (3) A great vulnerability is that of open-source software, as it is widely used across genomics, and acquired from several online code repositories [35]. (4) A specialized niche industry such as genomics and bioinformatics is generally not built with security in mind nor assessed for vulnerabilities. All these lead to numerous opportunities to manipulate data inputs and parameter settings or perform other modifications throughout the bioinformatics pipeline that are not easily visible to lab workers, leading to non-integrous outputs.

Alarmingly, ref. [44] suggests that genetic data could be manipulated globally, even during the early stages of the analysis. This could have profound consequences, especially in the context of new pathogens when the integrity of genomic data has been compromised in clandestine. It is hard to tell, if, when, and how such unknown discrepancies related to sensitive genetic information could be detected.

Just prior to the pandemic, Peccoud and coworkers [23] illustrated the largely unrecognized problem of sequence verification by whole-genome sequencing

(WGS). Although WGS is a popular method to obtain the entire genomic DNA of a cell, even the most foundational steps, including assembly, variant calling, and strain verification, are highly vulnerable. To make this point explicit, [23] demonstrated major flaws during strain validation of haploid yeast strains with an unexpected phenotype derived from a mutant collection. Their findings are sobering: even though the strains analyzed are commonly used in laboratories, there was no finished reference genome available. Disturbingly, when using a closely related reference genome, this resulted in a number of unexpected mutations.

Peccoud et al. [23] warn that the magnitude of the problem of identification of unexpected mutations is underappreciated. In this context, the critical shortcomings are summarized as follows:

- The software tools that are currently available are not well suited for verification workflows.
- In-depth analysis requires ad-hoc or heuristic decision points that mandate an advanced understanding of the software tools used.
- The results still needed to be manually validated by visualizing the reads, reliant on expert decisions requiring detailed knowledge about the function of the individual gene in question.

2.3.2 The First SARS-CoV-2 Reference Genome

The first reference genome for SARS-CoV-2 was obtained by Wang and collaborators [56]. It is based on 95 full-length genomic sequences of early strains of this virus, as published up to February 14, 2020 in NCBI and GISAID databases, obtained via multiple sequence alignment and phylogenetic analyzes.

For verification of the reference sequence, Wang et al. [56] referred to their previous work. Previously, their team had been able to build specific hepatitis B virus (HBV) sequence subtypes from a total of 3000 reported sequences that had been available from different countries. In that case, infectious plasmids that were constructed based on selected subtype-specific reference sequences "confirmed complete biological functions of these reference sequences." However, for SARS-CoV-2, a biological/physical characterization, as done for HBV, was omitted (in fact, at that point, comprehension and diagnostics of the disease caused by SARS-CoV-2 was still grossly incomplete). As acknowledged by Wang et al. [56], a shortcoming of their reference genome is that all the underlying sequences to build the reference (by selecting the most common nucleotide in each position) were retrieved from databases. They admit that "the accuracy of sequences could not be verified." Nonetheless, the authors argue they were able to confirm the validity of their constructed reference via several additional steps—albeit, with all of them via computer-based techniques.

In [56], Wang and collaborators validated the correctness of their calculated reference sequence in that they, among others, showed that it was identical to the genomic sequence of 15 strains isolated from clinical samples. In fact, they also found that compared to the reference sequence, the homology of the vast majority among all SARS-CoV-2 strains used in the study was extremely high (99.99%), both at the nucleotide and the amino acid level. For the latter, homology among full-length sequences was 99.99%, with homology among most isolates in each region being 100%.

The study authors, apparently surprised by the overwhelming homology, stress that they were also able to find (some) mutations, which to them suggested "that the virus in this epidemic might originate from the same animal species, and caused widespread infection in a short period of time."

Others argue this sequencing information is indicative of a lab origin: the high degree of genetic purity is unexpected especially for RNA viruses in general, and more so, for a postulated natural spillover, as animal viruses need time to get better adapted to their human host [41].

2.3.3 The problem of early sequence errors

A reference genome of high enough resolution to differentiate between a natural and viral origin requires precise early viral genetics data obtained from accurately diagnosed patients. As stated, both of these components are technically tricky, and especially so during a new disease outbreak.

Disturbingly, the correct identification of nucleotide sequences has proven even problematic in well-studied contexts such as cancer research. An analysis[3] of high-impact journals revealed that in 2020, 38% of all papers contained errors in their nucleotide-targeting reagents, with many of them likely caused by honest mistakes in part due to a computer interface (e.g. auto-correction errors in spreadsheets). The study highlights both the technical difficulty of error identification, especially in a context (medicine) where such mistakes previously were never thought to be so widespread, as well as evidence of targeted fraud.

The complexity and reality of these issues in context of the pandemic is demonstrated by inconsistencies of early SARS-CoV-2 data.

In May 2020, an independent study [1] from Colombia raised concerns regarding the published sequences of SARS-CoV-2 as "consolidated by the WHO during the early stages of the pandemic." Relying on these public sequences, and utilizing primer/probe sequences from 13 target regions for SARS-CoV-2 detection, all following official recommendations, including the Corman-Drosten test for RT-PCR, this study utilized Next Generation Sequencing (NGS) technologies to determine the whole genome of SARS-CoV-2 strains in Colombia.

However, their in-house molecular assays revealed significant differences from what officially had been reported. In particular, some of the "oligonucleotides displayed mismatches that were considered of minor or major importance for the test

[3] Highly cited genetics studies found to contain sequence errors. Kwon Diana Nature, 10 Feb 2023, DOI: https://doi.org/10.1038/d41586-023-00385-7

performance," including a mismatch located at the $3'$ end of the primer—a known critical vulnerability for molecular detection which is expected to severely affect primer hybridization and subsequent extension.

2.3.4 Exploiting the Knowledge Gap Related to New Pathogens

Especially with new pathogens, the availability and correctness of early genetic information are of paramount importance. However, all the processes and routines used in this context, ranging from sample identification, sequencing, uploading, sequence alignment, maximum-likelihood tree construction and other methods employed to infer phylogenetic relationships are subject to noise and inherent error, and could, as stressed before, intentionally be exploited. For example:

- Sequencing itself is a highly stochastic process. As such, it cannot readily be concluded that a sequence in question is authentic and accurate. For instance, in the previously mentioned analysis of WGS by Peccoud et al. [23], about 95% of the yeast genome were covered with at least 30 reads in all samples. Despite high sequencing depths, there were significant errors nonetheless. A correct answer is difficult to obtain and requires, in addition to extensive knowledge in bioinformatics, detailed manual curating.
- The resolution limit of WGS of viruses of this size (approx. 30,000 bp) does not seem to have been investigated. This in itself may lead to concerns under adversarial pressure, allowing the infiltration of mutations that cannot be identified for mere technical reasons alone.
- Genetic information is frequently transmitted between life science organizations across various networks. In terms of DNA sequencing, the task is often delegated to international third-party organizations, leading to potential network compromise and the danger that genetic data could be aggregated globally by nation-states and other actors even *during* the analysis phase [44, 49].

2.3.5 Mysterious Attacks on DNA Sequencers

In June 2019, some mysterious attacks on very unique devices installed in scientific, academic, and medical institutions puzzled cyber-security experts [4]. The attacks targeted web-based DNA sequencing applications using a still unpatched vulnerability that first became known in 2017. In particular, this vulnerability allowed the attackers to control the underlying web server from remote locations. The motives for the attacks are unknown. Although ZDNET [12] describes some scenarios, they do not seem to do justice to the event.

As it has been impossible to attribute a clear motive, the event seems to have been dismissed. "With the vendor refusing to patch the security flaw back in 2017, these systems remain open for attacks," the analysis of these intrusions concluded [12]. It is not apparent that the bug meanwhile has been patched. In fact, [4] describes a very funny response by the vendor, indicating that the implied vulnerability is not being taken seriously.

Yet, a critical concern is that all these DNA sequencers had come under the attacker's control. Surprisingly, nobody seems to have asked to what extent the attacks can lead to the disruption of sensitive genomic information. Disturbingly, if left unpatched, such cyber intrusions can corrupt rare sequencing data, such as those of early SARS-CoV-2. As discussed, such error infiltration can radically confound early genomic studies, all the while without anyone knowing such alterations have happened.

2.4 *The Challenge of Obtaining Accurate Genetic Information and Identification Issues*

The necessity of knowing "who is who" cannot be over-emphasized but in a biological context is difficult to be realized. In contrast to the implied singularity of notions like "the genome" or "the viral sequence," in reality, there is neither uniqueness nor a simple approach to comparing and measuring genetic differences. As a result, the identification of a pathogen may not be as easy as it first appears to be.

- Even after years of most devoted efforts related to the molecular diagnosis of SARS-CoV-2 subvariants, there are still enormous challenges. This once again demonstrates the great advantage malicious actors could have after an intended release of a new pathogen. For example, on December 20, 2021, the European Centre for Disease Prevention and Control (eCDC) and the World Health Organization (WHO) recommended partial Sanger sequencing of two specific SARS-CoV-2 genes on PCR positive samples [30]. Surprisingly, as pointed out by Lee [30], more than 2 years into the pandemic, this protocol still lacked actual test data validating outcome performance in diagnostic laboratories. Albeit, Sanger sequencing of PCR amplicons needs well-designed PCR primer sets. In [30], Lee showed this is very challenging for the highly mutating Omicron variants which he further demonstrated via clinical samples harboring mutations that affected the PCR primer binding site. Moreover, clinical samples with a high level of co-existing minor subvariant sequences—as can be present in an individual patient—could not be adequately distinguished by the eCDC/WHO protocol [19]. Further analysis revealed that multi-allelic single-nucleotide polymorphisms (SNPs) or recombinant viral subvariants impair automated base-calling accuracy, causing the computer to make various errors during base calling [30].
- The problem of estimating phylogenetic distances is most foundational to inferring evolutionary close relatives. For SARS-CoV-2, [16] showed that a critical insertion—which was previously believed to point to the natural origin

of SARS-CoV-2— is actually a deletion in that genetic region; this confirms that the methods used of how sequences are aligned, critically impact the outcome.

- The findings of [16] is but the tip of the iceberg of how results in biology can be shaped by inappropriate modeling. In fact, the widespread approach via minimal (Hamming) distance is not the most optimal measure to estimate genetic distances as it cannot account for the actual chain of mutations that occurred in a living context. This was first recognized by Buschmann and Bystrykh [8] who developed an alternative distance metric for DNA barcodes (tagging) used for high-throughput sequencing. In addition to inherent security vulnerabilities (see below), there is no complete solution to this distance problem, not even applied to the specially constructed synthetic barcodes, let alone in the context of phylogenetic analysis of rapidly mutating microorganisms.

2.4.1 Multiplex Sequencing as Potential Attack Vectors

The last two decades have seen a greatly increased reliance on sample multiplexing to effectively address the large number of sequences generated. Widely used multiplexing strategies, such as with the Illumina Genome Analyzer, utilize sample-specific index sequences which are attached to the sample molecules during sequence library preparation. Through such "barcoding," multiple samples can be pooled and sequenced in parallel. The samples are later de-multiplexed after sequencing by computationally identifying and partitioning them through their specific index sequences.

Sample multiplexing greatly increases experimental scalability but also brings the risk of sample mis-identification when sequences are incorrectly assigned to their original samples. Detailed experiments [28] have shown that multiplex sequencing on the Illumina platform has much higher rates of such inaccuracies than expected. With roughly 0.3%, such high ratios may severely confound applications that require highly accurate genotyping, such as when dealing with rare sequence variants, or when conclusions are drawn from a single sequence, as e.g., in ancient DNA research. Analogous concerns apply to early genetic data of a novel pathogen, which, as discussed, could bias towards either a lab or a natural origin, and more.

2.4.2 Adversarial Potentials Involving Index Cross-Talk and Sample Mis-assignment

Several sources and mechanisms that impair multiplex sequencing have been identified [28, 36]. These include contaminants during ultramer synthesis or library preparation, cluster overlaps (with random cluster amplification), and the presence of residual free index primers. It can be shown that these and related issues result in molecular errors in the form of "index cross-talk" during the computerized output.

As with other gaps that are inherent in complex protocols, these vulnerabilities—giving rise to unintended errors—could intentionally be exploited too. This was first demonstrated in [36]. The basis of the attack rests on the observation that DNA can be synthesized in a way to mimic those particular DNA strands which are known to be vulnerable to protocol errors and sample mis-assignment. Thus, in [36], when the malicious DNA library was constructed accordingly, it was able to exploit index cross-talking vulnerabilities when called in other multiplexed samples.

The attack was particularly concerning as it led to false variant calling of the targeted sample (which was attacked), even though the latter was aligned to a reliable reference genome. In fact, the maliciously designed sequences induced a high-quality false variant call despite high coverage and read depth.

Such a carefully designed attack makes it also possible to evade updated sequencing applications, including those that previously were thought to be rather robust to index cross-talk, simply because known protocol errors on the molecular level can effectively be mimicked by a maliciously synthesized strand. Remarkably, samples can thereby be compromised in clandestine, without anyone knowing that such manipulations were made.

Such attacks in high-throughput DNA sequencers can also be used to manipulate sequence results in a targeted way, e.g. to cause the incorrect genetic interpretation of concurrently sequenced genomic samples. In [36], this led to a sickle-cell disease-causing variant to appear in the wild-type human genome. Sobering downstream consequences of such alterations are manifold, ranging from false diagnoses to faulty treatments.

2.4.3 Sequence Contamination Not Ruled Out in the Context of SARS-CoV-2

These same attack potentials of DNA multiplexing are particularly relevant when dealing with new and highly variable pathogens. For instance, with SARS-CoV-2, the reliance on early sequences and variants confounded by contaminants has complicated critical analyses. The problem of genetic contamination is part of the ongoing discussion about the origin of SARS-CoV-2. Notably, some studies that reported the discovery of specific coronavirus strains—which purportedly demonstrated the natural origin of some of SARS-CoV-2's sequence insertions—were later found to be based on metagenomic datasets with unexpected reads indicating significant contamination [16] of the relevant pangolin dataset which may have been introduced during purification from cell culture experiments or during sequencing [16, 59].

During a bioterrorist attack, integration of contamination (whether targeted or not), could severely derail research on viral detection, epidemiological investigations, vaccine design, evaluation of drug effectiveness, and more.

2.5 *Attacks via Automation and Machine Learning*

In [20], Finlayson et al. describe types of attacks that pose enormous risks to the entire health sector. These so-called adversarial attacks rely on very subtle changes in how input data are presented to a computerized system. Such carefully designed changes have the potential to completely alter the system's output, "causing it to confidently arrive at manifestly wrong conclusions." These types of attacks first emerged with machine learning, deep learning algorithms, and pattern recognition algorithms [6]. Soberingly, even in the context of computer-science research, comprehending these vulnerabilities has proven challenging, in part due to inherent challenges and misconceptions related to the security evaluation of machine-learning algorithms [6]. Comprehending the main threat models and attacks is an ongoing major open problem in the design of more secure learning algorithms.

Machine learning has been widely applied to medical diagnostics and decision support, where they are deemed to have achieved diagnostic parity with physicians on certain tasks in radiology, pathology, dermatology, and ophthalmology. The current trend is to further extend those applications, including the use of artificial intelligence (AI) diagnostic systems for regulatory decisions. Nonetheless, it seems that the life science and medical community is largely unaware of adversarial attack potentials. Specifically, adversarial examples have been demonstrated for essentially every type of machine-learning model ever studied and across a wide range of data types, including images, audio, text, and other inputs [6] making numerous applications relied upon in the biological sciences highly susceptible to malicious exploitation.

2.5.1 Adversarial Attacks Are Known to Be Successful Even in Known Contexts

As a proof of concept of the deep—albeit largely unrecognized—vulnerabilities of adversarial attacks on medical machine learning, Finlayson et al. [20] demonstrated specific adversarial examples against highly accurate medical image classifiers. In one example, they applied some adversarial noise to a dermatoscopic image of a benign mole: notably, the original image was correctly flagged as benign with a confidence of $> 99\%$; yet, adding a carefully calculated perturbation to this image, so small that it was invisible to human beings, fooled the model into classifying the mole as malignant with 100% confidence.

Another example of an adversarial attack exploits natural learning processing and shows that substitutions of carefully selected synonyms for medical coding can be sufficient to hijack the underlying computer algorithms. In this case, the original data for billing codes, leading to the conclusion "reimbursement denied," could successfully be altered to "reimbursement approved." Furthermore, deep neural networks, as used in medical image classification for automated support for clinical diagnosis, are highly vulnerable to several types of adversarial perturbations [26].

2.5.2 An Existing Knowledge Gap Increases The Vulnerability

Conspicuously, the above types of attacks would be even easier in a situation with incomplete knowledge of a disease. As the pandemic has shown, especially with a new pathogen, the initial knowledge gap can be immense, despite international efforts and collaborations. For example:

- The disease spectrum of Covid-19 has been difficult to capture, most notably as it has changed considerably with the emergence of the Omicron variants [51].
- Covid-19 is very different than infections implied with common cold coronaviruses. Comprehending the main culprit for severe disease has taken years. It was not until 2021, that the spike protein itself has been shown to damage heart muscle cells [3], been implied with thrombosis [24, 40, 60, 60], and been shown to cause mitochondrial damage [13].
- The immune response triggered by SARS-CoV-2 infection is atypical in that it causes a suppression of T cells [17] and may trigger autoimmunity [25].

Weaponized pathogens may impair a natural immune response, enable asymptomatic transmission, and cause harmful symptoms radically different from their natural counterparts, allowing the wide and rapid spread of a bioweapon. Additionally, as biolabs are increasingly automated, it is possible that the release of a pathogen could even be facilitated remotely, making attribution even more challenging.

2.5.3 Inherent Gaps in the Life Science Fields Increase Susceptibility

The full spectrum of vulnerabilities fostered by the increased automation of the health science fields has been far from grasped, even though several factors greatly increase susceptibility in this context:

Virtually all of the most important processes and tools in the bioscience fields are highly stochastic in nature. As a result, rather small changes to input parameters or data may result in a substantially different output. Combined with the computerization of processes and equipment, and the lack of expertise in the handling and underpinnings of cyber-based applications, there is little control over technical mishaps, human error, and deliberate misuse.

Complete reliance on a computer output that gives flawed results could have detrimental consequences on human health, society, and the environment, both in a well-intended but also an adversarial context. As is known in general [6, 20, 26], fooling a deep learning algorithm is rather easy, even in a situation where the key parameters and relationships are known and understood. Albeit, when the reliability of the algorithm is difficult to be validated, e.g., when dealing with new pathogens, then an intended manipulation could take a long time to be identified as such.

3 Recommendations

This work has described major problems resulting from the computerization of biotechnologies, with a special focus on the Covid pandemic. Fully addressing the related security concerns will require open international dialogue and cooperation. To promote these, the following steps are recommended.

3.1 Acknowledging the Gap Between the Digital and the Biological/Physical

As a first step, it will be critical to acknowledge the inherent digital—biological/physical gap that results from the digitization of biology and the reliance on computers. Notably, we cannot recognize a virus, for example, other than through some computer interface. As the above has shown, this same gap is at the core of ongoing controversies related to the Covid pandemic and creates a significant knowledge gap that could effectively be exploited.

3.2 The Notion of a Signature Sequence as a Unique Identifier Needs to Be Revisited

In light of the problems identified above, a genetic sequence alone is not always sufficient for identification purposes. For instance, the assertion that SARS-CoV-2 jumped from animals to humans because of some genetic similarities found in some related coronaviruses [32, 37, 54] continues to be controversial and unresolved, essentially because the gap between the digital and the actual physical/biological has not been closed. While on the one hand, various studies (see, e.g., [32]) argue that specific genetic sequences point to some natural origin of the virus, these arguments are all based on contestable computer models and analyzes. Open questions in this regard include (1) the fact that no animal has been found that harbored the virus before it spread to humans ("the coronavirus did not spontaneously generate itself in market stalls in Wuhan" [37]), (2) the timing and location of occurrence of the most ancestral version of SARS-CoV-2, (3) lack of animals in China with SARS-CoV-2 antibodies or infection, and others [37, 41, 54].

On the other hand, the presence of a genetic signature does not automatically prove a lab origin either. A curious example with Covid is the finding of a proprietary sequence in SARS-CoV-2 of human origin which is not contained in any other coronavirus [2]. The insert at the furin cleavage site (or rather, its reverse complement—which biologically does make sense [2]), is a 100% match to a nucleotide-optimizes sequence that was patented in 2016 by Moderna.

Here, while the sequence itself is unique, it is unclear how it could have ended up in the viral genome. Ambati and coauthors [2] explain why it would have been very unlikely this happened by chance and offer a biological mechanism of how (the reverse complement of) the new sequence could have been incorporated in a lab setting. Albeit, the rationale for conducting the particular experiments is somewhat unclear.

Thus, genetic sequences alone cannot serve as identifiers, and biological plausibility, research propositions, goals, and context are just as relevant. In addition, it is important to note that alternative metrics and models can result in totally different interpretations as well.

As another example, the linear genetic sequence is regarded as the most basic information to estimate the phenotype of an organism. More realistic and complex information can be obtained by, e.g., structural modeling and protein interaction studies. Albeit, fully predicting 3-d structure and higher-degree organization in a biologically relevant context from amino acid sequence data remains one of the hardest problems in biology.

A key example is given by Piplani et al. [39] who investigated SARS-CoV-2 species susceptibility via in silico comparisons using different algorithms and metrics. This study aimed to rank the ability of this virus's spike protein to bind ACE2 from relevant species. Doing so led to some surprising and unexpected findings. With regard to key species believed to be possible progenitors of SARS-CoV-2, structure-based distance measures revealed the opposite of what had been suggested from sequence-based approaches: species that seemed to be most closely related to SARS-CoV-2 in terms of a high ACE2 sequence homology were shown to exhibit low structural similarities and vice versa.

Securing genetic information by, e.g., encrypting pathogen databases alone would not be enough, as it is unclear how exactly microbes will mutate and evolve—especially under selective pressure—and how this could be unequivocally modeled and differentiated.

Therefore, attribution of any possible future form of bioterrorism based on genetic information alone may be incomplete and ought to include circumstantial evidence as used in forensic analysis more generally. Specifically, processes that rely on automation and AI ought to be carefully scrutinized to make sure that gaps stemming from computer-assisted methods have sufficiently been closed.

3.3 Acknowledging the Fact That in Biology There Is No Clear Binary Yes-No Paradigm

Due to the flexibility of all of life and the inherent noise and stochasticity of biomedical processes, our comprehension of biological phenomena is essentially non-binary. The same extends to potential attacks, so even the line between intentional and unintentional release from a lab may be difficult to draw. Specifically, the attack landscape itself resembles a spectrum. For example:

- Most directly, bioterrorists may perform genetic manipulations aiming for a specific outcome. Alarmingly, the technology for the targeted genesis of pathogens with pandemic potential is in place, as was also described in [50] via the reconstruction of SARS-CoV-2 from purely digital information.
- Less directly, and fostered by the digital-biological/physical gap, details about the design and construction of pathogens can be hidden in various ways. Nefarious programs may even be camouflaged by benevolent research projects, and dangerous modifications of microorganisms through lab experiments may not be sufficiently documented or made publicly available.
- Lastly, threat actors may indirectly trigger the creation/release of a pathogen by establishing an environment where evolution is accelerated and accidents are prone to happen. Without necessarily knowing specific biologic outcomes, such opportunistic actors would resemble something like sending a kid with matches into a dry forest. A known method to mimic and utilize natural processes to covertly enhance pathogenic evolution has long been realized via recombination or serial passaging whereby zoonosis between species can be enforced within a laboratory in a much shorter time than required by a natural jump while leaving a genome behind that appears natural [45].

4 Conclusion

Some would say that projects that, say, enhance the infectivity of a pathogen can help provide valuable information to test a new vaccine, for example. Others argue that such "gain-of-function" is extremely dangerous as there is no guarantee that the "enhanced" pathogens can be sufficiently contained. This inherent *dual-use* research of synthetic biology triggered alarm already years ago [46] when the danger of misusing this type of research was clearly understood.

The computerization of biotechnology has added yet another layer, imperiling safety and security of more traditional research and created a sheer insurmountable gap between appearance and reality. Actual biological/clinical reality may effectively be hidden behind computerized methods, processes, and automated interpretations.

Through a computer interface, dual-use research in effect also becomes *dual-appearance* (Fig. 1). As synthetic biology relies on and mimics nature, this inherent camouflaging feature could enable dangerous research to be hidden behind benign biological characteristics and natural phenomena. As not all the modifications of pathogens are predictable, and consequently, as people may not be looking for them in a targeted way, covet research may therefore not easily be identified. The computer interface also enables the dual-appearance of forbidden dual-use research,

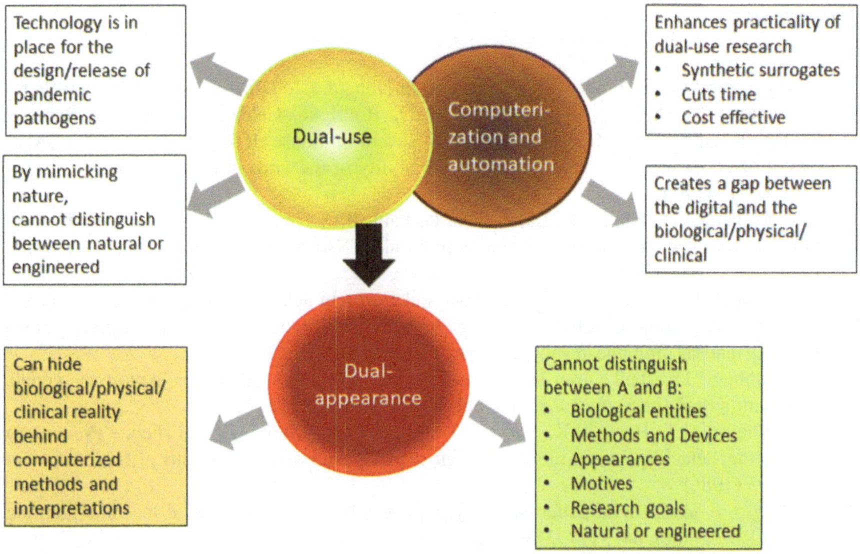

Fig. 1 The computerization and automation of dual-use research. The reliance on computers, web-based applications, and automation endanger dual-use ("gain-of-function") research not only in terms of (unintentional) safety issues, but also related to security. The marriage between synthetic biology and computers has created a gap between a digital presentation of a purported thing and the actual biological/physical/clinical entity in question. This gap in and of itself enhances the dangers of dual-use research and can further aid bad actors to effectively perform even more perilous research, by exploiting not only dual-use but also "dual-appearance" (camouflaging) related to basic identification issues (involving pathogens, processes, and machines), as well as what is being done, why, and how. In the figure, the gray arrows indicate some of the main cyberbiosecurity concerns of pathogen research

so it can be hidden behind well-motivated or beneficially looking projects, with computerization making it all look safe and secure.

In summary, if dual-use ("gain-of-function") research has been recognized as dangerous, then via a cyber-interface, automation, and AI, the gap between the digital and the biological/clinical enables the development and release of bioweapons that can become an existential threat to humanity.

Just as addressing cyber-security issues rely on new ideas, sophisticated research and targeted support for their mitigation, it is imperative to carefully scrutinize cyberbiosecurity dangers related to the weaponization of dual-use research (Fig. 1). As the workings of nature remain incompletely appreciated—and remain susceptible to malicious exploitation—independent critical research, international and open discussions, and a commitment to the divinity of humanity and nature cannot be an option.

Acknowledgments I would like to thank Guy Halevi for his careful reading of an earlier version of this work and his helpful comments.

References

1. D.A. Álvarez-Díaz, C. Franco-Muñoz, K. Laiton-Donato, J.A. Usme-Ciro, N.D. Franco-Sierra, A.C. Flórez-Sánchez, S. Gómez-Rangel, L.D. Rodríguez-Calderon, J. Barbosa-Ramirez, E. Ospitia-Baez, et al., Molecular analysis of several in-house rRT-PCR protocols for SARS-CoV-2 detection in the context of genetic variability of the virus in colombia. Infect. Genet. Evol. **84**, 104390 (2020)
2. B.K. Ambati, A. Varshney, K. Lundstrom, G. Palú, B.D. Uhal, V.N. Uversky, A.M. Brufsky, MSH3 homology and potential recombination link to SARS-CoV-2 furin cleavage site. Front. Virol., 10 (2022)
3. American Heart Association, Coronavirus spike protein activated natural immune response, damaged heart muscle cells. https://newsroom.heart.org/news/coronavirus-spike-protein-activated-natural-immune-response-damaged-heart-muscle-cells (2022)
4. A. Anubhav, Dnashell : Iran based attackers utilize a DNA sequencer exploit for genetic information theft. https://www.ankitanubhav.info/post/dnashell (2019)
5. N. Arnheim, G.T. Horn, R.K. Saiki, S.J. Scharf, K.B. Mullis, H.A. Erlich, Process for amplifying, detecting, and/or-cloning nucleic acid sequences. https://patents.justia.com/patent/4683195 (1986)
6. B. Biggio, F. Roli, Wild patterns: Ten years after the rise of adversarial machine learning. Pattern Recognition **84**, 317–331 (2018)
7. J.D. Bloom, Recovery of deleted deep sequencing data sheds more light on the early Wuhan SARS-CoV-2 epidemic. Mol. Biol. Evol. **38**(12), 5211–5224 (2021)
8. T. Buschmann, L.V. Bystrykh, Levenshtein error-correcting barcodes for multiplexed DNA sequencing. BMC Bioinformatics **14**(1), 1–10 (2013)
9. E. Callaway, Deleted coronavirus genome sequences trigger scientific intrigue (2021)
10. J. Caswell, J.D. Gans, N. Generous, C.M. Hudson, E. Merkley, C. Johnson, C. Oehmen, K. Omberg, E. Purvine, K. Taylor, C.L. Ting, M. Wolinsky, G. Xie, Defending our public biological databases as a global critical infrastructure. Front. Bioeng. Biotechnol. **7**, 58 (2019)
11. Centers for Disease Control and Prevention, 07/21/2021: Lab alert: Changes to CDC RT-PCR for SARS-CoV-2 testing. https://web.archive.org/web/20220822080253/https://www.cdc.gov/csels/dls/locs/2021/07-21-2021-lab-alert-Changes_CDC_RT-PCR_SARS-CoV-2_Testing_1.html (2021)
12. C. Cimpanu, Mysterious iranian group is hacking into DNA sequencers hackers are scanning the internet and planting shells on web-based dna sequencing apps. https://www.zdnet.com/article/mysterious-iranian-group-is-hacking-into-dna-sequencers/ (2019)
13. E. Clough, J. Inigo, D. Chandra, L. Chaves, J.L. Reynolds, R. Aalinkeel, S.A. Schwartz, A. Khmaladze, S.D. Mahajan, Mitochondrial dynamics in SARS-COV2 spike protein treated human microglia: Implications for neuro-covid. J. Neuroimmune Pharmacol. **16**(4), 770–784 (2021)
14. J. Cohen, Anywhere but here. Science (New York, NY) **377**(6608), 805–809 (2022)
15. V.M. Corman, O. Landt, M. Kaiser, R. Molenkamp, A. Meijer, D.K. Chu, T. Bleicker, S. Brünink, J. Schneider, M.L. Schmidt, et al., Detection of 2019 novel coronavirus (2019-ncov) by real-time RT-PCR. Eurosurveillance **25**(3), 2000045 (2020)
16. Y. Deigin, R. Segreto, SARS-CoV-2's claimed natural origin is undermined by issues with genome sequences of its relative strains: Coronavirus sequences RaTG13, MP789 and RmYN02 raise multiple questions to be critically addressed by the scientific community. Bioessays **43**(7), 2100015 (2021)
17. B. Diao, C. Wang, Y. Tan, X. Chen, Y. Liu, L. Ning, L. Chen, M. Li, Y. Liu, G. Wang, et al., Reduction and functional exhaustion of t cells in patients with coronavirus disease 2019 (Covid-19). Front. Immunol., 827 (2020)

18. C. Drosten, S. Günther, W. Preiser, S. van der Werf, H.-R. Brodt, S. Becker, H. Rabenau, M. Panning, L. Kolesnikova, R.A. Fouchier, A. Berger, A.-M. Burguière, J. Cinatl, M. Eickmann, N. Escriou, K. Grywna, S. Kramme, J.-C. Manuguerra, S. Müller, V. Rickerts, M. Stürmer, S. Vieth, H.-D. Klenk, A.D. Osterhaus, H. Schmitz, H.W. Doerr, Identification of a novel coronavirus in patients with severe acute respiratory syndrome. New Engl. J. Med. **348**(20), 1967–1976 (2003). PMID: 12690091
19. European Centre for Disease Prevention and Control, Methods for the detection and characterisation of SARS-CoV-2 variants – first update. 20 December 2021. https://www.ecdc.europa.eu/en/publications-data/methods-detection-and-characterisation-sars-cov-2-variants-first-update (2021)
20. S.G. Finlayson, J.D. Bowers, J. Ito, J.L. Zittrain, A.L. Beam, I.S. Kohane, Adversarial attacks on medical machine learning. Science **363**(6433), 1287–1289 (2019)
21. G. Fongaro, P.H. Stoco, D.S.M. Souza, E.C. Grisard, M.E. Magri, P. Rogovski, M.A. Schörner, F.H. Barazzetti, A.P. Christoff, L.F.V. de Oliveira, M.L. Bazzo, G. Wagner, M. Hernández, D. Rodríguez-Lázaro, The presence of SARS-CoV-2 RNA in human sewage in Santa Catarina, Brazil, November 2019. Sci. Total Environ. **778**, 146198 (2021)
22. U. Food, D. Administration, FDA facts: Biomarkers and surrogate endpoints. https://www.fda.gov/about-fda/innovation-fda/fda-facts-biomarkers-and-surrogate-endpoints (2017)
23. J.E. Gallegos, S. Hayrynen, N.R. Adames, J. Peccoud, Challenges and opportunities for strain verification by whole-genome sequencing. Scientific Reports **10**(1), 1–9 (2020)
24. L.M. Grobbelaar, C. Venter, M. Vlok, M. Ngoepe, G.J. Laubscher, P.J. Lourens, J. Steenkamp, D.B. Kell, E. Pretorius, SARS-CoV-2 spike protein S1 induces fibrin (ogen) resistant to fibrinolysis: implications for microclot formation in Covid-19. Bioscience Reports **41**(8), BSR20210611 (2021)
25. K. Hagemann, K. Riecken, J.M. Jung, H. Hildebrandt, S. Menzel, M.J. Bunders, B. Fehse, F. Koch-Nolte, F. Heinrich, S. Peine, et al., Natural killer cell-mediated adcc in SARS-CoV-2-infected individuals and vaccine recipients. Eur. J. Immunol. **52**(8), 1297–1307 (2022)
26. H. Hirano, A. Minagi, K. Takemoto, Universal adversarial attacks on deep neural networks for medical image classification. BMC Med. Imag. **21**(1), 1–13 (2021)
27. International Consortium of Scientists in Life Sciences (ICSLS), Review report corman-drosten et al. eurosurveillance 2020. https://cormandrostenreview.com/ (2020)
28. M. Kircher, S. Sawyer, M. Meyer, Double indexing overcomes inaccuracies in multiplex sequencing on the illumina platform. Nucleic Acids Res. **40**(1), e3–e3 (2012)
29. E. Kopp, Another missing database? Ecohealth project in southeast Asia is under construction. https://usrtk.org/biohazards-blog/another-missing-database-ecohealth-project-in-southeast-asia-is-under-construction/?mc_cid=ce6fb1f44f&mc_eid=6cb23747ff (2022)
30. S.H. Lee, Implementation of the ECDC/WHO recommendation for molecular diagnosis of SARS-CoV-2 Omicron subvariants and its challenges (2022)
31. A. Mandavilli, Your coronavirus test is positive. Maybe it shouldn't be. https://www.nytimes.com/2020/08/29/health/coronavirus-testing.html (2020)
32. A. Maxmen, Wuhan market was epicentre of pandemic's start, studies suggest (2022)
33. MSNBC, Covid tests don't do what you think they do, dr. fauci explains. https://www.youtube.com/watch?v=bAICMQ1D5F8&t=5s (2021)
34. R.S. Murch, W.K. So, W.G. Buchholz, S. Raman, J. Peccoud, Cyberbiosecurity: An emerging new discipline to help safeguard the bioeconomy. Front. Bioeng. Biotechnol. **6**, 39 (2018)
35. National Academies of Sciences, Engineering, and Medicine, *Safeguarding the Bioeconomy* (The National Academies Press, Washington, DC, 2020)
36. P.M. Ney, Securing the future of biotechnology: A study of emerging bio-cyber security threats to DNA-information systems (Doctoral dissertation). PhD thesis, 2019
37. A. Noymer, Four things i want to know about the origin of Covid. https://www.washingtonexaminer.com/opinion/op-eds/four-things-i-want-to-know-about-the-origin-of-covid?utm_source=substack&utm_medium=email (2022)

38. J. Peccoud, J.E. Gallegos, R. Murch, W.G. Buchholz, S. Raman, Cyberbiosecurity: from naive trust to risk awareness. Trends Biotechnol. **36**(1), 4–7 (2018)
39. S. Piplani, P.K. Singh, D.A. Winkler, N. Petrovsky, In silico comparison of SARS-CoV-2 spike protein-ace2 binding affinities across species and implications for virus origin. Scientific Reports **11**(1), 1–13 (2021)
40. E. Pretorius, M. Vlok, C. Venter, J.A. Bezuidenhout, G.J. Laubscher, J. Steenkamp, D.B. Kell, Persistent clotting protein pathology in long covid/post-acute sequelae of Covid-19 (PASC) is accompanied by increased levels of antiplasmin. Cardiovascular Diabetology **20**(1), 1–18 (2021)
41. S.C. Quay, A Bayesian analysis concludes beyond a reasonable doubt that SARS-CoV-2 is not a natural zoonosis but instead is laboratory derived (2021)
42. W. Randazzo, P. Truchado, E. Cuevas-Ferrando, P. Simón, A. Allende, G. Sánchez, SARS-CoV-2 RNA in wastewater anticipated Covid-19 occurrence in a low prevalence area. Water Research **181**, 115942 (2020)
43. W. Ren, X. Qu, W. Li, Z. Han, M. Yu, P. Zhou, S.-Y. Zhang, L.-F. Wang, H. Deng, Z. Shi, Difference in receptor usage between severe acute respiratory syndrome (SARS) coronavirus and SARS-like coronavirus of bat origin. J. Virol. **82**(4), 1899–1907 (2008)
44. G.J. Schumacher, S. Sawaya, D. Nelson, A.J. Hansen, Genetic information insecurity as state of the art. Front. Bioeng. Biotechnol. **8**, 591980 (2020)
45. K. Sirotkinm, D. Sirotkin, Might SARS-CoV-2 have arisen via serial passage through an animal host or cell culture? A potential explanation for much of the novel coronavirus' distinctive genome. BioEssays **42**(10), 2000091 (2020)
46. The Cambridge Working Group, Cambridge working group consensus statement on the creation of potential pandemic pathogens (PPPS). http://www.cambridgeworkinggroup.org/ (2014)
47. The Centers for Disease Control and Prevention (CDC), Cdc 2019-novel coronavirus (2019-NCoV) real-time RT-PCR diagnostic panel. https://www.fda.gov/media/134922/download
48. The Epoch Times, U.S. university concedes it may have broken law in contract with Wuhan lab. https://www.theepochtimes.com/us-university-says-its-poorly-drafted-contract-with-wuhan-lab-may-have-broken-law_4644724.html (2022)
49. The Federal Bureau of Investigation (FBI), FBI and CISA warn against chinese targeting of Covid-19 research organizations. https://www.fbi.gov/news/press-releases/press-releases/fbi-and-cisa-warn-against-chinese-targeting-of-covid-19-research-organizations (2020)
50. T. Thi Nhu Thao, F. Labroussaa, N. Ebert, P. V'kovski, H. Stalder, J. Portmann, J. Kelly, S. Steiner, M. Holwerda, A. Kratzel, et al., Rapid reconstruction of SARS-CoV-2 using a synthetic genomics platform. Nature **582**(7813), 561–565 (2020)
51. Tim Spector and Collaborators, The Zoe health study. https://health-study.joinzoe.com/, n.d.
52. U.B.C. on Biodefense, Biologia et machina: Cyberbiosecurity for today's hybrid evolution - virtual meeting. https://biodefensecommission.org/events/biologia-et-machina-cyberbiosecurity-for-todays-hybrid-evolution/ (2021)
53. U.B.C. on Biodefense, Revisiting gain of function research: What the pandemic taught us and where do we go from here - senate hearing. https://www.hsgac.senate.gov/subcommittees/etso/hearings/revisiting-gain-of-function-research-what-the-pandemic-taught-us-and-where-do-we-go-from-here (2022)
54. U.S. Senate Committee on Homeland Security & Governmental Affairs Hearings, Revisiting gain of function research: What the pandemic taught us and where do we go from here subcommittee on emerging threats and spending oversight. https://www.hsgac.senate.gov/subcommittees/etso/hearings/revisiting-gain-of-function-research-what-the-pandemic-taught-us-and-where-do-we-go-from-here (2022)
55. B.A. Vinatzer, L.S. Heath, H.M.J. Almohri, M.J. Stulberg, C. Lowe, S. Li, Cyberbiosecurity challenges of pathogen genome databases. Front. Bioeng. Biotechnol. **7**, 106 (2019)
56. C. Wang, Z. Liu, Z. Chen, X. Huang, M. Xu, T. He, Z. Zhang, The establishment of reference sequence for SARS-CoV-2 and variation analysis. J. Med. Virol. **92**(6), 667–674 (2020)

57. World Health Organization, Who information notice for IVD users 2020/05. https://www.who.int/news/item/20-01-2021-who-information-notice-for-ivd-users-2020-05 (2021)
58. M. Worobey, J.I. Levy, L.M. Serrano, A. Crits-Christoph, J.E. Pekar, S.A. Goldstein, A.L. Rasmussen, M.U.G. Kraemer, C. Newman, M.P.G. Koopmans, M.A. Suchard, J.O. Wertheim, P. Lemey, D.L. Robertson, R.F. Garry, E.C. Holmes, A. Rambaut, K.G. Andersen, The Huanan seafood wholesale market in Wuhan was the early epicenter of the Covid-19 pandemic. Science **377**(6609), 951–959 (2022)
59. D. Zhang, The Pan-SL-CoV/GD sequences may be from contamination (2020)
60. Y. Zheng, J. Zhao, J. Li, Z. Guo, J. Sheng, X. Ye, G. Jin, C. Wang, W. Chai, J. Yan, et al., SARS-CoV-2 spike protein causes blood coagulation and thrombosis by competitive binding to heparan sulfate. Int. J. Biol. Macromol. **193**, 1124–1129 (2021)

How to Protect Biotechnology and Biosecurity from Adversarial AI Attacks? A Global Governance Perspective

Eleonore Pauwels

Abstract The integration of genomics and biotechnology with AI is emerging as a geo-strategic, societal and welfare asset that can define a country's digital sovereignty and preserve national and international security. In the absence of a robust AI and cybersecurity framework, however, algorithms can be trained and misused to manipulate the integrity of genomic datasets, creating hybrid cybersecurity threats and potentially leading to widespread collective data harms, research and industrial sabotage, as well as compromised governance systems and data integrity crucial to health, food and civilian security. This chapter aims at analysing a new typology of AI-led cyberthreats that can manipulate and corrupt the integrity of genomic datasets and algorithmic models crucial to the global knowledge-production cycle in bio-medicine, biotechnology and biosecurity. It will also focus on a diagnosis of the current vulnerabilities inherent to genomics data security and demonstrate how adversarial data manipulation may not only produce lethal outcomes for populations and erode countries' digital sovereignty but also drastically undermine public trust in the bio-economy's critical information and governance infrastructures. In the discussion section, this contribution will cover the nascent regulatory debates that assess the adequacy and applicability of international law and governance mechanisms to cyberattacks and adversarial operations that target populations' genomic data.

Keywords Cyberbiosecurity · Genomics · Artificial intelligence · Adversarial attacks · Data targeting and manipulation · Governance

1 Introduction

In the last two decades, the world of biotechnology has moved from analogue to digital, converging with artificial intelligence (AI) as an innovation catalyst.

E. Pauwels (✉)
International Expert in the Security Implications of Converging Technologies and Senior Fellow, The Global Center on Cooperative Security, New York, NY, USA
e-mail: eleonore@eleonorepauwels.com

D. Greenbaum (ed.), *Cyberbiosecurity*, https://doi.org/10.1007/978-3-031-26034-6_11

New collaborations between AI, geneticists and bio-engineers have led to the field of functional genomics, a more precise understanding and optimisation of functional processes in genome biology [4]. Deep learning algorithms can help analyse and test genetic functions in silico and help predict the effect of a genetic mutation on an individual's overall genome. Such algorithms improve analysis of the combinatorial relationship between genotype and phenotype in genomic datasets related to humans, animals and pathogens. Other deep learning models aim to unveil important features of genome biology, from simulating RNA processing events to modelling the genetic regulatory code governing gene expression [10].

The new frontier of functional genomics is therefore increasingly happening in silico, producing important knowledge insights that build on synthetic datasets as well as algorithmic and advanced computing [25]. Substantial progress will also derive from digitising, processing and learning from genomics and other multimodal omics datasets that are part of comprehensive approaches to analysing complete genetic or molecular profiles of humans, animals and pathogens. Functional genomics, and biosciences in general, are becoming not only crucial and sensitive digital assets but also critical information infrastructure. Transformational opportunities range from improving trust in precision medicine diagnoses and therapies to ensuring reproducibility and efficiency in complex biotech supply chains and isolating potential harmful genetic functions in biosecurity screening.

The integration of genomics and biotechnology with AI is emerging as a geostrategic, societal and welfare asset that can define a country's digital sovereignty and preserve national and international security [2, 3]. In the absence of a robust AI and cybersecurity framework, however, algorithms can be trained and misused to manipulate the integrity of genomics datasets, creating hybrid cybersecurity threats and potentially leading to widespread collective data harms, research and industrial sabotage, as well as compromised governance systems and data integrity crucial to health, food and civilian security.

This chapter aims at analysing a new typology of AI-led cyberthreats that can manipulate and corrupt the integrity of genomic datasets and algorithmic models crucial to the global knowledge-production cycle in bio-medicine, biotechnology and biosecurity. It will also focus on a diagnosis of the current vulnerabilities inherent to genomics data security and demonstrate how adversarial data manipulation may not only produce lethal outcomes for populations and erode countries' digital sovereignty but also drastically undermine public trust in the bio-economy's critical information and governance infrastructures. In the discussion section, this contribution will close on the nascent regulatory debates that assess the adequacy and applicability of international law and governance mechanisms to cyberattacks and adversarial operations that target populations' genomics data.

2 Technical Considerations

2.1 A New Era of AI and Genomics Convergence

"Cyberbiosecurity" can be defined as the emerging discipline addressing the unique vulnerabilities and threats that occur at the confluence of AI, cybersecurity and biosciences [19]. Understanding the cyberbiosecurity threats' landscape requires mapping the scope and depth of the convergence between AI, cyber-physical systems and biological systems. AI computing has acted as a catalyst in two ways: first, by drastically advancing bio-informatic capacities, improving both in silico and synthetic biology; second, AI programmes are leading to an increase in automation, turning labs into smart, fully connected facilities, operating often with cloud services and decentralised networks of devices (from automated DNA assembly programmes to mobile DNA sequencers and synthesisers). In other words, AI computing, automation and decentralisation have made modern biotechnology, both more powerful and more vulnerable.

The current COVID-19 pandemic serves as a wake-up call across the world of our shared vulnerability to biothreats and the crucial importance of biomedicine and biodefense programmes for threat mitigation. The combination of genomics surveillance, AI and advanced bioinformatics has drastically bolstered the ability to track the spread of SARS-CoV-2 variants and decipher transmission dynamics in real time [6]. Technological convergence has also aided in the development of timely diagnostics tools and accelerated the synthesis of vaccines. Global disease control programmes, such as those for tuberculosis, malaria, HIV, foodborne pathogens and antibiotic resistance, now recommend genomics-based surveillance as a vital component [27].

The integration of AI computing within modern biomedicine allows researchers to rely on synthetic datasets and predictive methods to produce actionable knowledge in a genome's biology and assess its clinical value. AI computing also creates increased potential for monitoring and optimising data analytics across the multimodal datasets that constitute the complete genetic or molecular profiles of humans, animals and pathogens. A significant advantage that AI computing could bring to public health and clinical research is to process simultaneously massive amounts of genomic, physiological, health, ecosystem and lifestyle data about populations in their environment [15]. These approaches are crucial to improving our understanding of genomics and biological processes related to human and animal pathologies, including infectious diseases.

The convergence of AI with biotechnology could help identify which genetic functions are key to augmenting the capacity of a pathogen to infect a host, evade the immune system, spread among subpopulations or resist vaccines and antibiotics. Predictive modelling is important for real-time disease surveillance and for monitoring and preventing future zoonotic spill-overs using advanced bio-forensics and sensing capacity for detecting pathogens [26]. The fast production of medical countermeasures (such as immunoassay diagnostic tests for detecting the antigen or antibody properties of certain proteins, liquid biopsies and vaccines) also increasingly depends on advances at the intersection of genomics, AI and bioinformatics.

The COVID-19 pandemic serves as a prime example of how emerging in silico capacities in pathogen genomics are key for developing rapid medical countermeasures. In 2020, at the onset of the pandemic, scientists designed a platform to automate the synthesis of existing RNA viruses, which are estimated to make up 44% of all emerging infectious diseases [24]. Using this platform, they were able to synthesise clones of the SARS-CoV-2 virus a week after receiving the synthetic DNA fragments. Such technical advances enable both the real-time genotypic detection of viral traits and the modelling of the pathogen's mutational landscape.

2.2 *Rising Cyberthreats and Adversarial AI Attacks*

At the same time as biosciences are digitising, the field of cybersecurity is being challenged by a new type of adversarial attacks. AI malware can be designed to inject noise and manipulate the integrity of datasets and algorithmic models crucial to biomedicine, biotechnology and biosecurity [8, 16]. Such AI malware can increasingly evade detection and learn how to harness human and machine vulnerabilities. The field of cybersecurity is currently witnessing an explosion of new AI techniques to engineer behavioural vulnerabilities and scale up attacks, from precision spear phishing, audio/video forgeries (deepfakes), password spraying and biometrics theft. The ability of AI to decode and manipulate behaviours can target human weaknesses to the point of increasingly equipping external actors with insider and tacit knowledge. This is what can be framed a "human computation" problem where the distance between external and insider attacks is shrinking. Importantly, this shift is taking place while staff in hospitals and genomics laboratories, biotech firms, gene synthesis and biomanufacturing facilities are struggling to acquire skills and build internal capacity to prevent offensive cyberoperations.

At each stage of the information life cycle, the digital infrastructure that underpins biosciences is a target for AI-led cyberattacks, in particular adversarial data manipulation, that could sabotage and weaponise biomedical research, clinical trials, biotech facilities and supply chains. It follows that three types of potential threats are rising.

2.2.1 Manipulating Biomedical Research and Population Datasets

The digital interdependence of modern biosciences subjects our growing functional intelligence about genome biology to new information risks, particularly adversarial attacks that could corrupt the integrity of biological datasets and manipulate the functioning of deep learning analysis systems. Several studies in AI security have demonstrated how generative adversarial networks can be trained to drastically undermine the predictive ability of a wide range of medical image analysis systems that are based on deep learning [1, 13]. In 2018, researchers at Ben-Gurion University designed a malicious attack to manipulate cancer data in hospital CT scans, generating false lung tumours that conformed to a patient's unique anatomy, leading to a misdiagnosis rate in excess of 90% [11]. Furthermore, researchers at Harvard University tested adversarial attacks against algorithms used to diagnose skin cancer images, showing that such attacks only required modifying a few pixels in the original biopsy picture to corrupt a diagnosis [5]. As medical intelligence about the treatment of cancers, blood clots, brain lesions and infections could be manipulated, adversarial attacks on deep learning pose a substantial risk to our most critical medical and clinical infrastructures.

The attack surface extends far beyond medical diagnosis and clinical trials with adversarial malware that could target the integrity of genomics and other omic datasets related to humans and pathogens. Researchers at the Sandia National Laboratory have demonstrated how autonomous malware could be used to manipulate raw data within large curation of human genomes [14]. The malicious malware could be used to target the functioning of genetic analysis software and alter actual fragments of DNA sequences within individuals' genomes. Such malicious tampering could result in misdiagnosis with an impact on clinical decisions. This type of data poisoning could affect in silico predictive models in functional genomics, including how we diagnose and treat complex genetic diseases, how we analyse and study the pathogenicity of viral and microbial threats and how we develop adequate medical countermeasures for subgroups of patients. What is at stake is the global knowledge-production cycle in biomedicine.

2.2.2 Sabotaging Bio-Engineering and Bio-Manufacturing

New capacities in automation and remote manufacturing – including cloud laboratories and commercial DNA sequencing and synthesis – are accelerating the decentralisation of bio-engineering experiments [21]. Increasingly, biotech laboratories and bio-pharmaceutical manufacturing systems are automated, equipped with AI analytics software and connected to cloud services. On such platforms, technical skills and tacit knowledge are encoded into "automated protocols" that programme and standardise the instructions of a biotech experiment. Equipped with connected sensors to measure experimental variables, the AI operating system uses constant learning and iteration to augment the precision of automated protocols and may even lead to the in silico design of novel experiments with less outside guidance. Automated laboratories therefore offer advantages that are crucial to precision medicine as they allow for scalability, reproducibility and outsourcing to a broader and more diverse talent pool. The advent of autonomy provides an increasing potential to weaponise biotech laboratories and biomanufacturing supply chains through adversarial attacks waged in cyberoperations. Adversarial algorithms could target vulnerabilities in automated protocols to corrupt networks of sensors and duly impact control decisions related to important experimental parameters. Resulting harm could range from producing pharmaceutical products that do not match specification standards (leading to waste and shortage) to spoiling vital stocks of vaccines, antibiotics, cell or immune therapies for cancer treatment.

Cyber criminals and state actors have already mounted targeted cyberoperations against firms researching, producing and distributing COVID-19 vaccines. In December 2020, IBM researchers and the US Cybersecurity and Infrastructure Security Agency (CISA) unveiled global social engineering attacks "intended to steal the network log-in credentials of corporate executives and officials at global organizations involved in the refrigeration process necessary to protect vaccine doses" [20]. The underlying goal could be to access and manipulate shared information about how the vaccine is shipped, stored, kept cold and delivered.

Weaponising biotech laboratories could escalate into a strictly biothreat-based scenario while avoiding traditional screening and oversight. Automated bio-labs could be used to (1) produce toxins that can disrupt cellular metabolism, (2) synthesise a known lethal pathogen or (3) use gene editing to augment the capacity of a pathogen to infect a host, evade the immune system, spread among subpopulations or resist vaccines or antibiotics [12]. An area of near-term concern is the automated design of bacteria with multidrug resistance or the modification of commensal bacteria to become super-producers of toxins.

The convergence of AI and automation with biotechnology is increasingly challenging the compliance tools, verification methods and overall oversight that countries can rely on to ensure nonproliferation within the current disarmament regime, the Biological Weapons Convention (BWC) [17]. Importantly, no adequate guidance exists to prevent the adversarial use of biological data and algorithmic models to produce pathogens of concern or produce a biosecurity consequence by exploiting vulnerabilities in the cyberbiosecurity infrastructure.

2.2.3 Hacking and Corrupting Biosecurity Screening

By improving our knowledge of DNA functions, AI computing is becoming an integral part of biosecurity screening mechanisms [18]. In particular, algorithmic models are instrumental in preventing illicit gene synthesis and illicit experiments in gain-of-function research, a field that studies the potential to enhance the transmissibility or pathogenicity of potential pandemic pathogens. Government-funded programmes are already designing deep learning systems to predict how genetic sequences are meant to function, before being assembled, even if the combination is not seen in nature. Gene synthesis companies are developing computational threat models that can be applied to characterise the function of novel combinations of DNA sequences. Similar algorithmic tools play an increasing role in microbial forensics, using their capacity for anomaly detection to identify the specific signatures left in modified organisms [22].

Adversarial attacks could be designed to corrupt the predictive ability of screening algorithms to identify threats based on functional analysis of DNA sequences. By obfuscating functional data from sequences of pathogen and toxin DNA, generative adversarial networks could manipulate the integrity of the priority dataset shared by stakeholders to train screening algorithms. Such data manipulation could drastically undermine the confidence level of screening algorithms when they aim to ascribe threat potential to known and unknown genes, including genes responsible for the pathogenesis of viral threats, bacterial threats and toxins. Both human and algorithmic understanding of functional genetic data is still weak and fraught with complex unknowns. Adversarial attacks therefore have a very high potential to succeed in undermining stakeholders' trust in DNA screening.

3 Global Governance and Legal Discussion

The digital infrastructure that supports biomedicine and biotechnology is a global public good. Emergent hybrid threats that compromise AI and cybersecurity within the bio-economy are contributing to a new geopolitics of inequality and insecurity that cuts across societies and borders.

Adversarial information operations that target the biotechnology sector are a powerful type of hybrid threat. They may serve an array of offensive goals and involve broad coalitions of malicious actors, including states, non-state actors and surrogates. They target systemic vulnerabilities and different civilian and security interfaces, from population datasets and industry's clinical trials to biosecurity screening. They also interfere with diverse levels of strategic and emergency decision-making.

New forms of covert, adversarial data manipulation attacks are extremely hard to detect, creating new challenges for attribution. What is potentially under attack is the global knowledge-production cycle in biosciences. The aim is not only to seriously erode a country's digital sovereignty but also to undermine both global leadership crisis response and people's trust and resilience. Combinations of poisoning population datasets, falsifying biomedical research, sabotaging bio-manufacturing and corrupting biosecurity screening would have drastic economic costs and potentially lethal outcomes for populations. Yet the most damaging impact would be on citizens' trust in governing institutions, emergency data systems, industrial laboratories, food supply chains, hospitals and critical health infrastructures. This could have powerful, long-term implications for peace and security. As vulnerable states are unable to prevent and mitigate data poisoning attacks, they could become fertile operating grounds for cyber mercenaries, terrorist groups and other actors, increasingly compromising the data integrity and robustness of our globalised intelligence system.

3.1 International Legal Debates

Legal experts and multilateral governance processes have indicated how AI-led offensive cyberoperations do not take place in a legal vacuum. International law provides a comprehensive regulatory framework that can be applied to offensive cyberoperations [7]. The international legal regime affords protections to civilians and civilian objects and prohibit certain types of hostile cyberactivities, including inter alia: direct attacks against civilians and civilian objects; indiscriminate attacks that do not distinguish between military objectives and civilians or civilian objects; disproportionate attacks that may cause incidental loss of civilian life, injury to civilians, damage to civilian objects or a combination thereof; and attacks that would destroy, remove or render useless objects indispensable to the survival of the population. In a 2019 position paper, the International Committee of the Red Cross (ICRC) emphasises two legal ambiguities about how the international legal framework applies to offensive cyberoperations [9]:

- **First**, there is no internationally agreed definition on what constitutes a cyber-attack or cyberhostilities within international law. Important technical questions persist as how to define and qualify offensive cyberoperations or technical terms such as "attack" when they rely exclusively on cyber means. In particular, legal ambiguity remains as whether cyberoperations that would not cause physical damage but result in limited disruption of essential services' functionality, or in erosion of public trust into critical systems, would qualify as an attack and therefore violate international legal regimes. For instance, it is likely that an AI-led cyberoperation could qualify as an attack if it is designed to disable automated protocols and drastically increase levels of sodium hydroxide in urban water supplies with direct implications for human health.

Now, consider a scenario in a private sector's biotech facility, while avoiding detection, an AI-based malware target vulnerability in automated data protocols to manipulate networks of sensors and impact quality control processes. Resulting harms extend from making pharmaceutical products that do not match specification standards (leading to waste and shortage) to undermining trust in stocks of vaccines and therapeutics. In this case, forensic analysis to characterise the nature and threshold of the attack and determine technical attribution would be complex and could remain highly contested. Would this scenario – where an AI malware influences automated data processing required for biotech quality controls – qualify as an attack under international law? Extrapolation from the Second Oxford Statement by international experts in the context of the COVID-19 pandemic could help support such legal qualification: "International law prohibits cyberoperations by States that have significant adverse or harmful consequences for the research, trial, manufacture, and distribution of a COVID-19 vaccine, including by means that damage the content or impair the use of sensitive research data, particularly trial results, or which impose significant costs on targeted facilities in the form of repair, shutdown, or related preventive activities" [23].

- **Another** salient question of legal interpretation is whether datasets can be considered as "object" with the consequence that adversarial cyberoperations targeting essential civilian datasets for manipulation or destruction would then violate international law. While specific protection is afforded to digital medical records, a growing amount of biological data about civilians is not necessarily managed and stored in hospitals and medical facilities per se but transferred to universities, direct-to-consumer genomics databases and private sector platforms. Yet, even in such context of decentralised data management, the Second Oxford Statement may serve as a basis to extrapolate and infer that an AI-led cyberattack would be prohibited if the malware is designed to manipulate the integrity of human genomics data in a biotech research trial setting.

Interestingly, the same statement would not apply to important datasets, data analytics and algorithmic processes used in a parallel sector, **biosecurity**. Consider a case where malicious actors conduct an adversarial attack to poison the integrity of pathogen's genomic datasets critical to screening of biosecurity threats by universities and the private sector. Pathogens' genomic data do not necessarily rest in centralised high-security databases but can be held in public and open-source repositories. Biosecurity screening efforts are complex and relatively fragmented and increasingly integrate predictive algorithms. Assessing cyberthreats and gaps in legal protection in the biosecurity sector would therefore gain from being considered by technical and legal experts in the field.

3.2 Policy Recommendations

The convergence of AI with cybersecurity and biotechnology is changing regional security threats' landscapes and posing complex transnational challenges requiring multilateral policy and governance responses. Targeting biomedical datasets and the digital infrastructure of the bio-economy could increasingly being used by state and non-state actors alike for adversarial or commodification purposes, with the potential to sabotage or weaponise biomedical research, biotech facilities and biomanufacturing supply chains. Motivations behind adversarial attacks on the biotechnology sector range from falsifying clinical trials and research, holding the integrity of biomedical data hostage, undermining trust in precision medicine diagnoses and treatments and sabotaging critical infrastructure for health, food and biosecurity.

In this context of multipolar tensions, where boundaries are blurred between national and corporate responsibilities, legal and technical experts, civil society, states and private sector actors urgently need to work together to better understand, mitigate and regulate the harmful impact of adversarial data manipulation on biotechnology and the biosciences. Beyond strengthening legal and normative frameworks, public and private sector actors should collaborate on techniques to help secure biological and genomics data integrity. Policymakers need to start working with technologists to better understand the security risks emerging from the combination of AI and biotechnology and the implications for the bio-economy and its critical information systems. Preventing and mitigating such threats require a substantial departure from legacy approaches conceived to contain biological weapons by strictly controlling physical access to biotechnological equipment, materials and listed bio-agents. Importantly, no adequate guidance exists to prevent the adversarial use of biological data and algorithmic models to produce pathogens of concern or produce a biosecurity consequence by exploiting vulnerabilities in the cyberbiosecurity infrastructure:

- First, policymakers and technologists should use foresight to anticipate and more clearly determine the functional definitions of dual use that are emerging across AI, cyber and biosecurity domains.
- Second, they should collaborate to identify and assess the potential vulnerabilities that can be exploited in the convergence of AI, cyber and biotechnologies to cause extensive civilian harm, produce biosafety and biosecurity incidents and compromise the knowledge-production cycle of the bio-economy.
- Third, the new digital vulnerabilities emerging in biomedicine and biotechnology will increasingly require updated standards and practices that do not exist in our legacy policy, cyber- and biosecurity frameworks. For instance, encryption can be used to protect data-at-rest. Secure multiparty computation can help protect data-in-motion. Data authentication and verification mechanisms, such as cryptographic checksums and digital watermarking, may become critical to ensure data integrity. Importantly, modelling and simulating the ways in which sensitive civilian datasets are stored, accessed and retrieved for analysis are useful methods for testing such data systems, forecasting potential threats, identifying systemic vulnerabilities and building mitigation plans to address them. Diverse sectors facing information security risks already rely on these forms of sandboxing or operational foresight.

Ultimately, what is at stake is how we frame corporate responsibility and accountability in offensive cyberoperations that target biotechnology and the biosciences. Most efficient and scaled-up technological capabilities in AI and cybersecurity are the intellectual property of private companies. These companies are in a race to develop cybersecurity programmes that can detect the behavioural strategies used by AI-enabled malware to propagate across systems and avoid detection. As AI is enhancing speed, stealth and autonomy of cyberattacks, public sectors and civilian protection actors will become increasingly dependent on the cutting-edge expertise of AI and cybersecurity companies. This asymmetry gives private sector actors across the globe unprecedented power and a potential role to play in the protection of the bio-economy.

References

1. J. Allyn et al., Adversarial attack on deep learning-based dermatoscopic image recognition systems. Medicine **99**(50) (2020, December). PubMed Central). https://doi.org/10.1097/MD.0000000000023568
2. Bipartisan Commission on Biodefense. 'Cyberbio Convergence: Characterizing the Multiplicative Threat', 17 September 2019. https://biodefensecommission.org/events/cyberbio-convergence-characterizing-the-multiplicative-threat/
3. Bipartisan Commission on Biodefense. Biodefense in crisis: immediate action required to address national vulnerabilities. March 2021. https://biodefensecommission.org/reports/biodefense-in-crisis-immediate-action-needed-to-address-national-vulnerabilities/
4. Caudai C, Galizia A, Geraci F, Le Pera L, Morea V, Salerno E, Via A, Colombo T. AI applications in functional genomics. Comput. Struct. Biotechnol. J. 2021, October 11 195762–5790. doi: https://doi.org/10.1016/j.csbj.2021.10.009. PMID: 34765093; PMCID: PMC8566780
5. S.G. Finlayson et al., Adversarial attacks on medical machine learning. Science **363**(6433) (2019, March). PubMed). https://doi.org/10.1126/science.aaw4399
6. J. Giandhari, S. Pillay, E. Wilkinson, et al., Early transmission of SARS-CoV-2 in South Africa: An epidemiological and phylogenetic report. medRxiv **2020.05.29.20116376** (2020). https://doi.org/10.1101/2020.05.29.20116376
7. L. Gisel, T. Rodenhauser, K. Dormann, Twenty years on: International humanitarian law and the protection of civilians against the effects of cyber operations during armed conflicts. International Review of the Red Cross (2020) http://international-review.icrc.org/articles/twenty-years-international-humanitarian-law-and-protection-civilians-against-effects-cyber-913
8. H. Hirano, A. Minagi, K. Takemoto, Universal adversarial attacks on deep neural networks for medical image classification. BMC Med. Imaging **21**, 9 (2021). https://doi.org/10.1186/s12880-020-00530-y
9. ICRC, International Humanitarian Law and Cyber Operations during Armed Conflicts, Position Paper submitted to the Open-Ended Working Group on Developments in the Field of Information and Telecommunications in the Context of International Security and the Group of Governmental Experts on Advancing Responsible State Behaviour in Cyberspace in the Context of International Security, 2019. Available at: www.icrc.org/en/document/international-humanitarian-law-and-cyber-operations-during-armed-conflicts
10. S. Konur, L. Mierla, H. Fellermann, C. Ladroue, B. Brown, A. Wipat, J. Twycross, B.P. Dun, S. Kalvala, M. Gheorghe, N. Krasnogor, Toward full-stack in silico synthetic biology: Integrating model specification, simulation, verification, and biological compilation. ACS Synth. Biol **10**(8), 1931–1945 (2021, August 2). https://doi.org/10.1021/acssynbio.1c00143
11. Y. Mirsky et al., *CT-GAN: Malicious Tampering of 3D Medical Imagery Using Deep Learning* (Cornell University, 2019, January). https://arxiv.org/abs/1901.03597v3
12. National Academies of Sciences, Engineering, and Medicine, *Biodefense in the Age of Synthetic Biology* (The National Academies Press, Washington, DC, 2018). https://doi.org/10.17226/24890
13. A.K.M.I. Newaz et al., Adversarial attacks to machine learning-based smart healthcare systems. arXiv (2020, Octobre 7. arXiv.org). https://doi.org/10.48550/arXiv.2010.03671
14. Sandia National Laboratories, Personalized medicine software vulnerability uncovered by Sandia researchers. Sandia Labs News Releases (2019, July 1) https://share-ng.sandia.gov/news/resources/news_ releases/genomic_cybersecurity/
15. E. Pauwels, The New Geopolitics of Converging Risks. United Nations University, (2019). https://collections.unu.edu/eserv/UNU:7308/PauwelsAIGeopolitics.pdf
16. E. Pauwels, Hybrid CoE strategic analysis 26: Cyber-biosecurity: How to protect biotechnology from adversarial AI attacks. Hybrid CoE - Eur. Cent. Excell. Countering Hybrid Threats (2021) https://www.hybridcoe.fi/publications/cyber-biosecurity-how-to-protect-biotechnology-from-adversarial-ai-attacks/

17. E. Pauwels, G. Dunlap, *The Intelligent and Connected Bio-Labs of the Future: Promise and Peril in the Fourth Industrial Revolution* (Wilson Center, 2017) https://www.wilsoncenter.org/publication/the-intelligent-and-connected-bio-labs-the-future-promise-and-peril-the-fourth
18. S. Reardon, How machine learning could keep dangerous DNA out of terrorists' hands. Nature **566**(7742), 19–19 (2019, Janvier). www.nature.com, https://doi.org/10.1038/d41586-019-00277-9
19. L.C. Richardson, N.D. Connell, S.M. Lewis, E. Pauwels, R.S. Murch, Cyberbiosecurity: A call for cooperation in a new threat landscape. Front. Bioeng. Biotechnol. **7**, 99 (2019). https://doi.org/10.3389/fbioe.2019.00099
20. D.E. Sanger, S. LaFraniere, Cyberattacks discovered on vaccine distribution operations. The New York Times (2020, December 3) https://www.nytimes.com/2020/12/03/us/politics/vaccine-cyberattacks.htm
21. M. Segal, An operating system for the biology lab. Nature **573**(7775), S112–S113 (2019, September). www.nature.com, https://doi.org/10.1038/d41586-019-02875-z
22. E. Tegler, IARPA's bioweapon detection tools have difficulty finding what they're not looking for. Forbes (2021) https://www.forbes.com/sites/erictegler/2021/01/27/iarpas-bioweapon-detection-tools-have-difficulty-finding-what-theyre-not-looking-for/
23. The Second Oxford Statement on International Law Protections of the Healthcare Sector During Covid-19: Safeguarding Vaccine Research. Oxford Institute for Ethics, Law and Armed Conflict. https://www.elac.ox.ac.uk/the-oxford-process/the-statements-overview/the-second-oxford-statement/
24. T. Thi Nhu Thao, F. Labroussaa, N. Ebert, et al., Rapid reconstruction of SARS-CoV-2 using a synthetic genomics platform. Nature **582**, 561–565 (2020). https://doi.org/10.1038/s41586-020-2294-9
25. G. Zampieri, S. Vijayakumar, E. Yaneske, C. Angione, Machine and deep learning meet genome scale metabolic modeling. PLoS Comput. Biol. **15**(7), e1007084 (2019). https://doi.org/10.1371/journal.pcbi.1007084
26. M. Wardeh, M.S.C. Blagrove, K.J. Sharkey, et al., Divide-and-conquer: Machine-learning integrates mammalian and viral traits with network features to predict virus-mammal associations. Nat. Commun. **12**, 3954 (2021). https://doi.org/10.1038/s41467-021-24085-w
27. WHO. HIV Drug Resistance Report 2019. (2019). https://www.who.int/publications-detail-redirect/WHO-CDS-HIV-19.21

Eleonore Pauwels is an international expert in the security, societal and governance implications generated by the convergence of artificial intelligence with other dual-use technologies, including cybersecurity, genomics and neurotechnologies. Pauwels provides expertise to the World Bank, the United Nations and the Global Center on Cooperative Security in New York. She also works closely with governments and private sector actors on AI-cyberthreats prevention, the changing nature of conflict, foresight and global security. In 2018 and 2019, Pauwels served as Research Fellow on Emerging Cybertechnologies for the United Nations University's Centre for Policy Research. At the Woodrow Wilson International Center for Scholars, she spent 10 years within the Science and Technology Innovation Program, leading the Anticipatory Intelligence Lab. She is also part of the Scientific Committee of the International Association for Responsible Research and Innovation in Genome-Editing (ARRIGE). Pauwels regularly testifies before US and European authorities including the US Department of State, NAS, NIH, NCI, FDA, the National Intelligence Council, the European Commission and the UN. She writes for *Nature*, *The New York Times*, *The Guardian*, *Scientific American*, *Le Monde*, *Slate*, *UN News*, *The UN Chronicle* and *The World Economic Forum*.

Safeguarding the Guardians to Safeguard the Bio-economy and Mitigate Social Injustices

Roba Abbas, Katina Michael, M. G. Michael, Christine Perakslis, and Jeremy Pitt

Abstract Cyberbiosecurity (CBS) is essential to humanity due to dangers arising from digitalization of information, processes, and materials of various branches of biology. Humans are threatened by intensifying potential for malicious destruction, misuse, and exploitation of our biological data and information. As society seeks to identify and mitigate CBS risks, we must also work to ensure the absence of avoidable or rectifiable disparities among groups of people, whether those groups are defined socially, economically, demographically/psychographically, or geographically. In this chapter, we identified behaviorally, currently recognized risks at the demographically/psychographically, or interface of the life sciences and the digital world that lead to a failure "to protect opportunity or capability of people to function as free and equal citizens." We then identify and explore the more imperceptible uses of technology relative to the life sciences that negatively impact freedom and equity for humans. We considered these technologies against the backdrop of such social justice principles as inequality of outcomes, inequality of process, and inequality of autonomy.

Keywords Social justice · Economic justice · CBS · Cyberbiosecurity · Equity · Inequality of autonomy · Inequality of outcomes · Inequality of process · Allostatic load · A-Load · Physiological freedom · Vigilance fatigue · Biohacking · Bioengineering biochemistry · HCI · HCD

R. Abbas · M. G. Michael
University of Wollongong, Wollongong, NSW, Australia

K. Michael · C. Perakslis (✉)
Arizona State University, Tempe, AZ, USA

J. Pitt
Imperial College, London, UK

D. Greenbaum (ed.), *Cyberbiosecurity*, https://doi.org/10.1007/978-3-031-26034-6_12

1 Introduction

Cyberbiosecurity, which requires the protection of networks, devices, and data [1], is essential for humanity due to dangers arising from the digitalization of information, processes, and materials within, and back-and-forth from, the various branches of the life sciences. When considering the bio-economy, there is intensifying potential for malicious destruction, misuse, and exploitation [1, 2]. Technological advancements have yielded enhanced research and innovation through such emerging technologies as artificial intelligence/machine learning (e.g., artificial neural networks (ANN), cognitive computing, predictive analytics, cloud/big data, robotic process automation (RPA)). Yet, as organizations across the bio-economy are increasingly adopting digital transformation, agents of the companies utilizing, or interfacing with, the bio-economy can become a nexus of societal information as biodata are sent, stored, queried, analyzed, applied, revised, merged, shared, and/or archived. Data are moving not only within but also from the bio-economy to and through the products and services provided by companies serving the bio-economy. Unintended consequences are likely to arise from this confluence. Calamitous consequences emerge if actors interfacing with the bio-economy are permitted incomprehensible or imperceptible concentrations of influence and power [3, 4]. Thus, this chapter explores societal contexts and the consequential impacts on freedom and equity for humans.

1.1 Transforming to a Digital Society: Context Matters

Context matters; we recognize that the global society is rapidly moving toward the digital society. Therefore, we must not only address aspects of security of the bio-economy relative to external threats but also internal threats ("inside threats").

As a result of the digital transformation, there will be increasingly pervasive use of robotics, automation, artificial intelligence (AI) and information and communication technologies (ICT) in all areas of human enterprise: hence references to (and expectations of) Health 4.0, Industry 4.0, Smart Cities, platform economies, and so on. This vision and these expectations present a fundamental challenge: to engineer ever more complex socio-technical systems to support and enhance the full spectrum of human endeavor across social activities, business enterprises, and networked infrastructure. These human endeavors are most effective and beneficial if human values and issues of public interest are at the core [5].

However, as the old saying has it, "the invention of the ship was also the invention of the shipwreck" [6]. The opportunity presented by technological innovation also exposes the risk of unintended consequences, from the good (e.g., generative technology in the form of tools that enable the unsupervised creation of new tools that had not been imagined or intended by the designer of the original tool) [7], to the bad and downright ugly (e.g., degenerative technology). In the rush to deploy

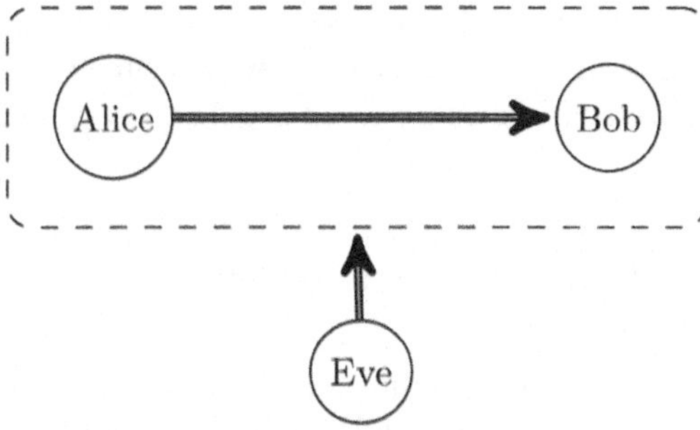

Fig. 1 The simplest abstraction of computer security

new technologies in the digital transformation, there is a primary – and deeply underestimated – risk: Society might not create robust safety and security measures.

The traditional and simplest abstraction of security is illustrated in Fig. 1. Alice, the source, wants to convey information to an information receiver, Bob. To prevent an eavesdropper, Eve, from accessing this information, a security perimeter is established: We may imagine everything inside this dotted line to be our socio-technical system for some application. Then in establishing this perimeter, there are some standard questions that need to be considered in terms of security, namely: What are we trying to secure? What are we trying to secure it for? Who are we trying to secure it against?

In the context of cyberbiosecurity, we must consider some *nonstandard* answers to these questions, and in particular:

- *What are we trying to secure?*

  We answer: Information, people, processes, and materials yielding biodata, or data indirectly leading to biodata, generated consciously or unconsciously, captured within materials or recorded by whatever device, and transmitted to an information receiver. In this chapter, we consider the bio-economy-proper, as well as what flows to and from companies interfacing with the bio-economy.
- *What are we trying to secure it for?*

  We answer: To ethically achieve or maintain some personal or collective value, with particular mind to serving the public interest.
- *Who are we trying to secure it against?*

  There is the possibility that the information receiver(s) will not act in good faith. We answer: We are also concerned with those who create and exploit inside threats, as well as those information receiver(s) who enjoy concentration of power through the convergence of digitalization.

One way to think about this is to expand our view of cybersecurity from the idea of preventing harm *from coming to you*, to include conditions of "cybersafety," which would involve protecting *you from coming to harm*. In other words, we are not only concerned with an outsider trying to break through the security perimeter to the bio-economy and/or what moves *out from* the bio-economy, but also with what goes on *inside* the perimeter among humans who are supposed and expected to safeguard the bio-economy, and the potential misuse and abuses of biodata, especially in the context of asymmetric power relationships [5, 8, 9]. We also recognize that vulnerable and underserved populations are often more at risk in these scenarios.

2 Conceptual Framework

Figure 2 is a diagrammatic representation of the foci of the chapter. The depiction when considered from left to right demonstrates external factors (societal factors) seeping into the internal, or inner, sanctum of the human (i.e., "inside threats"), thereby leading to diminished humanity. With diminished humanity, humans suffer the erosion of freedom and equity [10]; society is likely to face far more risks relative to inequalities of autonomy, inequalities of outcomes, and inequality of processes.

In the following sections, we will present a more detailed interpretation of Fig. 2. We will also clearly note the respective chapter sections for the reader's convenience.

2.1 *The External: Societal Factors*

As shown in Fig. 2, the key societal factors are as follows: The first societal factor is *predatory goods and services* (e.g., food, gambling, etc., as mentioned in Sect. 3). These predatory goods and services often disproportionately and negatively impact vulnerable and underserved members of society [11]. The second societal factor is

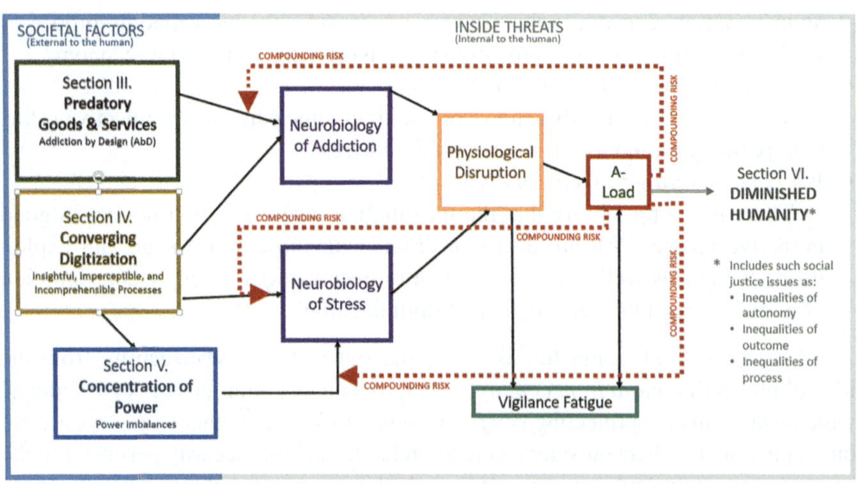

Fig. 2 Societal factors leading to inside threats leading to diminished humanity

converging digitalization, operatively defined in this chapter as diverse technologies integrated by an array of actors across the micro-, meso-, and macro-levels of society leading to potent capabilities to harvest and utilize vast amounts of amalgamated data (as discussed in Sect. 4). This convergence often disproportionately and negatively impacts those vulnerable members of society with lower levels of digital maturity [12, 13]. The third societal factor is *concentration(s) of power* as a few elite actors (aka the Powerful Elite [8]) are afforded imperceptible power due to their control of vast amounts of data about humans individually and collectively (as discussed in Sect. 5). This concentration often leads to such social justice issues as greater societal divides [8, 10–12].

2.2 The Internal or Inner Sanctum of the Human: Inside Threats

We also identify in Fig. 2 two significant inside threats which are as follows: first, the *neurobiology of addiction* (as discussed in Sects. 3 and 4), and second, the *neurobiology of stress* (as discussed in Sect. 3). Notably, these inside threats often disproportionately affect vulnerable and underserved populations [11, 14–17]. Then, as inside threats lead to physiologically enervated humans (e.g., *allostatic load* or the physiological wear and tear on humans as will be described in Sects. 4 and 5), humans are likely to face cyclical and compounding risk. Enervated humans are likely to become more vulnerable to the aforementioned societal factors and face further exacerbation of the impacts of inside threats (as denoted by dotted lines in Fig. 2).

2.3 Consequences: Diminished Humanity

The resulting *diminished humanity* in Fig. 2 is likely to be a significant risk to social justice. Degenerated humans are less likely to identify well, prevent, and/or robustly remediate disparities among groups of people, whether those groups are defined socially, behaviorally, economically, demographically/psychographically (as discussed in Sect. 6) [10, 16, 18, 19].

3 Predatory Goods and Services Leading to Inside Threats

In recent years, there has been an explosion of research and applications accomplished through the vast array of actors in the bio-economy for the benefit of non-bio market sectors. For example, the integrative applications of bioengineering work in tandem with various disciplines to design, deliver, deploy, and maintain a plethora of goods and services. At the intersection of materials science, bioengineers positively impact our world through such applications as biomimetics and biomaterials. Furthermore, at the T-junctions of chemical, mechanical, and electrical engineering, those in the bio-economy create and/or improve biofuels, prosthetics, and robotics, respectively. Lastly, in concert with neuroscience, those in the bio-industries interface with the fields of neurobiology. This includes the exploration of biological mechanisms by which the nervous system is measured and mediates behavior. Other examples include agricultural (e.g., food growth, food production), medical (e.g., population genetics, pharma, DNA synthesis), environmental (e.g., bio-energy, bioremediation), and industrial (e.g., bio-machine interfaces), to name a few [1, 2, 20, 21]. Thus, the bio-economy has increasing impact on, and involvement with, more societal domains.

In the next sections, we offer a representative sampling of these domains leading to inside threats through predatory products and services that trigger the *neurobiology of addiction.* We ask the following questions: What happens when the world around us is teeming with products and services that are negligently or intentionally designed, developed, and/or deployed to trigger dependency, or even unintentionally, addiction, in humans? What happens when vulnerable or underserved members of society are likely to be more susceptible to this predation? We can work, eat, drink, watch, listen, play, shop, interact, and then risk intense dependency and even enslavement to the very goods and services we purchase and use to improve or enhance our lives and the organizations for which we work. We are not likely to remain vigilant to safeguard information, people, processes, and materials if we are physiologically enervated by dependency and/or addiction in our day-to-day existence. We begin with an exploration of the *neurobiology of addiction* triggered by optional activities (i.e., gambling, gaming, apps [22], e-commerce, marketing, and mobile interfaces) and conclude this section with triggers found through life's essential activities (e.g., nourishment through eating).

3.1 Laundered Lives and Indentured Playtime

The British colloquial verb "to rinse" is to completely deplete someone or something until there is nothing left [23]; in particular, this can apply to "rinsing" someone of their money and assets. Such "rinsing" has a temporal dimension: one way to take a lot of money is to try to rinse someone for a large amount in one "go," for example, through a con trick. This does, however, raise the problem of "burning" (as in "once burned, twice shy"). Therefore, another way of rinsing someone is to extract a small amount of money, but over many "goes," without ever revealing any subterfuge. This can eventually yield a greater return over the long term, but to make a large amount over the long term, the small rinse has to be done at scale. In pursuit of such scale, the casino and the lottery were invented.

Dow-Schüll [24] charts how the gambling industry has leveraged many features of psychology and neurobiology to get gamblers (aka punters) "into the zone," that is, a engaged in condition, where the outcome, in the form of winning or losing money, is secondary to the point of being immaterial to the process itself. For example, a slot machine's repetitious behavior produces in a punter's brain the necessary chemical and neurological stimuli to accomplish a "coins in/quids in result" for the casino owners.

The physical environment of a casino is also carefully crafted and controlled to promote this zone, from the density and proximity of the machines (e.g., gamblers want privacy in their own bubble but to still be dimly aware of others in theirs), through to the regulated oxygenation levels, free drinks, acoustics, and lighting (e.g., a sense of the passage of time is deliberately made opaque). To further maintain the zone, casinos increasingly use biometrics (e.g., facial recognition, facial emotion recognition, fingerprint recognition, etc.), so that gamblers can move freely, fluidly and uninterrupted through the space, enjoying instant authorization to access areas on-demand. Casino operators also increasingly use bio-surveillance to capture, understand, and anticipate the desires of their gamblers. With this, gamblers can remain focused on nonstop play at tables/machines due to effortless reloading of monies in real time and can be inveigled by highly customized and personalized goods and services. Therefore, biometrics allow for not only rapid identification, enhanced security, and tracking of behavioral patterns (e.g., entry/exit/usage) at casinos, but also help create augmented and unbroken experiences for gamblers to stay in the zone.

The casino environment is complemented by other cognitive exploits, in particular the "near-miss effect." This effect causes a near-miss to be misinterpreted by players as a sign that they should keep playing [25]. In the UK, this could have been exploited by fixed odds betting terminals, in which the percentage of actual wins was regulated by law, but the percentage of "nearly wins" was not; consequently, the diminished margin of return according to the regulations could have been compensated for by an increased volume of input derived from exploiting the near-miss effect.

3.2 *From Gambling to Gaming*

It is not a significant shift, orthographically speaking, from "gambling industry" to "gaming industry," nor is it such a significant difference in commercial practice to use similar tactics and techniques in order to get gamers (aka players) so consumed by game-playing that it amounts to an addiction. Indeed, internet gaming disorder (IGD) has been recognized as a potential mental disorder and is therefore included in the fifth edition of *DSM-5* (the *Diagnostic and Statistical Manual for Mental Disorders*), the standard reference handbook for psychiatric diagnosis [26, 27]. A meta-analysis and literature review have demonstrated that there are significant neurobiological variances between individuals presenting with IGD and control subjects (healthy "normal" individuals) [28]. Additional risks surface when considering the advent and development of affective gaming; players can be easily exploited as their emotional states are evaluated in real time and as they share their physiological signals during gaming (e.g., heart rate, muscle tension, brain wave activity, temperature, respiration, facial recordings) [29].

Addiction by design [24] refers to tech developers designing with the objective of instigating chronic behavior and dependency on a particular machine, device, and/or system, through encouraging entry into a flow state or trance, where that state can be maintained over time. In Schüll's [24] book, *Addiction by Design,* flow is examined from an anthropological perspective with reference to machine gambling and gamblers' entry into the "machine zone" (flow state) through "machine play," highlighting the industry's desire to keep individuals in this state for purposes of profitability, while claiming that they are simply responding to the desires of the individual. The tech industry also exploits humans by creating apps to be addictive [30–33]. Internet addiction disorder (IAD) is also believed to be an emerging mental health issue with evidence of measurable changes in the brain structure and function of humans (e.g., abnormal white matter structure) as well as behavioral impairments [34]. Thus, these predatory goods and services might be better demarcated as creating a state that can be termed *dark flow*.

3.3 Concept of Flow: Positive and Dark

The concept of flow was originally defined by positive psychology Professor Mihaly Csikszentmihalyi [35] as an optimal experience and absorbed mental state whereby an individual is engrossed in a particular activity. The state has been described as one of "dynamic equilibrium" shaped by the interaction between an individual and their environment [36]. Attaining a state of flow requires numerous components, including concentration, the presence of a challenge, a sense of control, and some degree of skill, among other elements [36]. When individuals enter a flow state or are in the flow channel, they are completely engrossed in a given environment and experience [35] and, as such, seek cues in the form of feedback that enable actions to be modified [37]. Flow is typically equated with a positive experience, leading to fulfilment, productivity, and happiness.

Flow, in a technology-mediated environment, refers to a subjective user experience resulting from the interaction of the individual and the technology, focusing on the four dimensions of control, attention focus, curiosity, and intrinsic interest [38, 39]. In the human-computer interaction (HCI) domain more specifically, flow theory has been adopted, researched, and applied in the context of user-centered design to enhance interactive user experiences, particularly in the gaming industry. The objective here is offsetting the concepts of challenge and ability or skill to encourage entry into the "flow zone," recognizing that adaptability is required due to end users having divergent flow zones [40]. A primary motivation of related research is to develop frameworks or instruments that enhance both enjoyment and game design [41].

Furthermore, flow theory has been featured in studies exploring customer satisfaction with respect to e-commerce applications, in order to determine precursors to an optimal shopping experience, based on a stimulus-organism-response (S-O-R) paradigm [42]. Lee et al. [42] apply a five-factor measure of flow for e-commerce, and state the importance of product- and service-related cues and the inverse relationship between quality and flow as major findings. Flow in an online environment has also been studied from a range of disciplinary perspectives, such as marketing (see [43] for an overview).

More recently, the integration of flow in the design of mobile interfaces was identified as an area requiring further research and attention, from the perspective of deliberately inducing the flow state by design, and recognizing when compulsive usage is evident in order to prevent "dark flow" [44]. While such examples arguably point to the positive applications of flow theory, the concept of dark flow must also be explored as it relates to addiction by design (AbD), cyberbiosecurity (CBS), and the potential for the exploitation and manipulation of the neurobiological systems of individuals.

Interestingly, in the language of gambling machine development designed for flow, slot machines are also referred to as fruit machines (rather than cash-extracting machines). Unfortunately, it is not just *associations* with food that can be leveraged in pursuit of dependency and addiction, but also food itself. In the next section, we will explore how researchers reveal such methods as are executed by those companies who design and develop products for nourishment for humans (e.g., food giants within agribusiness or BigAgro [30]).

3.4 Nourishment: Enslaved Through Ingestion

Researchers have shown how food manufacturers first manipulated processed foods in order to make them alluring as well as cheap [30]. Subsequently, the manufacturers manipulated those foods to make them addictive [45]. Manipulation techniques tended to focus on a neuroscientific understanding of how the brain's reward system works, that is, on the roles of both chemicals in the brain (mainly, dopamine) and processing regions of the brain (primarily the prefrontal cortex), and how these are activated or deactivated by specific additives to food products [30]. Even seemingly innocuous wellness supplements could be exploited for neurohacking if genome editing tools were used to design formulas for over-the-counter (OTC) gut therapies (e.g., probiotics) to covertly impact the brain negatively, such as by altering mood or productivity once ingested [46].

We could also explore medical marketing and how DTC (direct-to-consumer) advertising for marketing prescription drugs, health services, disease awareness, and laboratory tests becomes a big business designed not to *help* consumers but rather to *sell* products and services [47]. We could also easily extend our evaluations to such chemical addictions as found through pharmaceuticals and/or to such behavioral addictions as found through entertainment and social media site usage [34, 45–51]. Although some might contend these goods and services create compulsion rather than addiction, researchers identify dopamine-inducing effects on the brains of users of social media sites. These platforms can cause the brain's reward area to trigger neural circuitry with the same kind of chemical reactions as those who are hooked on gambling or using recreational drugs such as cocaine [30–33].

4 In a World of Converging Digitalization, Addiction, and Stress

In the previous section, we asked: What happens when the world around us is teeming with products and services that are designed, developed, and/or deployed to create dependency or addiction in humans? Now, we ask: What happens when there is also a convergence across various micro-, meso-, and macro-segments of the bio-economy resulting in digitized processes becoming progressively more imperceptible, incomprehensible, and/or also insightful? Adding to this, what happens when work environs for humans within the life sciences are increasingly pressurized due to such issues as high-pressure tasks and/or an accelerated sense of time for speed-to-market? In intense work environments, humans can suffer such consequences as vigilance fatigue at a minimum but also physiological wear and tear leading to allostatic load (A-Load) which often results in brain fog, preoccupation, distraction, ambiguity, confusion, and unconstructive vulnerability [52–56]. Humans operating in A-Load are therefore more likely to be more susceptible to inside threats, as well as more likely to be debilitated in their efforts to safeguard the bio-economy.

4.1 In a World of Addiction by Design: The Neurobiology of Addiction

As we identified and explored risks related to neurobiology in the digital age, we also recognized the ramifications of addiction by design on humans. If such aforementioned products and services as gaming, gambling, apps, e-commerce, marketing, mobile interfaces, food, social media sites, and pharmaceuticals have been developed without responsible regard to, or intentionally to instigate, *the dependence syndrome*, the human is likely to move through three stages of *dependence development*: from the binge-intoxication stage to the withdrawal-negative affect stage, and into the preoccupation-anticipation ("craving") stage. The consequences for humans in these states include outcomes such as a strong sense of desire or compulsion relative to the behavior around the product or service, decreased concentration and focus, progressive neglect of former interests, difficulties with self-regulation (e.g., around a technology), and often recurrent upsurges of usage thereby creating further intensification of negative consequences. Additionally, a physiological withdrawal state often causes the human to compulsively seek relief by reengaging in the addictive behavior to avoid the discomfort of withdrawal [57–59].

When products and services induce addiction, or even maladaptive dependence, humans can suffer psychosocial stressors such as social segregation or isolation and psychological stressors such as impaired cognitive functionality leading to abandonment of necessary responsibilities and/or degradation of previously fulfilling activities. Furthermore, humans may suffer such physical stressors as circadian disruption. Increased stress and fatigue, can result and those factors subsequently can exacerbate negative impacts on a person's physical, psychological, emotional, social, and spiritual well-being [16, 17, 56, 58–60].

Vulnerable or underserved members of society are often most at risk in these scenarios. Social justice issues often correlate to lower levels of socioeconomic status (SES) or to lower measures of such components as income, occupation, and education [61–63]. Additionally, SES often correlates significantly with addiction patterns. For example, while young adults in wealthier families may be more likely to experiment with illegal substances at younger ages, segments of society with lower SES are far more vulnerable to goods and services leading to addiction. For example, inexpensive, yet nutritionally deficient, food options are often the most economically feasible. Additionally, within communities with lower SES, substances leading to addiction are often more readily accessible, while support and care options for combating addiction are often far less accessible [16, 17, 64, 65].

Research [14, 65–67] also reveals higher levels of chronic stress within communities with lower SES. Chronic stress is a risk factor for not only almost all mental illnesses but also for high susceptibility to addiction. For example, addiction can be a familial system dysfunction where one individual with maladaptive dependence archetypically creates physical and psychological disruption, economic constraints, and relational conflicts for the other members of the household, thereby creating higher levels of environmental stress [21, 62, 68]. Thus, the context within which populations with lower SES live and work may lead to more exposure and defenselessness to the aforementioned dependence, or addiction, of predatory products and services [16, 18, 65].

4.2 In a World of Converging Digitalization: The Neurobiology of Stress

Converging digitalization can be defined as diverse technologies integrated by an array of actors across the micro-, meso-, and macro-levels of society leading to the potent capabilities to harvest and utilize vast amounts of amalgamated data.

Undoubtedly, this mass confluence of digitalization can be of great benefit to humanity in certain contexts. Companies across the globe can pool resources to increase speed to market, reduce costs, and increase revenues. Additionally, global convergence can alleviate labor exploitation in that improved job prospects allow people the freedom to move to where opportunities exist. In these scenarios, impoverished nations with lower standards of living can benefit from more competitive wages. Higher wages often yield improved living standards [69]; optimistically, research [70] correlates higher incomes with better health and lower maternal and infant mortality. With digitalization, product access can improve; citizens in the global periphery can suffer less resource scarcity [20, 71]. Yet, as technological progress accelerates, and digitized interdependencies integrate in the life sciences, our ability to perceive risks becomes more complicated and convoluted. As technologies converge within (and back and forth from) the global bio-economy, there are significant risks to consider such as insightfulness, imperceptibility, and incomprehensibility of these processes.

Converged digitalization yields *insightfulness* [72]. Security risks emerge when aggregated data gleaned across a variety of technological sources in the bio-economy can assess humans in multiple contexts, capacities, and times, allowing the system to have a precise and profound understanding of a human in their past, present, and proposed future states. Insightfulness becomes even more formidable when genetic and other biological data are analyzed with social and environmental factors; researchers can understand gene-environment interactions and create bio-social models. These risks exacerbate as non-bio actors collect, store, analyze biodata. In one example, biosocial surveys allowed for social scientists to gather, and disseminate bio-specimens (e.g., blood, urine, saliva), bio-markers (i.e., measurable factors derived from a specimen and associated with a current or probable medical condition), and biodata (e.g., digital data derived from bio-specimens) in nonclinical settings [73]. Positively, such surveys could allow society to better understand the interplay between health and social inequities to alleviate injustices. These noninvasive quantification measurements of the human body can be correlated with social and environmental factors to yield rich models and solutions to better address the disproportionately negative effects on marginalized populations (e.g., anthropometry to assess nutritional adequacy for children in areas with lower SES) [16, 68]. However, these data are often not sufficiently managed to rightfully protect the data gleaned, analyzed, and warehoused about their human subjects. Research [73] reveals that social scientists acting as information receivers inadvertently may not act in good faith due to ignorance about essential protocols and protections needed around biodata that is moving within, or outside of, the bio-economy.

Genomic surveillance and DNA profiling have also been used potently in conjunction with social engineering (e.g., digital DNA, or D-DNA, such as mining posting behavior on social media) as well as facial recognition to yield invasive and pervasive insightfulness. Predictive analytics through AI and ML further galvanize the ability to accurately predict human behavior. Such DNA profiling as done through law enforcement, immigration, or the Combined DNA Index System (CODIS) can lead to discriminatory behaviors, violations of privacy and due process, and even such reprehensible human rights violations as ethnic cleansing. Algorithms can be used advantageously to detect the often humanly undetectable discrimination within; yet algorithms can also inadvertently yield direct or indirect discrimination and/or statistical discrimination [11, 15, 74].

Converged digitalization also yields *imperceptibility* [72]. Technology often happens behind lines of visibility. As those within the life sciences design, develop, deploy, and warehouse data, they may not realize or perceive the far-reaching effects of how such information, processes, people, and materials can be used in conjunction with other data and/or technologies within or outside of the bio-economy. Often, good actors do not perceive what is happening behind these lines of visibility, such as what is being collected by whom, for how long, how it is, or could be, synthesized with other data, and who ends up owning the data, now or in the future. Corporations have sold biotechnologies to bad actors unknowingly or negligently. Research published in the bio-sector has yielded unintended unethical consequences as bad actors have used published biodata for nefarious purposes [75, 76].

Finally, converged digitalization results in *incomprehensibility* [72], as technology can outpace our ability to comprehend existing or emerging processes that require new or revised policies and procedures to safeguard people, places, and processes. Aging populations, as well as communities with lower SES (socio-economic status), often struggle disproportionately with lower levels of digital literacy and digital maturity [12]. Interestingly, even leading international corporations report ongoing struggles to achieve and sustain digital maturity [12, 71]. Highly educated members of society can fall prey to nefarious practices of biohackers who obtain biodata under false pretenses. Technological advancement also continues to outpace legal and ethical frameworks that are best created early so as to address and homogenize appropriate practices [77]. With integrated digitalization, we often do not even fully comprehend a simple opt-in to (often murky and mutable) terms and conditions of technologies.

Additionally, those in the bio-economy have traditionally subscribed to a culture of openness and information sharing in order to advance scientific discovery and the practical application of these discoveries. Reliance on self-governance has also been an acceptable practice in the past [75]. With converging digitalization, these previously acceptable norms of collegiality create significant risks to the emerging bio-economy that may not be comprehendible to experts in the bio-economy domains. Encouragingly, policy makers are beginning to address such issues. In the USA in 2020, the government focused more effort on better defining, and more robustly clarifying, classifications and demarcations within the bio-economy, as well as creating well-integrated strategies to better address national security challenges of the fast-emerging and converging digitalization of the bio-economy [21].

4.3 Context of Lived Experiences: The Neurobiology of Stress

We will now explore the *neurobiology of stress* taking into account two key contexts. The first context is the workplace with a particular focus on those employed in the life sciences because they are the primary guardians of the bio-economy. The second context is the community (e.g., community-level stress or community allostatic load or A-Load). Significantly, Community A-Load, or the aggregated measure of physiological wear and tear on humans living in the same area, does appear to disproportionately affect the vulnerable or underserved members of society [55, 63, 65]. To explore categories of stressors, we will choose the four broad categories of stressors, which are as follows: physical stressors, psychosocial stressors, psychological stressors, and psycho-spiritual stressors [56, 78, 79].

4.4 Stress in the 9–5

Societal stress is a growing global crisis regardless of income. The World Stress Index (WSI), represents 95% of the global adult population. The index purports an average of 35% stress globally, which continues to increase each year [80]. Global studies [81, 82] reveal that the push for excellence and speed to market has led to intense pressures faced by researchers working in such fields as biology. Such twitter hashtags as #MedSciLife and #takebreaksmakebreakthroughs, created by molecular biologists and researchers, seek to alleviate intensifying burnout among those working in the life sciences. Those surveyed articulated a high personal toll on well-being due to an accelerated sense of time, bullying, long hours, and a focus on such metrics as impact factors of journals. Surveys from the lab and field reveal chronic stress in the working lives of scientists contending with damaging cultures of competition, leading to hostility, mean or aggressive working conditions, and persistent stress and anxiety [82, 83].

The Internet of Behavior (IoB), which represents a meso-level form of converging digitalization in workplace environments, is likely to yield such benefits as safer work spaces, as sensors and tags monitor employee key strokes and voice behavior or whether an employee is certified to operate a certain machine or is properly washing up after using the lavatory. Yet, as human behaviors are continuously surveilled and analyzed in the workplace, researchers [84, 85] report that employees often suffer a sharp rise in stress levels.

Stressors can also combine to create a prolonged effect on humans in the various spheres of their life [56]. Workplace stress often yields the four aforementioned stressors, as do patterns of addiction in humans. Additionally, the convergence of processes in the life sciences, which yields insightfulness, incomprehensibility, and imperceptibility, leads to stressors due to information overload and also as a result of the threats that operate behind the line of visibility [72, 86].

4.5 Prolonged Stress: Allostatic Load

The mind and body interact as a complicated, interconnected whole. The interplay can be both positive and negative. Allostatic load (A-Load), as previously mentioned, is index of the biological "wear and tear" on the physiology of the human. Prolonged stress can lead to chronic over-activation of the sympathetic nervous system (i.e., fight, flight, freeze response). When demands outweigh our ability to cope, we experience stress. When we suffer acute stress or allow ourselves to endure chronic exposure to stressors leading to the persistent over-activation of the sympathetic nervous system, our mind-body system works overtime. Therefore, without robust buffers to offset the stressors, we move into maladaptive states of A-Load in which our sympathetic nervous system is cast into an extended state of heightened alert. Without strategic, consistent, and intentional activation of the parasympathetic nervous system (e.g., eliciting the relaxation response for resilience building) to offset the risk of this maladaptive state [52, 54, 56, 87], humans suffer harmful physiological consequences including nearly all the identified consequences of addiction. Another adverse outcome is *vigilance fatigue* or the failure to accurately perceive, identify, or analyze bona fide threats. This can lead to serious negative consequences or even to a life-threatening state of affairs [88, 89]. Innovation is then usually impeded, and our resilience and ability to adapt are significantly constrained [17, 53]. Researchers believe that this phenomenon can be brought about by four factors as follows: (1) prolonged exposure to ambiguous, unspecified, and ubiquitous threat information, (2) information overload, (3) overwhelming pressure to maintain exceptional, error-free performance, and (4) faulty strategies for structuring informed decision-making under conditions of uncertainty and stress [87, 88].

Research [16, 55, 65, 90] relative to individual A-Load has expanded to collective A-Load, yielding evidence of significant effects of the community on individual A-Load and vice versa. Thus, we can experience detrimental vicious cycles of collective stress, breeding more individual stress, thereby breeding more collective stress. There are believed to be several categories of factors leading to A-Load in humans or communities. Such stressors can be chronic or acute, as well as internal or external to the human. As previously mentioned, we chose to consider four broad categories of stressors: physical stressors, psychosocial stressors, psychological stressors, and psycho-spiritual stressors.

Physical stressors may include genetic or biological factors. They may also include resource deprivation for adequate survival, living and working conditions, trauma or illness, such dietary stress as nutritional deficiency and unhealthy eating habits, such issues of work-life balance as overexertion, and such environmental concerns as climate change, noise, or pollution. Chronic stress is often more prevalent in populations with lower socioeconomic status (SES) due to such factors as crowding, crime, noise pollution, discrimination, and other hazards [16, 65]. Community A-Load is believed to be exacerbated by such physical stressors as deteriorated natural or built environs, residential/population churn, lack of healthy or safe public spaces, and lack of access to aesthetically-pleasing resources or to nature [16, 18, 56, 65].

Psychosocial stressors may include lack of family and/or community social support, employment or housing issues, and other interpersonal difficulties. Community A-Load is believed to be intensified within societal enclaves of marginalization, low/unstable employment, intergenerational poverty, and low social efficacy [16–18, 56, 65].

Psychological stressors may include those of a perceptual nature such as attitudes and beliefs, as well as information overload, frustration, grief, fear, an accelerated sense of time, and other types of emotional and cognitive stress. Researchers studying Community A-Load identified such factors as slow and/or inequitable response times of municipalities, power imbalances, and isolation as chronic psychological stressors leading to hopelessness and despair. These factors are often correlated significantly with communities of lower SES [16, 17, 55, 56, 65].

Psycho-spiritual stressors are likely to include a misalignment or suppression of core spiritual beliefs or crises of meaning, purpose, or values. Spiritual capacity, which has decreased in many segments of society, often serves as essential sustenance to offset stress for humans when facing adversity, as well as a robust source of resilience and motivation [16, 54–56, 65, 91].

When these four broad categories of stressors are recurrently triggered or exacerbated by adverse internal and external environments, humans are further negatively impacted cognitively, emotionally, behaviorally, physiologically, and socially. Cognitively, humans can struggle with anxious thoughts, poor concentration, and difficulty with memory. Emotionally, humans often experience feelings of depression, disquiet, tension, irritability, restlessness, and an inability to relax. Behaviorally, humans may avoid tasks, develop maladaptive eating or substance consumption habits, and experience sleep deprivation or degradation [56, 78]. Physiologically, humans can suffer with tense muscles, myalgia, headaches, difficulty swallowing, stomach aches, nausea, digestive issues, sexual dysfunction, weight loss or gain, and heart malfunctions. Socially, humans can experience decreased quality of relationships, increased conflicts in interpersonal interactions, a need for withdrawal from social activities, and a loss of a sense of belonging [55, 56, 65, 92].

These contexts of working and living can accumulate and thrust humans into a state of A-Load [56] individually and propagate Community A-Load [56, 65]. Therefore, if we are an enervated society, we lack vigilance to safeguard the bioeconomy; we risk living in a zombie-like sleepwalking trance. We are also likely to lack the endurance or resilience to challenge asymmetries of power enjoyed by the Powerful Elite as they accumulate unchecked control and influence [10]. Most lamentable, we are far less likely to eradicate or even mitigate inequalities in society; in fact, we allow vulnerable members of society to be far more susceptible to harm.

5 Concentration of Power Leading to Inside Threats

We previously asked: What happens when a convergence of digitalization is likely to proliferate imperceptible or incomprehensible processes? Now, we ask: What happens when this convergence also leads to concentration of power that often exerts influence and control behind lines of visibility? What happens if wide swaths of society are suffering physiological disruption, vigilance fatigue, or worse physiological dysregulation leading to A-Load and thereby are unable to address these asymmetries of power? Humans are likely to be diminished in their ability to robustly recognize, resist, or rectify such risks. These risks are also likely to lead to negligence in appropriately identifying and/or addressing societal inequalities.

Corporate entities with centralized portals have the potential to bring together disparate islands of information about human activities and behaviors relative to not only the economic, political, spatiotemporal/geographical, and social but also biological data. Biorepositories can include large multinational collections of countless specimens. Powerful AI/ML applications can now analyze large troves of data, making judgments about the current or future states of individuals behaviorally and increasingly also physically and physiologically. These diverse processes present unique risks when facilitated by converging digitalization.

Convergence can occur at the individual or collective level, the device level, the application level, and even the data level, thereby providing an entity (or alliances of entities) with granular analytic capabilities that can be turned into potent intelligence [72]. Entities managing large data warehouses using powerful relational database management systems on the cloud are able to take advantage of big data capabilities. "Dark data" or data stored and left unused in the current state are in reality a form of deferred data that can be "lit up" to be used at a later time. This has led to organizations potentially knowing humans better than humans know themselves. In these contexts, members of society may also lose control over their own data, thereby making their data abstruse and ineradicable [5, 72]. Humans are thereby stripped of data ownership and divested of rights to delete or rights to be forgotten.

Converging digitalization can aid in driving product development through predictive analytics and data-driven innovation processes but can also have counter-effects as organizations can mine sensitive biodata and thereby combine it with other data sets for profound understanding to manipulate the masses. As previously mentioned, social engineering alone now allows for generating digital DNA (D-DNA) of individuals, potentially leading to bio-discrimination [15, 74]. Societal shifts and trends have already proved to be surreptitiously influenced by both Big Tech corporations that span nationally and transnationally as well as by publicly organized and administered information systems by government agencies. Society has observed companies such as Facebook (nka Meta) that have been afforded unhampered and remarkably broad societal reach even without certain convergences. Often these mega-scale platforms become potent surveillance mechanisms leading to *surveillance capitalism* [93] or the commodification of personal data being used for profit making [8, 94].

When humans cannot recognize, resist, or rectify the concentration of power to which they are exposed, we risk the rise of techno-feudalism [8, 94], which can be defined as "a socio-political economic system in which a Big Tech company holds sway over a particular domain of enterprise ... as granted by an elected government, in exchange for political and financial support, and post-parliamentary career support" [94, 95]. In essence, the Powerful Elite [8] relegate the "rest of us" to "info-peasantry ... obliged to exist within the asymmetric terms and conditions of the service: we provide the data in return for the service, but the aggregator (the platform owner) is the primary beneficiary" [95]. Cashless societies and the provision of digital currency by platform owners are likely to further subjugate humans. The subjects, or the "info-peasants," either comply "on the grid" under particular terms and conditions, or suffer the plight of being socially excluded or perhaps even being relegated to nonpersonhood [95]. Although we continue to contend that the vulnerable and underserved members of society are disproportionately impacted by such contexts, this concentration of power undoubtedly leads to a failure to protect opportunity or capability of *most of society* to equitably function as free and equal citizens [96].

In terms of the devices, the apps, the data, the converging digitalization that we describe above – much of this is owned privately and/or capable of being monitored by the state. Boundaries between private and public actors are often blurred. Nearly 64% of pharmaceutical lobbyists were former government employees, and often private sector actors end up in government positions and vice versa [97, 98]. Boundaries between academic medicine (e.g., teaching hospitals) and the pharmaceutical industry have also been eroding; pharma companies, with a fiduciary responsibility to shareholders, do not have education budgets but rather have marketing budgets [97]. Thus, there is already a precarious concentration of power, as well as clear motivations and operations based on profit motives rather than on benefits to humans.

Techno-feudalism [8, 94], generated by the Powerful Elite [8], exploits the end user in the name of innovation and economics. With rich and broad amalgamated data, society is exposed and vulnerable. There is likely to be transparency of each individual's current and predicted motivation, intent, sentiment, behavior, and actions [99]. With converging digitalization, the result is then an ability to find the consumer, like a needle in a haystack, with precision based on historical movements and current context. This creates extraordinary vulnerability for the exploitation of humans.

The outcome of biodata being involuntarily or voluntarily added into the amalgamation is that we become susceptible to Überveillance (i.e., watching of the inner sanctum of the human) [48, 99]. As these converged processes increasingly have the ability to mine thoughts and activate behaviors, will the Powerful Elite [8] exploit the space between our skull where we were once free to roam, think, feel, and reflect without scrutiny [48, 100]? The ultimate end would easily lead to furthering the concentration of power among the few with the many relegated to a form of modern indentured servitude [101, 102]. Human freedom would be the cataclysmic collateral damage.

6 Consequences of Diminished Humanity: Inequalities of Autonomy, Outcome, and Process

In previous sections, we explored societal factors (e.g., predatory goods and services, converging digitalization, and concentration of power) leading to humans suffering heightened neurobiological states of stress, as well as neurobiological states of addiction. We also explored possible ramifications of these neurobiological states because humans can become enervated due to allostatic load. We thus contended that humans become less likely to recognize, and far more susceptible to harm from, these factors and, in particular, concentration of power; the common human can be relegated to modern serfdom.

In this section, we revisit our original three questions: *What are we trying to secure? What are we trying to secure it for? Who are we trying to secure it against?* We will now explore possible consequences if the information receiver(s) who enjoy(s) concentration of power do not act in good faith to avoid inequalities of autonomy, inequalities of outcome, and inequalities of process. Although all humans (with the exception of the Powerful Elite) are at risk, vulnerable or underserved members of society often are disproportionately impacted. Therefore, we will now focus on risks to the more susceptible members of our global community: those members of society in the global periphery.

6.1 Inequalities Defined

Inequalities of autonomy can be defined as unfairness relative to self-reflection, active or delegated decision-making, and limitations relative to high-quality options. Resultant barriers can include conditioned expectations, coercion, and such structural constraints as lack of advice and support. Inequalities of outcomes can be defined as disparities in society relative to income, wealth, education, health, and nutrition. Inequalities of process can be defined as spatial and symbolic boundary maintenance, emotion management, othering or objectification, and subordinate adaptation such as obstructed socioeconomic mobility or trading autonomy for protection [103]. These inequalities are exacerbated, or possibly created, when technological advancements lead to a concentration of power in the hands of the few who have far too much impact on the well-being of those in the global periphery [10, 93]. One example in the field of industrial biotechnology (for the benefit of the pharmaceutical industry) that we will discuss next is the creation and development of synthetic artemisinin or semisynthetic artemisinin (SSA) to produce drugs and treatments to prevent and/or treat malaria, as well to effectively treat various types of cancer.

Prior to exploring mostly concentration of power in this example, however, we would be remiss if we did not bring attention to the fact that these same types of firms have been known to either negligently or intentionally create predatory goods [97, 98]. Some allege these companies not only manufacture drugs, but more so, they have often intentionally manufactured pharmacologically induced epidemic-level crises of addiction [104]. These dire consequences of addiction and death have pervaded all segments of society, yet the underserved members of our society have suffered disproportionately. For example, the Cherokee Nation in the USA, whose tribes have had the highest per capita rate of opioid overdoses, reached a landmark settlement with key players in the pharmaceutical industry in the amount of $150 million. When considering other tribes in the USA, Native Americans collectively have been awarded settlements in excess of $40 billion due to such predatory goods as are peddled by these companies [105].

6.2 Destabilization in the Global Periphery: Local Economies Suffer

We now return to the example of industrial biotechnology leading to synthetic or semisynthetic artemisinin for the benefit of Big Pharma. Unintended consequences of these technological advancements have destabilized the supply chain of agricultural production of wormwood for those in the global periphery. With SSA, pharmaceutical companies no longer require naturally produced artemisinin from the global periphery but rather are self-sufficient in the bio-production of these synthetic ingredients. These companies can enjoy faster and cheaper production but also enhanced centralized power and control over knowledge, pricing, processes, and distribution. In this scenario, the adverse effects on the global periphery are magnified as hundreds of thousands of local and regional farmers, extractors, and micro-producers of artemisinin lose their sources of income; these local economies lose fair rewards for productive activities [20, 106].

Notably, the ramifications are far worse for local and regional families if the business owners took on debt to launch or grow the now-defunct business or if families uprooted their lives to move closer to these now-curtailed employment options. Local communities that relied on these industries and once invested in infrastructure are now left with shuttered structures leading to brown fields, economic decline, and population loss or churn and urban decay, thereby exacerbating Community A-Load. With micro-businesses shuttered, local and regional societies suffer intense stressors, as well as inequalities of outcomes due to reduced (or annihilated) incomes, which in turn often create inequalities of autonomy as families suffer limitations relative to higher-quality options for housing, education, and health [17, 18, 55, 65].

6.3 Resultant Barriers: Cost Increases and Access Decreases

Because these advancements in biotechnology were not measured against the social impact on the global periphery (e.g., impacts on equality of autonomy, outcomes, and process), the micro-manufacturing of artemisinin-based combination therapy (ACT) has also shifted away from regions where malaria is prevalent and toward main production sites of Western pharmaceutical companies. Remarkably, ACT is currently no longer widely available in malaria-endemic areas. These companies with concentration of power now hold the authority to control the flow of knowledge and information, to preside over materials, to undercut vendor contracts, to dominate the balance of supply and demand, and to increase costs to consumers. Few companies in the global periphery own prequalified ACT; the result is increased cost for delivery, a more significant carbon footprint for distribution, and a retail price that is prohibitive for the majority of those in need who are exposed to the threat of malaria [20, 106]. This is likely to result in inequalities of autonomy as access decreases for people who need access the most, as well as inequalities of outcome as costs increase for those who are far more likely to be contending with lower SES.

6.4 Disparities of the Digital Divide

This concentration of power of Big Pharma can also hamper the crucial need for sponsored technological initiatives for a more equitable global digital society [12]. These partnerships prove unnecessary with on-the-spot, self-sufficient manufacturing of Big Pharma [20]. Without a role in an integrated value chain, societies in the global periphery no longer enjoy such subsidized or sponsored initiatives as operations security (OPSEC), cybersecurity awareness, education and training, supported distribution networks, and well-integrated enterprise management systems for inventory control across the value chain [20]. There is also far less underwritten innovation for new equipment and/or production methodologies, but rather new barriers of structural constraint arise due to an end to support, advice, and information sharing as these partnerships cease to exist. Thus, the digital divide expands, leading to such inequalities of autonomy as structural constraints and such inequalities of outcome as obstructed socioeconomic mobility for those most in need.

With such concentration of power in the hands of the Powerful Elite [8], power imbalances are further exacerbated due to concentration of knowledge as research becomes far more centralized within private organizations that have the economic resources to fund and exert influence on the researcher at the bench and often practice at the bedside. There is also far less impetus, and far fewer opportunities, for building scientific knowledge for the good of society, such as contributions to bio-digital enclaves for biotechs to share appropriate information to serve the underserved. Inequalities of autonomy emerge as local and regional communities now face the loss of previously enjoyed advice and/or support, as well as limitations relative to high-quality options [2, 75, 83].

6.5 Converging Digitalization to Exploit Those at the Global Periphery

As Big Pharma become the dominant global producer of drugs and treatments, they can more easily create opaque processes under their own unfettered and/or unregulated control. Such a closed system can undoubtedly diminish certain cybersecurity risks, barring the possibility of one insidiously successful cyberattack. Yet, concentration of power in the hands of the Powerful Elite [8] also allows for concentration of amalgamated data and processes that could further exploit humans as these companies are able to monopolize the balance of supply and demand, as well as control the types and timelines of drug launches. If Big Pharma has a fiduciary responsibility to achieve optimal financial performance for their shareholders, why would these companies research or launch life-saving drugs for underserved populations if new drugs will cannibalize other more successful drugs already performing well in the market?

Such companies or laboratories can also imperceptibly circumvent regulations in one country by conducting otherwise unauthorized experiments in another country with more lenient or lagging standards [46]. These types of machinations are likely to lead to harm of vulnerable communities in the global periphery in that this would lead to inequalities of process such as objectification and exacerbate inequalities of outcomes due to using unwitting members of these societies as test subjects, thereby risking damage to their health and well-being. Notably, these consequences further propagate the four stressors for humans as are identified in this chapter as follows: the physical, psychological, psychosocial, and psycho-spiritual aspects of humans. As previously mentioned, these tangible and intangible outcomes also lead to high levels of risk for individual A-Load, as well as Community A-Load [55, 65].

If segments of vulnerable or marginalized societies are facing decreases of necessary resources to thrive due to digitalization, and an increase of taxation on capacity to flourish due to digitalization, we must do better to protect equalities, as well as fight harder to identify and address inequalities. Yet, if we are enervated as a society, how well will we proactively recognize the long-term implications of these shifts? In a physiologically-disrupted state, will humans have the strength, tenacity, and resilience to champion such methodologies as responsible research and innovation (RRI) to ensure "a transparent, interactive process by which societal actors and innovators become mutually responsible to each other with a view to the ethical acceptability, sustainability, and societal desirability of the innovation process and its marketable products," to ensure generative technological advances [107]?

7 Discussion

The individuals, organizations, and communities affected by the bio-economy are diverse and vast. As we seek to steadfastly safeguard the bio-economy and robustly mitigate social injustices, we might do well to consider the following questions relative to accepting, altering, adapting, and avoiding to optimize impact of efforts.

7.1 *Accepting Strategically: What Is Within Our Locus of Control in Order to Affect Change?*

We recognize that such companies in our aforementioned examples have a fiduciary responsibility to shareholders to turn profits by increasing revenues and decreasing costs. Additionally, as a society, we have not only limited but also delayed avenues to make accountable such behemoth organizations who enjoy concentration of power. Yet encouragingly, there is swelling societal outcry for companies to embrace such prosocial practices as social and economic justice, especially within their established partnerships in the value chain. For example, society is increasingly imploring companies to embrace such accounting practices as triple bottom line (TBL) to make themselves accountable and transparent in quantifying not only economic but also social and environmental performance [108, 109]. Thus, we ask the communities of those working within (and with) the bio-economy to focus energies strategically by delineating between what is within their (and each stakeholder's) locus of control [110] and what is not.

7.2 *Altering: How Might We Shift These Societal Factors Within the External Environments?*

Once members of the life science communities have determined key loci of control, we now ask: How might those working with and within the bio-economy leverage prosocial shifts (such as the aforementioned) to motivate companies to ethically achieve or maintain collective value, with particular mind to the issue of public interest [111], ensuring those within the periphery of society are not disproportionately impacted? How might you motivate those companies who seek to do business with the bio-economy to also embrace such practices? Perhaps "pay to play" is now eclipsed by "equity to play" or "social justice to play." How might you embed a framework into decision-making for companies working in or within the bio-economy to assess the unintended consequences that create inequalities of access, outcomes, and process for the underserved or vulnerable members of our global society? How can you and your community better embed the principles of responsible research and innovation RRI [107] into your spheres of influence?

By embracing the principles of RRI [107], those in the life sciences would better ensure a robust array of societal actors (e.g., citizens, representatives of the underrepresented, third-sector organizations, those in the global periphery, etc.) are represented during all phases of designing, developing, delivering, and maintaining goods and services and/or sanctioning converging digitalization that is truly generative [72, 111]. With RRI, we would allow for the full spectrum of public engagement, so products, processes/protocols, and procedures within the life sciences align well with the known and need-to-be-known values, needs, and expectations of society. With this approach, our guardians of the bio-economy would not only allow for comprehensive and inclusive involvement but also robust trajectory thinking to anticipate and assess intended and unintended consequences that are likely to arise in the short term, midterm, and long term to ensure freedom and equity for humans [96].

7.3 Adapting and Avoiding: The Inner Sanctum as the First Line of Defense

Lastly, we ask: How might those working with and within the bio-economy change the way in which we interact with these sources of inside threats? Because humans are the nexus to safeguarding the bio-economy and mitigating inequalities of autonomy, inequalities of outcome, and inequalities of process, we contend that safeguarding the physiology of the human faculty might be an appropriate first line of defense.

Those working with or within the life sciences must insert buffers [56] to avoid significant harm to the bio-economy due to enervation of the guardians. How do we do better to create and improve the capacity and resilience of our guardians? How should we better create and maintain robust buffers for those working within or with the bio-economy [53, 54, 56]? Resilience, or the ability to survive, cope, recover, learn, and/or transform when confronted with volatility, uncertainty, complexity, ambiguity, and/or adverse circumstances, is an essential component for robust societies, as well as individuals to enjoy eudemonia, or optimal human well-being [21, 56, 65, 112]. We, as humans, do well when we prepare for and develop resilience within a variety of current or emerging contexts – even contexts riddled with acute and/or chronic disturbances – so as to arm ourselves against enervation. Humans must ensure they have the capacity (and adaptive capacity) to deal well with these adverse scenarios to guard well the bio-economy for freedom and equity for all members of society, but in particular the more vulnerable or underserved.

The previously mentioned four stressors (i.e., physical, psychosocial, psychological, and psycho-spiritual) can be utilized to mine out buffers [54, 56] that humans can tailor and apply at the individual level but also incorporate with intentionality into such community contexts [65] as workplaces within the bio-economy. Therefore, we ask the community to begin by exercising their autonomy to strengthen buffers, thereby designing and practicing methodologies to rigorously care for your physical, mental, psychological, social, and spiritual health. In doing so, you will gain resilience to offset the stressors and avoid enervation to better ensure the absence of avoidable or remediable disparities among groups of people, whether those groups are defined socially, behaviorally, economically, demographically/psychographically, or geographically [96, 113].

8 Conclusion

In this chapter, we considered the risks relative to enervated humans striving to safeguard the bio-economy while also endeavoring to protect human freedom and equity. We surveyed a sampling of commonplace yet predatory goods and services in society leading to such inside threats as the neurobiology of addiction; we also sought to touch upon unique vulnerabilities of marginalized populations in these contexts. As we explored converging digitalization leading to the likelihood of concentration of power, we sought to demonstrate the exacerbation of such inside threats leading to the neurobiology of addiction as well as the neurobiology of stress. We then explored degenerative consequences such as physiological dysregulation in humans resulting in allostatic load. We also highlighted the abundance of internal and external factors disproportionately impacting communities with lower SES. As we pursued such likely consequences as techno-feudalism due to concentration of power, we sought to emphasize the burgeoning risk to not only the individual but also to society. We thus emphasized the potential for a cataclysmic, yet somewhat indiscernible, persistently expanding societal divide between the Powerful Elite and the enervated common human who is likely to be relegated to serfdom. We then probed a real-world example of concentration of power to evaluate intended and unintended consequences for communities within the global periphery with special attention to such social justice issues as inequalities of autonomy, inequalities of outcome, and inequalities of process. In conclusion, we highlighted the importance of embracing such methodologies as RRI and of exercising autonomy with intentionality as an initial line of defense.

References

1. L.C. Richardson, N.D. Connell, S.M. Lewis, E. Pauwels, R.S. Murch, Cyberbiosecurity: A call for cooperation in a new threat landscape. Front. Bioeng. Biotechnol. **7**, 99 (2019). https://doi.org/10.3389/fbioe.2019.00099
2. R.S. Murch, W.K. So, W.G. Buchholz, S. Raman, J. Peccoud, Cyberbiosecurity: An emerging new discipline to help safeguard the bioeconomy. Front. Bioeng. Biotechnol. **6**, 39 (2018). https://doi.org/10.3389/fbioe.2018.00039
3. R. Abbas, K. Michael, M. G. Michael, C. Perakslis, J. Pitt, Machine learning, convergence digitalization, and the concentration of power: Enslavement by design using techno-

biological behaviors. IEEE Trans. Technol. Soc. **3**(2), 76–88 (2022). https://doi.org/10.1109/TTS.2022.3179756
4. T. Bonaci, K. Michael, P. Rivas, L. J. Robertson, M. Zimmer, Emerging technologies, evolving threats: Next-generation security challenges, IEEE Trans. Technol. Soc. **3**(3), 155–162 (2022). https://doi.org/10.1109/TTS.2022.3202323
5. R.J. Deibert, Toward a human-centric approach to cybersecurity. Ethics Int. Aff. **32**(4), 411–424 (2018). https://doi.org/10.1017/S0892679418000618
6. P. Virilio, *The Original Accident* (Polity, Cambridge/Malden, 2007)
7. J. Zittrain, Law and technology: The end of the generative Internet. Commun. ACM **52**(1), 18–20 (2009). https://doi.org/10.1145/1435417.1435426
8. D. Cuddy, *New World Order: The Rise of Techno-Feudalism* (Bible Belt Publishing, 2010)
9. P.R. Lewis, et al., A survey of self-awareness and its application in computing systems, in *2011 Fifth IEEE conference on self-adaptive and self-organizing systems workshops,* Ann Arbor, MI, USA (2011), pp. 102–107. https://doi.org/10.1109/SASOW.2011.25
10. W. Yang, S. Jin, S. He, Q. Fan, Y. Zhu, The impact of power on humanity: Self-dehumanization in powerlessness. PLOS ONE **10**(5), e0125721 (2015). https://doi.org/10.1371/journal.pone.0125721
11. D. Pager, H. Shepherd, The sociology of discrimination: Racial discrimination in employment, housing, credit, and consumer markets. Annu. Rev. Sociol. **34**, 181–209 (2008). https://doi.org/10.1146/annurev.soc.33.040406.131740
12. G. Kane, Digital maturity, not digital transformation. MIT Sloan Rev. (2017) [Online]. Available: http://sloanreview.mit.edu/article/digital-maturity-not-digital-transformation. Accessed 1 Oct 2021
13. A. Lyons, J. Kass-Hanna, A. Zucchetti, C. Cobo, Bridging the gap between digital skills and employability for economically vulnerable populations, in *Realizing Education for All in the Digital Age* (Asian Development Bank Institute, 2019), pp. 64–74 [Online]. Available: https://t20japan.org/wp-content/uploads/2019/03/t20-japan-tf7-9-bridging-gap-between-digital-skills-employability.pdf. Accessed 3 Jan 2022
14. D.R. Williams, Stress and the mental health of populations of color: Advancing our understanding of race-related stressors. J. Health Soc. Behav. **59**(4), 466–485 (2018). https://doi.org/10.1177/0022146518814251
15. J. Kleinberg, J. Ludwig, S. Mullainathan, C.R. Sunstein, Discrimination in the age of algorithms. J. Legal Anal. **10**, 113–174 (2018). https://doi.org/10.1093/jla/laz001
16. A. Baum, J.P. Garofalo, A.M. Yali, Socioeconomic status and chronic stress: Does stress account for SES effects on health? Ann. N. Y. Acad. Sci. **896**(1), 131–144 (1999). https://doi.org/10.1111/j.1749-6632.1999.tb08111.x
17. B. Gatersleben, I. Griffin, Environmental stress, in *Handbook of Environmental Psychology and Quality of Life Research*, ed. by G. Fleury-Bahi, E. Pol, O. Navarro, (Springer, Cham, 2017), pp. 469–485. https://doi.org/10.1007/978-3-319-31416-7_25
18. L. Prior, Allostatic load and exposure histories of disadvantage. Int. J. Environ. Res. Public. Health **18**(14), 7222 (2021). https://doi.org/10.3390/ijerph18147222
19. World Health Organization and Regional Office for the Western Pacific, *Health equity and Its Determinants in the Western Pacific Region* (2020) [Online]. Available: https://apps.who.int/iris/handle/10665/333944. Accessed 1 Mar 2022
20. L. Asveld, P. Osseweijer, J.A. Posada, Societal and ethical issues in industrial biotechnology. Adv. Biochem. Eng. Biotechnol. **173**, 121–141 (2020). https://doi.org/10.1007/10_2019_100
21. Committee on Safeguarding the Bioeconomy: Finding Strategies for Understanding, Evaluating, and Protecting the Bioeconomy While Sustaining Innovation and Growth et al., *Safeguarding the Bioeconomy* (National Academies Press, Washington, DC, 2020), p. 25525. https://doi.org/10.17226/25525.
22. K. Michael, Are you addicted to your smartphone, social media, and more?: The new AntiSocial app could help. IEEE Consum. Electron. Mag. **6**(4), 116–121 (2017). https://doi.org/10.1109/MCE.2017.2714421

23. J. Taylor, *British slang words & phrases* (Oxford International English Schools, 2019). https://www.oxfordinternationalenglish.com/dictionary-of-british-slang/. Accessed 3 Jan 2022
24. N.D. Schüll, *Addiction by Design: Machine Gambling in Las Vegas* (Princeton University Press, Princeton, 2014)
25. S. Sharman, L. Clark, Mixed emotions to near-miss outcomes: A psychophysiological study with facial electromyography. J. Gambl. Stud. **32**(3), 823–834 (2016). https://doi.org/10.1007/s10899-015-9578-2
26. F.A. Nasution, E. Effendy, M.M. Amin, Internet Gaming Disorder (IGD): A case report of social anxiety. Open Access Maced. J. Med. Sci. **7**(16), 2664–2666 (2019). https://doi.org/10.3889/oamjms.2019.398
27. K. Albrecht, K. Michael, M.G. Michael, The dark side of video games: Are you addicted? IEEE Consum. Electron. Mag. **5**(1), 107–113 (2016). https://doi.org/10.1109/MCE.2015.2484820
28. D.J. Kuss, H.M. Pontes, M.D. Griffiths, Neurobiological correlates in internet gaming disorder: A systematic literature review. Front. Psychiatry **9**, 166 (2018). https://doi.org/10.3389/fpsyt.2018.00166
29. W. Yang, M. Rifqi, C. Marsala, A. Pinna, Physiological-based emotion detection and recognition in a video game context, in *2018 International Joint Conference on Neural Networks (IJCNN)*, Rio de Janeiro (2018), pp. 1–8. https://doi.org/10.1109/IJCNN.2018.8489125
30. M. Moss, *Hooked: Food, Free Will, and How the Food Giants Exploit Our Addictions*, 1st edn. (Random House, New York, 2021)
31. N. Eyal, R. Hoover, *Hooked: how to build habit-forming products* (2014) [Online]. Available: http://sckans.axis360.baker-taylor.com/Title?itemId=0015269374. Accessed 28 Feb 2022
32. B.J. Fogg, *Persuasive Technology: Using Computers to Change What We Think and Do* (Morgan Kaufmann Publishers, Amsterdam/Boston, 2003)
33. J. Hilliard, T. Parisi, *Social Media Addiction* (Addiction Center, 2021). https://www.addictioncenter.com/drugs/social-media-addiction/. Accessed 6 Jan 2022
34. F. Lin et al., Abnormal white matter integrity in adolescents with internet addiction disorder: A tract-based spatial statistics study. PLoS ONE **7**(1), e30253 (2012). https://doi.org/10.1371/journal.pone.0030253
35. M. Csikszentmihalyi, *Beyond Boredom and Anxiety*, 1st edn. (Jossey-Bass Publishers, San Francisco, 1975)
36. M. Csikszentmihalyi, *Flow: The Psychology of Optimal Experience*, 1st edn. (Harper & Row, New York, 1990)
37. J. Nakamura, M. Csikszentmihalyi, Flow theory and research, in *The Oxford Handbook of Positive Psychology*, ed. by S.J. Lopez, C.R. Snyder, (Oxford University Press, New York, 2009), pp. 194–206. https://doi.org/10.1093/oxfordhb/9780195187243.013.0018
38. L.K. Trevino, J. Webster, Flow in computer-mediated communication: Electronic mail and voice mail evaluation and impacts. Commun. Res. **19**(5), 539–573 (1992). https://doi.org/10.1177/009365092019005001
39. J. Webster, L.K. Trevino, L. Ryan, The dimensionality and correlates of flow in human-computer interactions. Comput. Hum. Behav. **9**(4), 411–426 (1993). https://doi.org/10.1016/0747-5632(93)90032-N
40. J. Chen, Flow in games (and everything else). Commun. ACM **50**(4), 31–34 (2007). https://doi.org/10.1145/1232743.1232769
41. X. Fang, J. Zhang, S.S. Chan, Development of an instrument for studying flow in computer game play. Int. J. Hum.-Comput. Interact. **29**(7), 456–470 (2013). https://doi.org/10.1080/10447318.2012.715991
42. Y.J. Lee, S. Ha, Z. Johnson, Antecedents and consequences of flow state in e-commerce. J. Consum. Mark. **36**(2), 264–275 (2019). https://doi.org/10.1108/JCM-10-2015-1579
43. D.L. Hoffman, T.P. Novak, Flow online: Lessons learned and future prospects. J. Interact. Mark. **23**(1), 23–34 (2009). https://doi.org/10.1016/j.intmar.2008.10.003
44. C.D. Johnson, B.C. Bauer, N. Singh, Exploring flow in the mobile interface context. J. Retail. Consum. Serv. **53**, 101744 (2020). https://doi.org/10.1016/j.jretconser.2019.01.013

45. M. Moss, *Salt, Sugar, Fat: How the Food Giants Hooked Us*, 1st edn. (Random House, New York, 2013)
46. M. Elgabry, S. Johnson, *Synthetic Biology and Future Crime* (Dawes Centre for Future Crime at UCL, 2021), p. 4
47. *Do Not Get Sold on Drug Advertising* (Harvard Health Publishing, 2017) [Online]. Available: https://www.health.harvard.edu/drugs-and-medications/do-not-get-sold-on-drug-advertising. Accessed 3 Dec 2021
48. M.G. Michael, K. Michael, C. Perakslis, Überveillance, the web of things, and people: What is the culmination of all this surveillance? IEEE Consum. Electron. Mag. **4**(2), 107–113 (2015). https://doi.org/10.1109/MCE.2015.2393007
49. V. Insley, D. Nunan, Gamification and the online retail experience. Int. J. Retail Distrib. Manag. **42**(5), 340–351 (2014). https://doi.org/10.1108/IJRDM-01-2013-0030
50. A.L. Alter, *Irresistible: The Rise of Addictive Technology and the Business of Keeping Us Hooked* (Penguin Press, New York, 2017)
51. B. Verplanken, A. Herabadi, Individual differences in impulse buying tendency: Feeling and no thinking. Eur. J. Personal. **15**(1_suppl), S71–S83 (2001). https://doi.org/10.1002/per.423
52. H. Benson, *The Mind/Body Effect: How Behavioral Medicine Can Show You the Way to Better Health* (2019) [Online]. Available: http://search.ebscohost.com/login.aspx?direct=true&scope=site&db=nlebk&db=nlabk&AN=2145500. Accessed 1 Mar 2022
53. H. Benson, M.Z. Klipper, *The relaxation response*, Updated & Expanded edn. (Quill, New York, 2001)
54. H. Benson, W. Proctor, *Beyond the Relaxation Response: How to Harness the Healing Power of Your Personal Beliefs* (Berkley Books, New York, 1985)
55. D.E. Saxbe, L. Beckes, S.A. Stoycos, J.A. Coan, Social allostasis and social allostatic load: A new model for research in social dynamics, stress, and health. Perspect. Psychol. Sci. **15**(2), 469–482 (2020). https://doi.org/10.1177/1745691619876528
56. E. Park, P. Baim, L. Kagan, *Stress Management and Resiliency Training: The Relaxation Response Resiliency Program®* (Benson-Henry Institute and Harvard Medical School, 2021)
57. G.F. Koob, N.D. Volkow, Neurocircuitry of addiction. Neuropsychopharmacology **35**(1), 217–238 (2010). https://doi.org/10.1038/npp.2009.110
58. G.F. Koob, N.D. Volkow, Neurobiology of addiction: A neurocircuitry analysis. Lancet Psychiatry **3**(8), 760–773 (2016). https://doi.org/10.1016/S2215-0366(16)00104-8
59. The neurobiology of substance use, misuse, and addiction, in *Facing Addiction in America: The Surgeon General's Report on Alcohol, Drugs, and Health* (Office of the Surgeon General (US); Substance Abuse and Mental Health Services Administration (US)) [Online]. Available: https://www.ncbi.nlm.nih.gov/books/NBK424849/. Accessed 3 Nov 2021
60. H. Yaribeygi, Y. Panahi, H. Sahraei, T.P. Johnston, A. Sahebkar, The impact of stress on body function: A review. EXCLI J. 16Doc1057 ISSN 1611-2156 (2017). https://doi.org/10.17179/EXCLI2017-480
61. P. Chen, D.R. Voisin, K.C. Jacobson, Community violence exposure and adolescent delinquency: examining a spectrum of promotive factors. Youth Soc. **48**(1), 33–57 (2016). https://doi.org/10.1177/0044118X13475827
62. H.C. Covey, S. Menard, R.J. Franzese, Effects of adolescent physical abuse, exposure to neighborhood violence, and witnessing parental violence on adult socioeconomic status. Child Maltreat. **18**(2), 85–97 (2013). https://doi.org/10.1177/1077559513477914
63. A.B. Eisman, S.A. Stoddard, J. Heinze, C.H. Caldwell, M.A. Zimmerman, Depressive symptoms, social support, and violence exposure among urban youth: A longitudinal study of resilience. Dev. Psychol. **51**(9), 1307–1316 (2015). https://doi.org/10.1037/a0039501
64. M.E. Patrick, P. Wightman, R.F. Schoeni, J.E. Schulenberg, Socioeconomic status and substance use among young adults: A comparison across constructs and drugs. J. Stud. Alcohol Drugs **73**(5), 772–782 (2012). https://doi.org/10.15288/jsad.2012.73.772

65. A. Chandra, M. Cahill, D. Yeung, R. Ross, *Toward an Initial Conceptual Framework to Assess Community Allostatic Load: Early Themes from Literature Review and Community Analyses on the Role of Cumulative Community Stress* (RAND Corporation, 2018). https://doi.org/10.7249/RR2559
66. A. Tawakol et al., Stress-associated neurobiological pathway linking socioeconomic disparities to cardiovascular disease. J. Am. Coll. Cardiol. **73**(25), 3243–3255 (2019). https://doi.org/10.1016/j.jacc.2019.04.042
67. G. Evans, J. Brooks-Gunn, P. Kato-Klebanov, Stressing out the poor: Chronic physiological stress and the income-achievement gap. Community Invest. **23**(Fall), 22–27 (2011)
68. L. Lander, J. Howsare, M. Byrne, The impact of substance use disorders on families and children: From theory to practice. Soc. Work Public Health **28**(3–4), 194–205 (2013). https://doi.org/10.1080/19371918.2013.759005
69. N.D. Rao, J. Min, Decent living standards: Material prerequisites for human wellbeing. Soc. Indic. Res. **138**(1), 225–244 (2018). https://doi.org/10.1007/s11205-017-1650-0
70. D. Vilda, M. Wallace, L. Dyer, E. Harville, K. Theall, Income inequality and racial disparities in pregnancy-related mortality in the US. SSM Popul. Health **9**, 100477 (2019). https://doi.org/10.1016/j.ssmph.2019.100477
71. K. Sabbagh, R. Friedrich, B. El-Darwiche, M. Singh, S. Ganediwalla, R. Katz, *Maximizing the Impact of Digitization*, World Economic Forum 2012, The Global Information Technology Report 2012 (2012)
72. C. Perakslis, K. Michael, M.G. Michael, The converging veillances: Border crossings in an interconnected world. IEEE Potentials **35**(5), 23–25 (2016). https://doi.org/10.1109/MPOT.2016.2569724
73. *Conducting Biosocial Surveys: Collecting, Storing, Accessing, and Protecting Biospecimens and Biodata* (National Academies Press, Washington, DC, 2010), p. 12942. https://doi.org/10.17226/12942
74. C. Orwat, *Risks of Discrimination Through the Use of Algorithms. A Study Compiled with a Grant from the Federal Anti-Discrimination Agency* (Federal Anti-Discrimination Agency, 2020). https://doi.org/10.5445/IR/1000123477
75. Committee on Understanding the Global Public Health Implications of Substandard, Falsified, and Counterfeit Medical Products, Board on Global Health, and Institute of Medicine, *Countering the Problem of Falsified and Substandard Drugs* (National Academies Press, Washington, DC, 2013), p. 18272. https://doi.org/10.17226/18272
76. I. van de Poel, Z. Robaey, Safe-by-design: From safety to responsibility. *NanoEthics* **11**(3), 297–306 (2017). https://doi.org/10.1007/s11569-017-0301-x
77. G.E. Marchant, Addressing the pacing problem, in *The Growing Gap Between Emerging Technologies and Legal-Ethical Oversight*, ed. by G.E. Marchant, B.R. Allenby, J.R. Herkert, vol. 7, (Springer, Dordrecht, 2011), pp. 199–205. https://doi.org/10.1007/978-94-007-1356-7_13
78. N. Schneiderman, G. Ironson, S.D. Siegel, Stress and health: Psychological, behavioral, and biological determinants. Annu. Rev. Clin. Psychol. **1**(1), 607–628 (2005). https://doi.org/10.1146/annurev.clinpsy.1.102803.144141
79. H.-R. Lin, S.M. Bauer-Wu, Psycho-spiritual well-being in patients with advanced cancer: An integrative review of the literature: Psycho-spiritual well-being in patients with advanced cancer. J. Adv. Nurs. **44**(1), 69–80 (2003). https://doi.org/10.1046/j.1365-2648.2003.02768.x
80. J. Clifton, *Gallup Global Emotions 2021*, World Stress Index and Lighthouse [Online]. Available: https://bluesyemre.files.wordpress.com/2022/02/gallup-global-emotions-2021-report.pdf. Accessed 29 Jan 2022
81. A. Abbott, Huge survey reveals pressures of scientists' lives. Springer Nature Limited **577**, 460–462 (2020)
82. H. Moran et al., Understanding research culture: What researchers think about the culture they work in. Wellcome Open Res. **5**, 201 (2020). https://doi.org/10.12688/wellcomeopenres.15832.1

83. Approaches to Risk and Benefit Assessment for Advances in the Life Sciences, United States of America, Geneva, Meeting of the States Parties to the Convention on the Prohibition of the Development, Production and Stockpiling of Bacteriological (Biological) and Toxin Weapons and on Their Destruction GE.19-11823(E) (2019) [Online]. Available: https://www.centerforhealthsecurity.org/our-work/pubs_archive/pubs-pdfs/2019/190729-bwc-msp-mx-2.pdf. Accessed 3 Oct 2021
84. T. Singh, A. Johnston, *How Much is Too Much: Employee Monitoring, Surveillance, and Strain?* Presented at the international conference on information systems (ICIS), Munich, Germany (2019) [Online]. Available: https://aisel.aisnet.org/icis2019/cyber_security_privacy_ethics_IS/cyber_security_privacy/31. Accessed 2 Sept 2021
85. L. Stark, A. Stanhaus, D.L. Anthony, 'I don't want someone to watch me while I'm working': Gendered views of facial recognition technology in workplace surveillance. J. Assoc. Inf. Sci. Technol. **71**(9), 1074–1088 (2020). https://doi.org/10.1002/asi.24342
86. P. Roberts, *Information Visualization for Stock Market Ticks: Toward a New Trading Interface*, Master's thesis (Massachusetts Institute of Technology, Boston, MA, 2004) [Online]. Available: https://dspace.mit.edu/bitstream/handle/1721.1/16668/56675083-MIT.pdf?sequence=2. . Accessed 2 Sept 2018
87. E.R. Park et al., The development of a patient-centered program based on the relaxation response: The Relaxation Response Resiliency Program (3RP). Psychosomatics **54**(2), 165–174 (2013). https://doi.org/10.1016/j.psym.2012.09.001
88. J. Eisert et al., Vigilance and fatigue: A double sided coin? Proc. Hum. Factors Ergon. Soc. Annu. Meet. **60**(1), 1563–1568 (2016). https://doi.org/10.1177/1541931213601361
89. K. Sagherian, M.E. Clinton, H. Abu-Saad Huijer, J. Geiger-Brown, Fatigue, work schedules, and perceived performance in bedside care nurses. Workplace Health Saf. **65**(7), 304–312 (2017). https://doi.org/10.1177/2165079916665398
90. D. Ropeik, The consequences of fear. EMBO Rep. **5**(S1) (2004). https://doi.org/10.1038/sj.embor.7400228
91. J. Loehr, T. Schwartz, The making of a corporate athlete. Harv. Bus. Rev. **79**(1), 120–128, 176 (2001)
92. D. Goleman, J. Gurin, *Mind/body medicine: How to use your mind for better health* (Consumer Reports Books, Yonkers, 1996)
93. S. Zuboff, *The Age of Surveillance Capitalism: The Fight for a Human Future at the New Frontier of Power*, 1st edn. (PublicAffairs, New York, 2019)
94. J. Pitt, The BigTech-Academia-Parliamentary Complex and Techno-Feudalism [Editorial]. IEEE Technol. Soc. Mag. **39**(3), 5–8 (2020). https://doi.org/10.1109/MTS.2020.3012257
95. J. Pitt, J. Dryzek, and J. Ober, "Algorithmic Reflexive Governance for Socio-Techno-Ecological Systems," IEEE Technol. Soc. Mag., vol. 39, no. 2, pp. 52–59, Jun. 2020, doi: https://doi.org/10.1109/MTS.2020.2991500.
96. N. Daniels, *Just Health: Meeting Health Needs Fairly* (Cambridge University Press, Cambridge/New York, 2008)
97. Big pharma, bad medicine, Boston Rev. (2012). https://bostonreview.net/forum/angell-big-pharma-bad-medicine/
98. Industry profile: Pharmaceuticals/health products, Open Secrets (2020) [Online]. Available: https://www.opensecrets.org/federal-lobbying/industries/summary?cycle=2020&id=H04. Accessed 1 Sept 2020
99. K. Michael, M. Michael, Uberveillance: Microchipping people and the assault on privacy. Quadrant **3**(53), 85–89 (2009)
100. G. Orwell, 1984 (2020) [Online]. Available: https://search.ebscohost.com/login.aspx?direct=true&scope=site&db=nlebk&db=nlabk&AN=2533966. Accessed 1 Mar 2022
101. J. Pitt, M. Tzanou, Special issue introduction: Against modern indentured servitude ("I'm Spartacus"). IEEE Technol. Soc. Mag. **41**(2), 20–23 (2022)
102. J. Pitt, The digital transformation and modern indentured servitude. IEEE Technol. Soc. Mag. **41**(2), 6–9 (2022)

103. S. Alkire, *Developing the Equality Measurement Framework: Selecting the Indicators* (Equality and Human Rights Commission, Manchester, 2009)
104. AG Shapiro announces $48 billion opioid epidemic deal with five companies, Pennsylvania Office of Attorney General Josh Shapiro, Harrisburg (2019) [Online]. Available: https://www.attorneygeneral.gov/taking-action/ag-shapiro-announces-48-billion-opioid-epidemic-deal-with-five-companies/. Accessed 30 Sept 2021
105. J. Francis-Smith, Drug companies, tribes reach landmark opioid agreement. J. Record (2022) [Online]. Available: https://journalrecord.com/2022/02/01/drug-companies-tribes-reach-landmark-opioid-agreement/. Accessed 9 Feb 2022
106. P. Sarasas, *Risks and Potential Rewards of Synthetic Biology*, United Nations environment programme (2019), https://www.unep.org/news-and-stories/story/risks-and-potential-rewards-synthetic-biology. Accessed 1 Dec 2021
107. R. Von Schomberg, Towards responsible research and innovation in the information and communication technologies and security technologies fields. SSRN Electron. J. (2011). https://doi.org/10.2139/ssrn.2436399
108. J. Elkington, Accounting for the triple bottom line. Meas. Bus. Excell. **2**(3), 18–22 (1998). https://doi.org/10.1108/eb025539
109. A. Geva, Three models of corporate social responsibility: Interrelationships between theory, research, and practice. Bus. Soc. Rev. **113**(1), 1–41 (2008). https://doi.org/10.1111/j.1467-8594.2008.00311.x
110. J.B. Rotter, Generalized expectancies for internal versus external control of reinforcement. Psychol. Monogr. Gen. Appl. **80**(1), 1–28 (1966). https://doi.org/10.1037/h0092976
111. R. Abbas, S. Hamdoun, J. Abu-Ghazaleh, N. Chhetri, N. Chhetri, K. Michael, Co-designing the future with public interest technology. IEEE Technol. Soc. Mag. **40**(3), 10–15 (2021). https://doi.org/10.1109/MTS.2021.3101825
112. Infrastructure Resilience Planning Framework (IRPF), Cybersecurity & Infrastructure Security Agency, Version 1.0 (2021) [Online]. Available: https://www.cisa.gov/sites/default/files/publications/Infrastructure_Resilience_Planning_Framework_IRPF.pdf. Accessed 1 Dec 2021
113. K. Michael, K. Albrecht, We've got to do better. IEEE Technol. Soc. Mag. **33**(1), 5–7 (2014). https://doi.org/10.1109/MTS.2014.2300948

AI for Cyberbiosecurity in Water Systems—A Survey

Daniel Sobien, Mehmet O. Yardimci, Minh B. T. Nguyen, Wan-Yi Mao, Vinita Fordham, Abdul Rahman, Susan Duncan, and Feras A. Batarseh

Abstract The use of Artificial Intelligence (AI) is growing in areas where decisions and consequences have high-stakes such as larger scale software, critical infrastructure, and real-time systems. This transition in recent years has been accompanied by the growth of research in AI assurance in fields such as ethical, explainable, and trustworthy AI. In this work, we survey the literature to find the state of AI assurance for cyberbiosecurity systems as they exist now, particularly for water and agricultural supply systems; future directions are also presented. We focus on papers at the intersection of cyberbiosecurity, AI assurance, and water/agricultural supply systems, discuss how assurance techniques improve these systems, and provide pointers for future research into the application of AI for the cyberbiosecurity field. Current cyberbiosecurity solutions do not focus much on AI, but existing AI solutions for water supply and cyber or Cyber-Physical Systems (CPS) exist and can

D. Sobien
Hume Center for National Security and Technology, Virginia Tech, Arlington, VA, USA
e-mail: sdan8@vt.edu

M. O. Yardimci · W.-Y. Mao
Department of Computer Science, Virginia Tech, Blacksburg, VA, USA
e-mail: oguzy@vt.edu; wanyi@vt.edu

M. B. T. Nguyen
Bradley Department of Electrical and Computer Engineering, Virginia Tech, Blacksburg, VA, USA
e-mail: mnguyen0226@vt.edu

V. Fordham · A. Rahman
Deloitte Touche Tohmatsu Limited, Arlington, VA, USA
e-mail: vfordham@deloitte.com; abdulrahman@deloitte.com

S. Duncan
College of Agriculture and Life Sciences, Virginia Tech, Blacksburg, VA, USA
e-mail: duncans@vt.edu

F. A. Batarseh (✉)
Department of Biological Systems Engineering, Virginia Tech, Arlington, VA, USA
e-mail: batarseh@vt.edu

D. Greenbaum (ed.), *Cyberbiosecurity*, https://doi.org/10.1007/978-3-031-26034-6_13

be applied to benefit cyberbiosecurity. The inclusion of AI assurances help alleviate issues of applying AI to high-stakes human-centered infrastructure.

CCS Concepts Computing methodologies → Artificial intelligence, Security and privacy, General and reference → Cross-computing tools and techniques, Computer systems organization → Embedded and Cyber-physical systems

Keywords Cyberbiosecurity · AI assurance · Water supply systems

1 Introduction

The deployment of AI is outpacing the adoption of assurances that commit to its responsible use as policies and regulations lag behind. Assurances validate AI systems to assess the risk of failure, misuse, and even abuse, helping establish the trust needed for the adoption of AI. The risks of AI in infrastructure (e.g., agricultural supply chains, biological systems, and water supply systems) are significant, potentially affecting millions of citizens and resulting in loss of life, well-being, and economic opportunity.

For example, take a city-wide water distribution system that pumps in water from a reservoir and ensures every citizen has equal access to drinkable water. Imagine the city adopts an AI system that predicts demand and supplies regions of the system as needed. The system works fine to start, but years later it is not properly validated after new pumps are installed, so the sensor data changes and no longer predicts accurately. As a result there are large swaths of the city that are no longer receiving drinking water because the system forecasts are off. Or, maybe the system was trained with bias data because poorer neighborhoods had less data collected, so the system favors keeping the water supply greater for affluent regions resulting in poorer regions having intermittent supply issues.

For this water supply AI system to work properly assurances must validate outcomes are correct, fair, and that users can understand why the system has made its decisions. These concepts form the basis of AI assurance, which details the broad ways of verifying and validating AI systems, much the same way that traditional programming software (i.e., not machine learning) is verified and validated during its development process [1]. AI assurance applied during development would help avoid the mentioned issues of robustness and bias.

Water supply systems are a form of CPS, as physical sensors, pumps, and tanks act as data collectors to track the flow of water and relay data to a central computer. This data processing exposes the water supply to cyber-attacks. Additionally, water supply systems are part of the bioeconomy (the supply chain infrastructure that is tied to critical commodities like food, water, and medicine) meaning any impact to the system can have an effect on the livelihood of thousands or millions of people. The imagined *water supply AI* not only ensures proper water distribution, but there are additional security concerns, moving it into the relatively new realm of cyberbiosecurity, which is a discipline at the intersection of life science and information technology (IT) [2]. Cyberbiosecurity is defined in greater detail in Sect. 1.1.

Existing cyberbiosecurity research mostly focuses on the IT side of biology, or cybersecurity for biology labs and databases is a succinct way to put it. The cyberbiosecurity field, however, is lacking much research in applied AI for supply chain infrastructure, as most papers only identify vulnerabilities and propose high-level frameworks for addressing them. Our goal for this survey is to find papers at the intersections of cyberbiosecurity, AI assurance, and water and food supply systems and connect that to the bioeconomy. Our work searches for and discusses the applications of AI assurance to existing solutions within the cybersecurity and CPS to help ensure the proper function of cyberbiosecurity-related systems.

1.1 Relevant Terminology and Definitions

Proper use of AI assurances verifies and validates the outputs of those systems, convincing users that they are reliable. AI assurance codifies the process, so when changes occur to the water supply system, validation can be re-run to satisfy the AI is working properly or needs to be retrained. Definitions are intentionally broad in order to apply them to a wider range of applications. From Batarseh et al. [1], AI assurance is defined as:

> A process that is applied at all stages of the AI engineering lifecycle ensuring that any intelligent system is producing outcomes that are valid, verified, data-driven, trustworthy and explainable to a layman, ethical in the context of its deployment, unbiased in its learning, and fair to its users.

The importance of AI assurance is that it applies a process to all stages of the AI lifecycle, from the start of development all the way through deployment. Assurances are not merely tests of AI to check some boxes that it is okay to use. In order to trust the AI is working properly engineers need to validate it meets all the criteria of assurance:

- Ethical—the AI system can make "right" decisions that benefit the people impacted and not just the people in power of the technology [3].
- Fair—the AI system makes decisions without considering demographics, backgrounds, affiliations, or individual preferences (i.e., does not inherently value some citizens over others).
- Safe—the AI system ensures the life and well-being of those who are using it and impacted by it.
- Explainable—the AI system can explain, or be interpreted, to understand why it came to a decision or how the algorithm works.
- Secure—the AI system can prevent or mitigate attacks or other threats to the proper operation of the system.
- Trustworthy—users have confidence the AI system works properly.

For infrastructure systems in the bioeconomy, AI must be ethical to make the right decisions, safe to protect users it potentially impacts, explainable so humans can understand it, fair in the decisions it makes, trustworthy so we have confidence in its abilities, and secure to prevent cyber-attacks and threats.

The bioeconomy refers to the sector of the economy that relates to research or innovations in the life and biological sciences and fields related to biotechnology [2, 4–6]. This sector grows as progress continues in technology relating to computing and information sciences [7], including most crop production, especially as big data, AI, and machine learning become more involved for enhancing land use and water management via precision farming [4]. As the bioeconomy grows, cyber threats against it increase and require mitigation to safeguard investments in the bioeconomy [8].

Richardson et al. [2] described cyberbiosecurity as the intersection of IT and life sciences, but Duncan et al. [9] specified it further as the intersection of cybersecurity, cyber-physical security, and biosecurity. Each discipline with its own existing challenges and new vulnerabilities appearing where they overlap.

By its nature, cyberbiosecurity is grounded in IT and with that brings the risk of cyber-attacks. This is the traditional realm of cybersecurity, or the shielding of computer networks and information from damage, exploitation, and unauthorized use [10–15]. Linking any computer system to a network increases risk. This is compounded in the bioeconomy as more remote monitoring and controlling is added to existing physical infrastructure, because of this interaction of cyber and physical the security needs "safety and reliability requirements qualitatively different from those in general-purpose computing." [16]. A CPS integrates digital computing and physical processes, where a network monitors and controls a physical system via sensors and actuators, to interact with the real world [16, 17]. Communication and networking multiple devices is important because the components are often disparate and there is a back and forth of physical processes affecting the computer and vice versa, but this opens new vulnerabilities [16, 17].

The third aspect of cyberbiosecurity moves fully into the physical space for securing biological systems. Biosecurity is the protection of any form of life from the threat of disease and pests, including the protection of agriculture and food, or simply put the "re-branding of the centuries-old battle with disease" [18–20]. This includes threats that are natural, such as livestock and crop diseases, or intentional attacks, such as the deliberate use of smallpox and anthrax weapons [18]. The incorporation of biosecurity in the realm of cybersecurity and cyber-physical security is what sets cyberbiosecurity apart.

Traditional cyber-attacks are not necessary to impact biological systems, because there are physical, biological interactions outside the computer systems. We need to ensure that the biological aspects are operating properly, be it from natural causes (diseases, pests, etc.) or intentional cyber and physical attacks. There are three layers of interactions to protect: the cyber, the interactions of cyber and physical, and the biological.

Included in these biological systems are water supply systems, which can refer to distribution, treatment, agricultural, or storm water systems. Distribution systems control the transport and delivery of water through a network of pipes and pumps to ensure consistent supply, they are focused on the logistics of water transportation and storage. Treatment systems take raw or wastewater, unsafe for humans or the environment, and through a series of chemical and biological processing, filtering, and sanitizing produce either safe drinking water or water that can be released into the environment. Agricultural water systems focus on the distribution of water to crops and livestock. Unlike distribution systems, this water does not have to be safe for human drinking, but it must ensure the production of food for human use. This also closely ties agricultural water systems to food supply systems. Finally, storm water systems deal with the drainage of runoff water to prevent flooding or contamination of other water systems from the pollutants that it picks up.

These systems allow for the automation of critical infrastructure by adding more technology for monitoring and controlling human and agricultural water use. These water and food systems are not only cyber-physical but also biological as well. Their proper functioning is required for human livelihood, either through the supply of safe water or the growth of adequate food supplies. Water and food systems are cyber-physical and bio-infrastructure systems that are open to attacks (cyber and physical) and anomalies (such as maintenance issues, severe weather, sensor or equipment breakdowns).

Going back to our hypothetical city-wide water distribution system. If it were attacked by a bad actor who wanted to poison the water, they could give commands to add too much of a chemical or too little of a cleaning agent that would result in undrinkable water. In fact, there was an attack in 2021 on a Tampa, Florida water supply system where attackers increased the levels of lye in the water by 110 times before they were stopped [21]. We discuss this example further in Sect. 5.4, but it serves as a great example of the cyberbiosecurity threats to water supply systems. Threats can combine unauthorized access of computer systems to control physical processes; in the Tampa case, the lye controllers pose a biological threat to everyone that relies on the system for safe drinking water. The next section introduces the inclusion and exclusion criteria of the papers surveyed.

1.2 Description of Included Articles

In this survey, we used multiple online repositories and research paper search engines to find relevant papers on the topics of cyberbiosecurity, AI assurance, and water supply systems. Our focus was to find peer-reviewed papers at the intersection of two or more topics. We include papers from journals, conference proceedings, dissertations, books and book chapters, and industry white papers published from **2000** through **April 2022**. A complete repository of papers included in this study can be found here: https://github.com/AI-VTRC/CyberbiosecuritySurveyPaper.

Key search terms included the following to find papers:

- Cyberbiosecurity; Cyber-Biosecurity; Biocybersecurity; Bio-Cybersecurity

- Water Supply System; Water Distribution System; Water Treatment System; Water System
- AI Assurance (see assurance list in Sect. 1.1)
- Artificial Intelligence

Because cyberbiosecurity is a new research field, we kept search criteria as broad as possible to include enough papers for a survey. Some focus on the medical fields, but we tried to find relevant discussions that could apply to AI assurance or water supply systems as much as possible. Some focus just on the concept of cyberbiosecurity in general, but we focus on how best to apply the concept to AI assurance and water supply systems.

2 Survey Landscape

The papers surveyed for this research included publications between 2000 and 2022 (as of April 2022), but most are from 2016 onward. Figure 1 shows a histogram by publication year, and until 2016 there was not more than three publications per year that covered cyberbiosecurity, water systems, and AI assurance. There is a steady trend upward for the count of publications, and as cyberbiosecurity and AI assurance

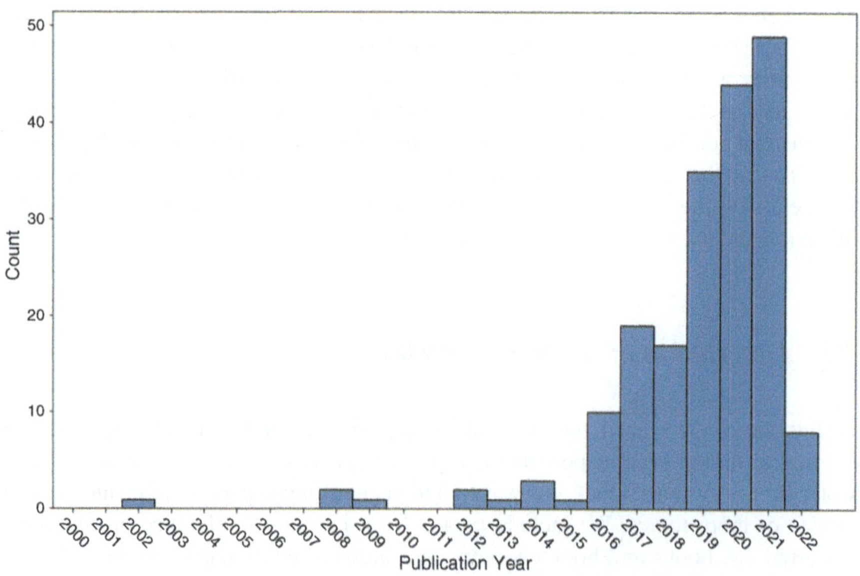

Fig. 1 Count of the number of publications by year that were used in this survey

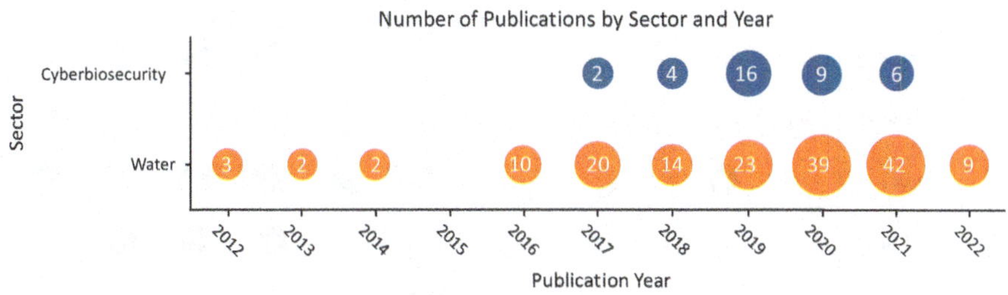

Fig. 2 The count of publications by year for the sectors of cyberbiosecurity and water supply (either water treatment or water distribution) systems. Papers are not confined to a single sector, and some are counted both as cyberbiosecurity and water supply papers. Most papers published since 2012, so older publications omitted from this figure

research continues to grow we expect the number of publications to continue to grow each year.

Figure 2 shows the breakdown of publications by cyberbiosecurity and water sectors. Publications on water systems had a low but steady trend from the early 2000s until about 2017 when they increased and held since. The year 2017 was also when the cyberbiosecurity term started showing in the scientific literature, and there is a sharp peak in 2019 before cyberbiosecurity publications return to a more steady pace.

We break down the AI assurance publications by assurance pillars in Fig. 3. Here, a majority of the papers deal with safe and trustworthy AI, especially just before the term of cyberbiosecurity starts showing in 2017. As AI becomes more popular, especially with deep learning (since 2015), we see an increase in publications for all the pillars of AI assurance.

Figure 4 shows a citation graph we created using Citation Gecko.[1] The yellow nodes are surveyed papers, gray nodes are other papers which cite our surveyed papers, and edges (lines that connect the nodes) are the citation link between two papers. The cyberbiosecurity literature is relatively disjointed from the literature on water supply systems and attack/anomaly detection. Most of the AI assurance papers remain independent in this view from each other and other sectors, with the exception of some trustworthy AI papers that form a small network. This graph shows the relative separation of the cyberbiosecurity literature from water system security and attack/anomaly detection (which includes secure AI). There is one citation chain from cyberbiosecurity to water system security via Mueller [22], Schmale III et al. [23], Moyer et al. [24], and Housh and Ohar [25]. (note that Moyer et al. [24] is the oldest link in that chain.)

[1] https://www.citationgecko.com/.

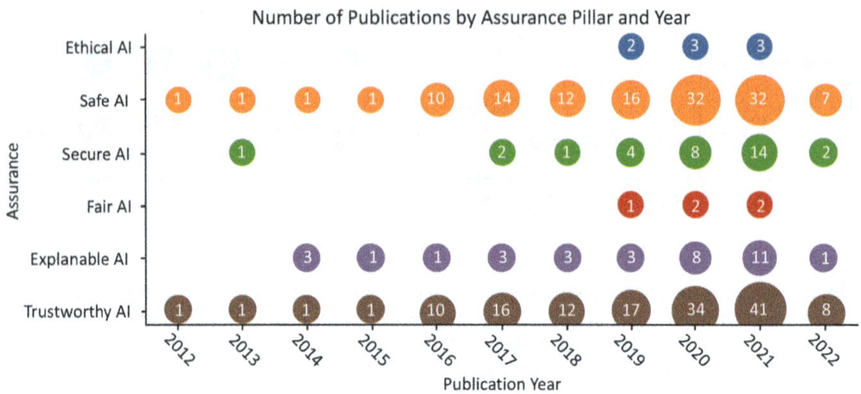

Fig. 3 The count of publications by year for the pillars of AI assurance. Papers are not confined to a single pillar, and some are counted for multiple. Most papers published since 2012, so older publications omitted from this figure

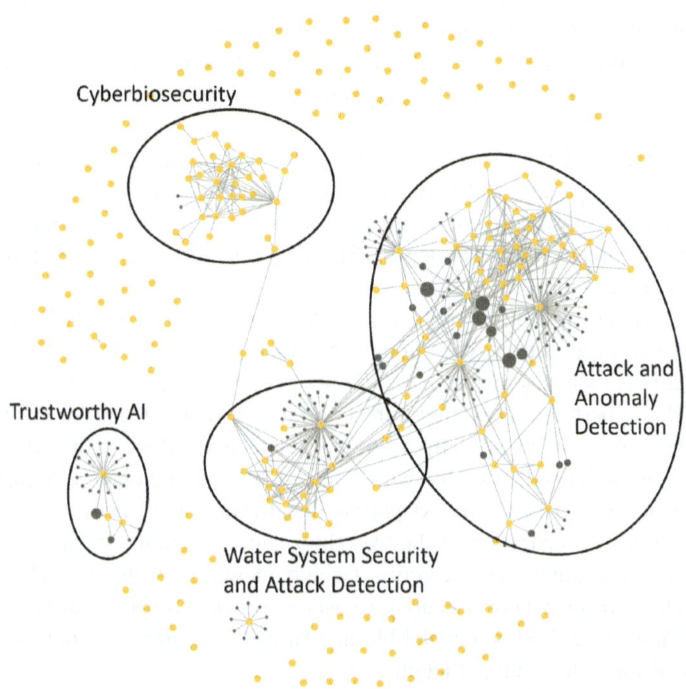

Fig. 4 Connected citation graph of the papers survey for this work. Yellow nodes are surveyed papers, gray nodes are other cited papers, and edges represent a citation between two papers. The cyberbiosecurity literature is relatively disjointed from the literature on water supply systems, AI assurance, and attack/anomaly detection. Graph generated using and courtesy of CitationGecko https://www.citationgecko.com/

3 AI Assurances for Cyberbiosecurity

In the introduction section, we described cyberbiosecurity as the intersection of life sciences and IT, and to be a little more specific it is the intersection of cybersecurity, cyber-physical security, and biosecurity [2, 9]. One of the best definitions we found is from Murch and DiEuliis [26], who defined cyberbiosecurity as the

> **understanding [of] the vulnerabilities** to unwanted surveillance, intrusions, and malicious and harmful activities which can occur within or **at the interfaces of commingled life and medical sciences, cyber, cyber-physical, supply chain and infrastructure systems**, and developing and instituting measures to prevent, protect against, mitigate, investigate and attribute such threats as it pertains to security, competitiveness, and resilience. (emphasis ours).

It is the vulnerabilities at the intersections of these cyber, physical, and biological systems that make cyberbiosecurity what it is, complex interactions between machines and biology that are open to disruption. This interaction creates unique vulnerabilities open to biological systems that make detection, attribution, and mitigation difficult in a timely manner [27]. Bernal et al. [28] recreated a Distributed Denial-of-Service (DDoS) attack using bacteria "engineered to act as biosensors" in a novel cyberbioattack, demonstrating the unique risks of the field and that traditional cybersecurity measures are not always adequate for cyberbiosecurity applications. The literature addresses these issues with a widespread call for action and collaboration—"We call for analyses and publications to fully scope cyberbiosecurity and identify a comprehensive strategy to establish the discipline's goals and objectives" [2] and others, as called out by [29] and seen in [26].

The purpose of our survey is to find how cyberbiosecurity intersects with AI assurance; there are applications that go beyond applying security to biological applications, and here we are interested in answering the question: what makes cyberbiosecurity different than cybersecurity for biology? It is the assurances a cyberbiosecurity system brings to the continuing function of the bioeconomy and relevant infrastructure. This is summed up well in the paper from Schmale III et al. [23], and while cyberbiosecurity is only mentioned briefly, the goal of the water supply system discussed is to ensure the safety of the drinking water from naturally occurring harmful algal blooms and cyber-attacks. Cyberbiosecurity "models must capture the physical dynamics of the system as well as the cyber-interconnections" [23].

Cyberbiosecurity systems that deal with supply chain and infrastructure systems have, or the potential to have, large impacts on the livelihood of people who rely on the system. All the residents of a city rely on its water distribution system to bring them water for drinking, cooking, and cleaning. A break down is not merely inconvenient but could be life-threatening, especially if the system is down for a long time or the water is contaminated. Even if AI is not considered for a cyberbiosecurity system, assurances are important to what cyberbiosecurity attempts to accomplish. AI brings an opportunity to add security or corrective actions in the event of any issues, and AI assurances validate their use for cyberbiosecurity applications. The end goal of any assurance (AI or not) is validating and verifying a system is working properly, so people have trust and adopt that system for use.

Turning back to the example of a water distribution in a city, suppose an AI monitors the system for cyber-attacks or natural anomalies (e.g., low levels from draught, bacterial growth, broken equipment, etc.) and takes corrective actions. If the hypothetical water distribution AI meets all the criteria listed in Sect. 1.1, then there is assurance that it behaves in a way that benefits everyone it impacts (people in the city who rely on the system providing drinkable water on demand) and minimizes unintended consequences. There is also some assurance the AI mitigates issues or threats to the system that would endanger city residents.

All these AI assurances are relevant to cyberbiosecurity, especially the secure assurance because the objective of cyberbiosecurity is "understanding the vulnerabilities" and developing "measures to prevent, protect against, mitigate, investigate and attribute such threats as it pertains to security…" [26]. There is also the human side of cyberbiosecurity, Perakslis [30] included the field in their list of public interest technologies, which are technologies that focus on public good. Further emphasizing the need for assurances to validate any AI systems involved with cyberbiosecurity and help promote their adoption in cyberbiosecurity. AI systems need to be trustworthy and explainable so people want to use them knowing they can rely on them to operate correctly, and because cyberbiosecurity systems focus on biological systems, safety is a big issue in order to ensure people impacted are not threatened by AI making a wrong decision. Ethics and fairness are a large part of the safety assurance too, as AI needs to ensure it does not favor some people over others, that it is not designed to favor its developers and investors over everyone else. Ethics and fairness are ensuring equal safety for everyone impacted.

4 AI Assurances for Open-Source Water Supply Testbeds

Open-source information engages more researchers allowing them to build better tools, frameworks, and operational systems such as Git, PyTorch, or Linux. Similarly, open-source testbeds allow the community to contribute, propose, test, and improve upon ideas. Lack of real-world water and CPS datasets prevented significant research in security of these systems [31]. Data from real facilities cannot be shared for both security concerns and lack of accurate ground truth, so the availability of reliable, open-source water testbeds is critical for research. Open-source datasets also allow hands-on experience and training scenarios needed for collaboration and understanding the security requirements of these systems [32].

Assurances for water systems closely match those of cyberbiosecurity systems discussed in Sect. 3. The two major assurances are the safety of the water quality and the security of the system's operations. Explainability is another key assurance for water systems, so we can understand how the water and AI systems operate in order to ensure consistent and safe water supplies. This emphasizes the importance of open-source datasets to help the AI research community better understand the operation of water systems and develop explainable and interpretable AI that is open to the water industry. Here we present some open-source water distribution and treatment system (as defined in Sect. 1.1) testbeds available to researchers across the world [33].

4.1 Secure Water Treatment (SWaT) Dataset

SWaT is a scaled down water treatment plant with real cyber and physical equipment to investigate cybersecurity research, which started in 2015 by Singapore University of Technology and Design [31]. The testbed consists of a six-stage water treatment process with modern-day components. The data collected from the testbed consists of eleven days of continuous operation, including seven days' worth of data under normal operation and four days' worth of data under attack. All network traffic, sensor, and actuator data was stored in the database.

4.2 Water Distribution (WADI) Dataset

Due to the success of the SWaT testbed, Singapore University of Technology and Design launched WADI in 2016 as an extension of SWaT to form a complete water treatment, storage, and distribution system [34]. Similar to SWaT, data collected for the WADI testbed consists of sixteen days of continuous operation, including fourteen days' worth of data under normal operation and three days with attack scenarios. All network traffic, sensor, and actuator data were collected.

4.3 Battle of the Attack Detection Algorithms (BATADAL) Dataset

The BATADAL dataset is not based on real-world data, though it is considered realistic since it was constructed using the de facto standard simulation tool for water distribution system modeling, namely the open-source Matlab software package EPANET [35]. EPANET is a Windows based software application for simulating and representing water distribution systems used world-wide by engineers and researches to design new water infrastructure, update existing water systems, and develop more efficient solutions to solve water quality problems. The BATADAL dataset was constructed for a competition to compare the performance of algorithms for the detection of cyber-attacks on water distribution systems. BATADAL simulates a fictional C-Town water distribution network, first introduced for the Battle of the Water Calibration Networks by Ostfeld et al. [36]. C-Town is based on a real-world, medium-size network which contains 388 nodes, 429 pipes, 7 tanks, 11 pumps, and one actionable valve.

4.4 Modbus Penetration Testing Framework (Smod) Dataset

Laso et al. [37] created the Smod dataset was produced in 2017 to investigate how data and information quality estimation can detect anomalies and malicious acts in a CPS. The data were acquired using a cyber-physical subsystem consisting of liquid fuel or water containers, along with its automated control and data acquisition infrastructure. The data consist of temporal series representing five operational scenarios—normal, anomalies, breakdown, sabotages, and cyber-attacks—corresponding to fifteen different situations. To acquire the data, Laso et al. [37] used two tanks of different volumes for storage, one ultrasound depth sensor, four discrete sensors, and two pumps.

4.5 Digital Hydraulic Simulation (DHALSIM) Framework

DHALSIM is an upgraded framework of the BATADAL Framework, which uses the Water Network Tool for Resilience (WNTR) EPANET wrapper to simulate the behavior of the water distribution systems [38]. DHALSIM uses Mininet and MiniCPS to emulate the behavior of the Industrial Control System (ICS) controlling a water distribution system. This means that in addition to physical data, DHALSIM also provides network captures of the Programmable Logic Controller (PLCs), Supervisory Control And Data Acquisition (SCADA) server, and other network and industrial devices present in the system. Similar to BATADAL, DHALSIM can be integrated into a C-Town Network, using a Mininet network that connects the C-Town PLCs and SCADA servers through Local and Wide Area Networks (LANs and WANs). In DHALSIM, each ICS equipment is a Mininet node running a script that represents the behavior of such equipment. In the C-Town network PLCs have private Internet Protocol (IP) addresses and NAT and port forwarding is used to connect the LANs.

4.6 Datasets Comparison

Figure 5 compares the number of total citations (labeled "General Citations") to the number of cyberbiosecurity citations (labeled "Cyberbiosecurity Citations") for the five datasets above. We obtained the number of cyberbiosecurity citations and general citations by counting the numbers of papers citing these datasets in our survey and by the count of citations from Google Scholar, respectively. We see the

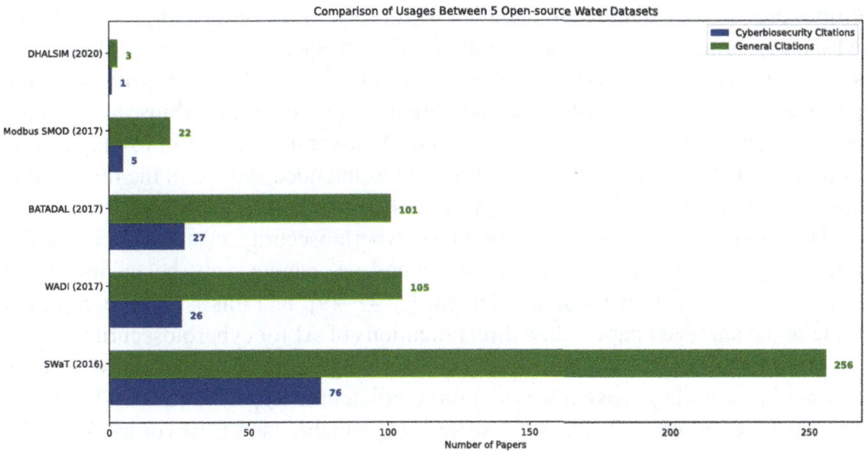

Fig. 5 Comparison between five open-sources water datasets in term of data usage

SWaT dataset is used the most, while DHALSIM dataset is used the least in both types of citations. This difference could be explained due to the early deployment of the SWaT dataset and the continuing collection and publishing of more data to that dataset by the University of Singapore in the years since its initial release. Although SMOD, BATADAL, and WADI are all water distribution systems published in 2017, the SMOD dataset is used significantly less. This could be explained by the scale of the datasets, specifically, both BATADAL and WADI simulate water distribution systems of large towns with multiple sensors, nodes, pipes, and a large recording time. On the other hand, SMOD only simulates a two-tank system, although SMOD is focused on different attack and anomaly scenarios than BATADAL and WADI. This shows that the research community prefers a dataset that can simulate a large scale, high quality real-world water distribution systems (WADI and BATADAL) and water treatment plant (SWaT) as benchmarks for model development.

5 AI Assurance Pillars

AI offers both opportunity and risk to cyberbiosecurity systems. It has the potential to detect and mitigate cybersecurity threats [2, 39–42], but at the same time offers an avenue for attacks [43–45], such as "poison" and "evasion" attacks on data or "inversion" attacks on AI models [43]. The current state of the cyberbiosecurity literature, however, focuses more on creating awareness and calls for collaboration to mitigate security threats rather than discussing the direct use of AI or AI assurance.

This supposition is not uniform, as Reed and Dunaway [40] praised the use of AI to "assist decision making… through the identification of cyberbiosecurity vulnerabilities and by providing recommendations for their elimination and/or mitigation." AI already brings a lot of benefit to the field of cyber and cyber-physical security, so the extension to cyberbiosecurity seems inevitable. However, with different physical, biological, and safety considerations required for cyberbiosecurity, there are no guarantees of success. This is where AI assurances come in to play a role, as they can help validate AI systems function as intended and aid in the responsible adoption of AI for the field of biology [1, 2, 46].

The multifaceted issues and solutions cyberbiosecurity systems face require interdisciplinary teams [47]. Solutions, therefore, cannot only be technical but require just as much of a human element [2, 47–49], and this is a more common topic in the surveyed papers than direct mentions of AI for cyberbiosecurity.

Assurances aid the adoption of AI by evaluating them for the benefit of humans and not because they make a solution more efficient, cheaper, or faster. The pillars of assurance are ethical, fair, safe, secure, explainable, and trustworthy. With the exception of secure, they are completely human focused. Clark et al. [48] claimed that cyber-defense is comprised of three aspects: technology, people, and physical protection and that these applications rely on people merging their knowledge rather than solely relying on automation. AI assurance is the way of merging the technological solutions of AI with the human values of the people within the cyberbiosecurity ecosystem. Aguilar et al. [49] argued a more holistic approach is required to solve the issues with the bioeconomy, one that includes "science, technology, economy, environmental issues, rural and industrial development, regulatory processes and social sciences."

5.1 Ethical and Fair AI

The most important question we can ask about AI is whether it works as intended or not. If not, how bad can the results be? And what kind of measures can we take in case of such a failure? In March of 2018, "an autonomous car operated by Uber—and with an emergency backup driver behind the wheel—struck and killed a woman on a street in Tempe, Arizona. It was believed to be the first pedestrian death associated with self-driving technology" [50]. This incident is a crucial example of when AI fails to make a safe decision. Although writing detailed contracts can legally reduce a manufacturer's liability, it might be morally unethical for the company to avoid legitimate liability.

With the growth of AI there are ethical and legal concerns regarding technology in areas, including how we can eliminate AI biases, ensure privacy, facilitate safety, and much more. AI should be made trustworthy, should be created and used with "an ethical purpose," and created to do good in society, but there are lots of questions that come up with AI and robots, such as if we "[assume that] the robots cannot be morally responsible—who will be responsible?" [51]. Furthermore, AI is already used in automated decision-making, and in high-stakes scenarios their decisions can be impactful. One issue with algorithmic decisions is bias, which can be "cognitive biases of programmers," "unrepresentative datasets used for training," or "bias in the data used to make the decision" [51]. It is just as important to start with ethical considerations before AI is designed, let alone deployed, to ensure it is making fair and ethical decisions [51].

The concerns of inclusive, equitable, and correct decisions from AI are not solely left to industry, in fact it is gaining more ground in research from large tech companies and academics. The ambiguity of "fairer" decision-making systems, however, leaves fair AI as a broad open ended question without a real solution. Besides defining what "fair" means, researchers must deal with how to train systems for fair decisions or the fact that systems made fairer for one group can result in bias against another.

One of the most common reasons for biased results is the under-representation of certain groups within a dataset. Increasing the representation of that group, for example, oversampling a certain demographic in certain areas predominantly held another, may be a solution to rectifying the data. When it is not possible to modify or edit data, the objectivity of the decision-making process can be resolved by adjusting the AI algorithm. For algorithms that learn from discriminatory practices it is possible to change the internal weights in a way that makes decisions more neutral. It is also possible to modify the decisions of AI algorithms directly to create more equitable outcomes.

In some instances, it is not the lack of representation, but rather, the over-representation on certain groups that can created biased results. In such fairness related cases, openness in the development and deployment of AI is required [52–55].

In short, it is possible for AI technologies to be more equitable, but this requires the cooperation of different stakeholders and a lot of work. Arnold et al. [56] pointed out the importance of ethical decision-making while raising critical questions for every AI developer. The authors also refer to relevant answers for these questions from the literature, making this article serve as a guidebook for comprehensive AI assurance deployment.

Laplante et al. [57] investigated the causes that lead to unethical AI and its potential results. The authors saw the main reason as unbalanced or underrepresented data. [57] also emphasized the importance of ethical considerations for AI over its importance for classical software.

Zicari et al. [58] provided a framework to assess the trustworthiness of AI systems. The parameters the authors investigated include, but are not limited to, ethical and fair AI. The article provided a lifecycle to ensure ethics in AI decision-making. The authors emphasized the required absence of conflict for a reasonable assessment of ethical AI.

Grady et al. [59] proposed an epistemic, ethical analysis framework; as the name suggests, the authors proposed ways to detect and analyze ethical issues in cyber-physical infrastructures including, but not limited to, water treatment and distribution systems. The article investigated the importance of ethical decision-making and the roots of the problems in this topic.

Freeman et al. [46] proposed a framework to investigate AI using AI assurance metrics. The authors brought together many AI measures on common ground in this work, challenged the readers, and provided answers to these AI assurance problems.

Calvo et al. [60] investigated the algorithmic, environmental, and human impact assessment of AI systems. They proposed a measurement algorithm called Human Impact Assessment for Technology (HIAT) and discussed ways to build trust into the algorithm using this method.

5.2 *Safe AI*

One goal of cyberbiosecurity is ensuring the safety and well-being of those impacted by the system. This stems from the biosecurity aspect of the field [61] but naturally extends to any form of safety ensured by systems like water and food supply chains (and agriculture [62] as an aspect of these supply changes). The goal of the safe AI assurance is for AI to guarantee some level of safety to ensure the life and well-being of anyone impacted by the AI. These two forms merge to, as Mueller [22] described cyberbiosecurity, develop, validate, and implement safety measures.

Physical consequences, including harm to humans, are what separates cyber-biosecurity from most forms of technological security. Walsh and Streilein [43] pointed out that "a successful cyber intrusion within the bioeconomy may yield a result that causes physical harm, something generally associated with biosafety and biosecurity but not cybersecurity." Any interference with the bioeconomy has potential to harm, and while Walsh and Streilein [43] focused on illicit interference, this extends to unintentional interference as well. It is the ability for any cyberbiosecurity system to cause physical harm, intentional or otherwise, that safe AI and safety assurances need fortifying.

Water and food supply systems are a prime example of a cyberbiosecurity systems where safety is a priority. Quality and supply from the system impact everyone in a service region, and both are affected by natural anomalies (algal blooms, weather, draughts, and floods) or cyber-attacks. Water supply systems require constant monitoring and threat mitigating to ensure safety of the water quality and supply [23, 63–79]. On the other hand, food supply relies less on technological innovations, whereas water systems have standardized the use of SCADA systems [48], food supply and agriculture have seen a more limited and hesitant adoption of technology, especially for small-scale farmers [9]. A more standardized approach to tech adoption helps by "securely sharing and interpreting data across sectors and identifying cyberbiosecurity risks," ultimately improving food supply chains by designing "agricultural and food systems to better meet consumers' need and protection of life science data" [80]. Data privacy is also a concern any time personal health information may be involved with genomic databases with the potential for cyber-attacks on lab automation [81, 82].

We found in the literature that water and wastewater sectors vary greatly in size, complexity, organization, security protocols, available resources, and even in imposed regulations [47, 48]. While the end goal of each water system is to supply clean water on demand, the approach each system takes is unique and requires different considerations, including adopting security measures specific to their organization [48]. This means that each system needs to take unique considerations to ensure to the quality of the water and consistency of the supply, posing a challenge to the field as a whole because standardized approaches to safety cannot be developed or relied on for all situations.

The bioeconomy, too, consists of large and complex systems that intertwine and connect, and it "harbors unique features that have to be more critically assessed for their potential to unintentionally cause harm to human health or environment" [22]. Water systems supply water to farms that impact agricultural production which in turn impacts food supplies to retails (grocery stores), prices, and the ag-economy. Any hiccup along the way can have unforeseen consequences. The complexity, however, makes it difficult for any one person, or even organization, to understand what consequences their actions have. This means that changes for the sake of mitigating external threats could lead to unintended consequences [39]. Cyberbiosecurity cannot focus solely on cybersecurity and attack detection or, as mentioned in the previous section, on monitoring natural phenomena as interference. We need to implement assurances to guarantee the safety of a system (e.g., quality of water or food for human consumption) at all times.

AI and other emerging technologies' reliance on data provides both benefit and potential harm. The concern of unintentional errors can arise in the data used for Safe AI. Caswell et al. [83] pointed out the potential issues of errors in biological databases, but the concern is applicable to any data-driven analysis in cyberbiosecurity. While referring to synthetic biology, Li et al. [84] emphasized that unintentional risks can lead to food scarcity despite the efforts of biosafety and biosecurity to provide more. Similar concerns for unintended consequences of dealing with biological data have been expressed in [84, 85]. As these technologies are implemented more into cyberbiosecurity systems (such as precision agriculture) more emphasis needs to be placed on quality assurance of the data and safety assurances for the final product.

5.3 Explainable AI

In the introduction section we defined explainable AI as AI that can "explain, or be interpreted, to understand why it came to a decision or how the algorithm works." Here, we expand this to include cyberbiosecurity systems in general because that is the environment the AI system operates in, the AI's behavior is dependent on the larger system, and the end user needs to understand both in order to operate the system correctly. Even if a cyberbiosecurity system does not incorporate AI, human understanding is crucial to its operation. Therefore, we expand the definition of explainability to include "the process of making complex systems human intelligible."

The literature surveyed often mentions the lack of training, understanding, and even awareness of cyberbiosecurity and cybersecurity risks as a vulnerability. This means a lack of knowledge and human understanding of threats, how to recognize them, and what to do about them is one of the biggest hurdles for the cyberbiosecurity field to overcome. Accordingly, a framework for making these complex systems understandable in order to avoid and mitigate risks is recommended. However, even in the biotechnology and cybersecurity realms "cyberbiosecurity is not well-known or understood" [86] and there is "a failure to recognize vulnerabilities" [40]. This lack of awareness is detrimental because cyberbiosecurity relies on understanding the vulnerabilities, threats, and risks to mitigate impacts [22, 26]. Even with the conventional cybersecurity approach, a "good cybersecurity plan is understanding the threat and establishing cybersecurity governance protocols" [47]. The mentioned approaches are not fully implemented or are done so inadequately resulting in "the failure of individuals to identify and address cybersecurity vulnerabilities" in cyberbiosecurity systems [40].

Part of this lack of awareness is from lack of education or training available in cyberbiosecurity [87]. Drape et al. [29] surveyed researchers from the agricultural sector attending a cyberbiosecurity workshop and found that no participants had cybersecurity training or resources, and attendees were uncertain about obtaining training or implementing solutions. Despite the research going into cyberbiosecurity vulnerabilities, there is no "one size fits all" solution, the difference in educational resources for agricultural security varies from county to county in the USA [29]. It is no stretch of the imagination to see that disparities exist country to country for agriculture, water supply, and food supply chains. These sectors are critical everywhere around the world, but the resources for cyberbiosecurity are not equally distributed, so a solution needs to be general and easy to implement and maintain. Authors in Duncan et al. [88], by focusing on the US food supply chain, stated that "this gap in education and training increases risks to the domestic [U.S.] food supply chain and the ultimate mission of securing the U.S. and global food supply."

Lack of understanding is a significant risk for any cyberbiosecurity system, but especially for small farms where available knowledge and resources are less than large infrastructure organizations (e.g., utility companies, and industrial farms). More needs to be done to explain cyberbiosecurity as a concept and raise awareness of the vulnerabilities it creates. Richardson et al. [2] point out that as agricultural becomes more reliant cyber-enabled systems the security of these systems is "unclear from a cyberbiosecurity perspective." This is at the same time that technology is increasingly incorporated into water supply and food supply systems, creating similar vulnerabilities [9, 34, 43, 48, 89–91]. Although, Reed and Dunaway [40] were optimistic that technology would bring solutions without any vulnerabilities.

As the size of an organization increases (e.g., industrial farms, utility water supplies, and the bioeconomy) so does complexity and difficulty in understanding how the system operates. Lack of understanding of minute details and interconnectedness are a vulnerability, as even changes to mitigate external threats can lead to unintended consequences [39]. Imagine updating security software and a bug prevents water tanks in a system form relaying fill levels to the central control. More effort needs to be placed on understanding how the system actually operates and how best to explain that operation to the people it matters most.

This approach needs to be done on a case by case basis, as the variability in each individual systems differs. Germano [47] and Clark et al. [48] both point out that differences among organizations and utilities in the water and wastewater sectors include size (employee count and water processed), management, available resources, regulatory oversight, and even security protocols. These differences make a unified approach to cyberbiosecurity in the water sector unfeasible, as each organization or utility needs to build their own approach to match their unique operation and threats. The water distribution system for a large city is going to vary in size, available resources, and security measures from that of a small rural county. This disparity exists in the other sectors of the cyberbiosecurity as well, no two farms, food supply chains, or any other large-scale infrastructure are going to be the same as the issues each one deals with greatly varies. Understanding the needs and shortcomings of each system is critical for cyberbiosecurity.

Awareness of threats and how cyberbiosecurity systems operates is a form of threat mitigation, and several papers make the case for simply making people aware of the risks [26, 44, 45, 47, 92, 93]. Even something as simple as "understanding the threat and establishing cybersecurity governance protocols" is all it can take to protect these systems [47]. That said, understanding these complex systems is no trivial tasks. Both cyberbiosecurity and AI can benefit from the explainability assurance to make them human intelligible. Explainable AI systems are easier to understand how they operate and therefore understand what might negatively impact the system cyberbiosecurity systems, on the other hand, could be explained via machine learning techniques like clustering or even learning a Directed Acyclic Graph (DAG) of the data like Lin et al. [94] did for the SWaT dataset.

The next step for building understanding of cyberbiosecurity systems is through education and training. Richardson et al. [87] call for a standardization of the training process, in the same manner as biosafety and cybersecurity, through credentialing. They also called for integrating training into existing programs or relying on existing programs, as did [29], while others merely made a call for increasing education and awareness [95]. Another theme that emerged in the literature was a need for training across sectors in the water and agricultural industries, so employers training employees [45, 47], cross-sector training [80, 96, 97], government or university curated resources and training, both formal and informal [48, 88, 97], and even war-gaming [98].

5.4 Secure AI

Undoubtedly, one of the most important factors in ensuring the security of water distribution systems is to detect anomalies that may occur in these systems or malicious attacks that may come from adversaries. Water treatment and distribution systems have been increasingly targeted by cyber-physical attacks in recent years [99]. This is partially due to the expansion of the Internet of Things (IoT) and proliferation of AI increasing the digitization of the decision-making processes and creating an adversarial attack opportunity following recent development in the machine learning field, which led to black-box adversarial methods that work well even with limited information [100].

The Kemuri Water Company (KWC) [101] attack in 2016 is a very important example of the risk these national infrastructures are under. The attack has resulted in more than 2.5 million records stolen, but more importantly, the attackers were able to change control data to manipulate the water supplied to the area. The attacks were halted before any public health damage occurred, nonetheless, it showed how vulnerable these infrastructures are and how important it is to ensure their safety.

Another recent, important incident was the Florida Water Supply hack in 2021 [21]. In this malicious attack, the hacker was able to gain remote access to the PLC (Programmable Logic Controller) unit that controls the sodium hydroxide level (also known as lye) of the water supplied to more than 15,000 residents in Tampa, Florida. The hacker was able to increase the amount of sodium hydroxide content of the water by 110 fold. Fortunately, the attack was mitigated before the poisonous levels of chemical diffused into the distribution network.

Both of these incidents show how important it is to detect any anomaly or malicious attacks early to mitigate, or hopefully prevent, any damage. Taormina et al. [35] investigated the vulnerabilities of these critical infrastructures in-depth in their research.

Pasqualetti et al. [102] investigated the detection and identification of CPS attacks from two different perspectives in their 2013 paper. They categorized the monitoring limitations from "graph-theoretic" and "system-theoretic" while proposing a mathematical framework for the problem's solution. The framework they proposed considers the CPS as a linear time-invariant descriptor system. They then defined a comprehensive set of assumptions and equation systems to measure and detect the corrupted signals in the system. They have also made a theoretical quantification of the limitations of both monitoring approaches to determine undetectable and unidentifiable attacks boundaries. Their paper is also one of the earliest attempts to formally describe the attack detection against CPSs and in this sense, its importance in the field is substantial.

Machine learning is a powerful and important tool for ensuring cyber-physical security. It is not surprising to see deep learning, more specifically Long-Short Term Memory Recurrent Neural Networks (LSTM-RNN), as efficient solutions to a problem with a time-dependent and high sequential relations such as attack detection [103]. Goh et al. [104] used the SWaT dataset [105] as a small-scale representation of a water treatment plant to detect anomalies and identify the sensors affected by this anomaly. They proposed to use the Cumulative Sum (CUSUM) method to mitigate the effects of an extremely unbalanced distribution of positive and negative classes (millions of negative samples to only thousands of positive samples with a sequential dataset). The SWaT dataset is a comprehensive and very important dataset for cyber-physical security research and the contributions of the authors and supporting organizations to the field should not be left unacknowledged.

Inoue et al. [106] applied another deep learning approach in their 2017 paper. The authors used a Deep Neural Network (DNN) to evaluate the Support Vector Machine (SVM) method's performance for anomaly detection problems. The paper also made a side-by-side comparison of the two models while discussing their advantages and disadvantages. Unlike Goh et al. [104], the authors did not address the data imbalance in the paper. The researchers used the SWaT dataset and the simulation to test the models.

BATADAL is a planning and management competition for Water Infrastructures and it takes place as part of the Water Distribution Systems Analysis Symposium. This competition presents an imaginary C-Town as a water distribution network dataset to detect the real-life size and real-time, simulated data from this town (SCADA) [107]. The paper includes seven well-performing solutions to the problem on this dataset from the competitors. Others (Aghashahi et al. [108]) used a two-stage approach to solve the anomaly detection problem. In the first stage they make a feature extraction, and in the second stage they use a supervised classification method, Random Forests, to detect attack instances.

Brentan et al. [109] proposed a statistical approach to the problem. They used the sectioned nature of the problem environment and trained Recurrent Neural Networks (RNNs) to learn each district's normal behaviors and then calculated the deviation from these expected normals to measure the anomaly levels on the system.

Chandy et al. [110] used a similar two-staged approach to Aghashahi et al. [108]. Chandy et al. [110], however, first make a detection of the anomaly and then confirm or reject this detection is with a second model, a Convolutional Neural Network (CNN) Auto-Encoder, by calculating reconstruction probabilities.

Giacomoni et al. [111] proposed another two-stage approach. In the first stage, the authors created a set of rules and calculated the integrity of the rules for each instance. In the second stage, they analyzed the dataset to calculate certain thresholds of normalcy. They also proposed using Principal Component Analysis (PCA) and convex optimization routine to perform this analysis [112].

Abokifa et al. [113] proposed a three-stage model and they classified different types of attacks on each stage of the process. In the first stage, the authors used statistical methods to detect local outlier events. In the second stage, they introduced a neural network to the process to detect operational outliers. In the third stage, they focused on the global scope to detect events that might affect more than one aspect of the system with PCA.

Pasha et al. [114] introduced another three-stage method for anomaly detection. The first stage checked the consistency of the underlying rules of the water distribution system. The second stage checked each component for behavioral patterns to see if the system is following the normal patterns it is supposed to do. If any anomaly is detected in the first two stages, the third stage confirms the detections by comparing the estimations of the system made by the method.

Housh and Ohar [25] used EPANET to create a simulation of a water distribution system's behavior to calculate the difference between the SCADA and the expected values from the simulation to detect and locate anomalies in the systems. Housh and Ohar [115] also used a similar approach to detect contamination attacks against water distribution systems with successful results.

Taormina et al. [107] have comparatively investigated all these proposed approaches, and many more, are discussed along with the advantages and disadvantages of the models. Even though the methods are very diverse, one common factor should not be unnoticed: each of the major competitors followed a direction of first discovering underlying behavioral principles of the system in some manner and then proposed ways to measure the diversion from these principles in anomalous scenarios.

The BATADAL competition provides immense contributions to the cyber-physical security field by providing a great dataset to the researchers as well as creating a valuable comparative environment for all the approaches to provide assurances methods for cyber-physical security [107], an approach (competitions) that proved successful in other areas of AI. Kravchik and Shabtai [116] investigated the attack detection problem from an ICS perspective in their 2018 paper. They used the SWaT dataset to train CNN and Long-Short Term Memory (LSTM) models to compare their effectiveness to detect anomalies. The experimental results showed that 1D CNNs can outperform RNN and LSTMs in more complex multivariate tasks.

Umer et al. [117] investigated attack detection from a distributed system. In their work, they separated the endeavor into two categories: "design-centric" and "data-centric," while proposing a model for each category. The research used the SWaT dataset [105] as a small-scale representation of a water treatment plant. The methods they proposed utilize Association Rule Mining (ARM). They also compare the advantages and disadvantages of the two approaches proposed in the paper.

Junejo and Goh [118] proposed a behavior-based machine learning approach for the detection and classification of cyber-physical attacks. Their approach promised a low false-positive rate, which some of the other approaches discussed earlier suffer from, and still provided high recall and precision. They used the SWaT dataset to evaluate the effectiveness of nine different algorithms from supervised machine learning literature ranging from Bayesian networks, naive Bayes, logistic regression, neural networks, SVM, and more while making comparisons between models for advantages and disadvantages.

Adepu and Mathur [119] proposed a Single-Stage Multi-Point (SSMP) type of attack with a distributed detection method. Even though they focus on single-stage attacks in their paper, the authors noted that they found it more effective to detect this kind of attack using the information from neighboring stages. The researchers used the SCADA dataset to create two invariants: State-Dependent (SD) and State-Agnostic (SA). Later the authors combined both invariants to create a more efficient tool for distributed detection problems.

In another paper, by Adepu and Mathur [120] authors used the SWaT dataset to investigate ways to improve cyber-physical security and attack detection problems by asking the following questions: "What attacker and attack models should be used to understand the behavior of a CPS?", "How do cyber-attacks impact a specific CPS with respect to the number of actuators affected, state of a CPS when the attack is launched, and duration of the attack?", and "Given the response of a CPS to one or more cyber-attacks, how does one design attack detection mechanisms using the physical properties of the system?". While trying to answer these questions with experimental results the authors disclaimed the generalizability of their findings and stated that this research only targets the SWaT testbed.

This disclaimer shows a very important direction that requires more attention in the field, which is the generalization of the proposed methods since almost all of the methods we discussed so far require prior knowledge of the attack samples to be effective in the first place. The need for generalizability of the proposed approach is the utmost importance since solutions cannot wait until the attacks happen on the real systems to collect the necessary data to train the models.

Adepu and Mathur [121] must have seen this problem as well, as they tried to address it in their next work with a case study of their earlier distributed attack detection proposal [119]. Adepu and Mathur replicated real-life scenarios to test their improved attack detection mechanism and shared their findings with the strength and weaknesses of the model with an in-depth discussion [121].

As we pointed out earlier, fast adoption of automation and networking technology does not come without drawbacks. Al-Abassi et al. [122] tried to remark these issues and address the vulnerabilities created by another attack detection method while promising generalizability on the way. The researchers propose a combined model of DNN and Decision Tree with results that outperformed most of the conventional machine learning models including DNN and Random Forest. The authors also addressed the imbalanced class distribution and effective performance of the proposed approach with experimental results.

5.5 Trustworthy AI

AI is used in an increasing number of different systems, for example, autonomous vehicles, search engines, recommendation systems, medical imaging [123], public health [124], and others. It appears well-developed, yet there are still a lot of issues that need to be addressed and discussed, especially when it comes to the question can AI be trusted in "these scenarios that have life-critical consequences?" [125]. The foundation of societies, economies, and sustainable development is based on trust. If there is no trust the whole societal system would not grow or be stable [126, 127], and the same applies to cyberbiosecurity applications. Inderwildi et al. [128] discussed the impact of intelligent CPSs in energy provision and gave policy recommendations to lower potential risks. The same applies to AI systems, the idea of trustworthy AI is to build trust between users, developers, and the system itself [129].

Trust is a concept that is difficult to build, and trust in AI is even harder to address. The "black-box" characteristic is one of the most important reasons of mistrusting AI [130]. It is hard to build trust without knowing why the system makes its decision. We need to be able to explain the results, and this leads to the importance of explainable AI (see Sect. 5.3). Another situation where trust in AI faces scrutiny is ethical decisions, such as the trolley problem. What is the priority that the system should follow? Are there any guidelines to follow? There are so many different questions to address in order to build trust.

In recent years, a significant amount of research on trustworthy AI has been conducted in different academic and industry areas (see Fig. 3). Each study focused on different aspects of trustworthy AI, for example, [131] focused on government guidelines, which advise how to establish a trustworthy AI system through rules and regulations, and other studies focused on the computational aspect of achieving trustworthy AI [132–137]. Most of the research agrees that trustworthy AI systems should include a set of properties: reliability, safety, security, privacy, availability, usability and can be extended to the following dimensions: accuracy, robustness, fairness, accountability, transparency, interpretability/explainability, and ethics [56, 125, 126, 129, 131–133, 138–141].

Trust is a complicated concept that combines numerous factors, and different researchers from various backgrounds would also see trustworthy AI from a diverging perspectives. Liu et al. [132] defined trustworthy AI from three perspectives: technical, user, and social. The system should focus on accuracy, robustness, and explainability from a technical perspective; while it should focus on availability, usability, safety, privacy, and autonomy from the user's perspective. Whereas from the social perspective, there should be a guideline or regulation regarding legality, ethics, fairness, accountability, and environmental-friendliness. To have more clear guidelines for accomplishing trustworthy AI, the EU established the High-Level Expert Group (HLEG) to provide ethical guidelines, not just principles to follow but also concrete operational steps that allow an AI developer to examine when building and deploying an AI system [131]. Zicari et al. [58] proposed a state-of-the-art process to evaluate the trustworthy AI based on applied ethics called "Z-Inspection," which is also first process in practice that HLEG defined to evaluate the trustworthiness of AI. Z-Inspection consists of three processes: set-up, access, and resolve, and each phase breaks down into different aspects to examine whether the AI systems are trustworthy.

Toreini et al. [133] pointed out that there are various AI policy frameworks to follow from different nations and organizations, and categorize those objectives into eight qualities: privacy, accountability, safety & security, transparency & explainability, fairness & nondiscrimination, human control of technology, professional responsibility and promotion of human values. They further mapped these eight qualities with four principles, including fairness, explainability, auditability, and safety. The authors separate two main technologies of trustworthiness: Data-Centric Trustworthiness and Model-Centric Trustworthiness.

Liu et al. [132] stated "Trustworthy AI are programs and systems built to solve problems like a human, which bring benefits and convenience to people with no threat or risk of harm." They focused on six dimensions in achieving trustworthy AI including safety & robustness, nondiscrimination & fairness, explainability, privacy, accountability & auditability, and environmental well-being. Instead of focusing on policy framework or guidelines, they worked on specific computational solutions for each dimension for realizing trustworthy AI.

Li et al. [138] mentioned AI practitioners, including researchers and developers, should focus on pursuing system performance as the main goal, whereas this is not sufficient to reflect the trustworthiness of an AI system. Therefore, they proposed a methodology that takes the entire lifecycle of AI systems into consideration, from data management to model development, deployment, and all the way to monitoring and governance. For the future research direction, while adopting this systematic approach, there are side-effects due to increased learning time and slowed development by using this new approach.

We mentioned that the trustworthiness of AI is essential when it comes to AI systems related to life-critical consequences. There were incidences where critical CPSs came under attack [142] and affected the overall trust in CPSs. For example, an attack happened on a water treatment plant in Florida in 2021 and the level of sodium hydroxide in the water supply was increased over 100 times higher than usual [143]. There were also numerous cyber-attacks on Israel's water system in 2020 [144]. That exposes how vulnerable those CPSs are and the importance of the security of those systems [145–155]. There has been no lack of related research done in the area of anomaly detection in water system or its security challenges using machine learning methods [33, 107, 116, 156–190], statistical methods [191–198], or other tangential methods [106, 199–213].

Wang et al. [214] applied probabilistic model learning to probabilistically validate a real-world CPS. MR and Mathur [215] proposed "AICrit" to effectively detect anomalies in real-time with low false alarms. Another factor contributing to the complication of evaluating trustworthiness is that most of the research or review that discusses how to achieve trustworthy AI focuses more on social science topics, such as ethics and policy [59, 139]. Most of the frameworks or guidelines they proposed, however, do focus on the human factor. Uslu et al. [216] proposed a decision-making framework to manage Food-Energy-Water (FEW) resources. While developing the optimal solutions under different scenarios, they included humans in the framework to make the solutions more trustworthy. They introduced two new metrics, trust sensitivity and trust pressure, in the framework and used a game-theoretical tool to explore the relationship between trust sensitivity and the distance of community-desirable solutions.

6 Discussion

6.1 Attack Detection Models for Water Systems

Cyberbiosecurity attack/anomaly detection research in the literature mainly focused on three datasets SWaT, WADI, and BATADAL which have been introduced in Sect. 4. These three datasets have become field leading benchmarks. As a part of the survey, we have created tables for each dataset. In order to make a fair comparison, we have used the most commonly reported statistical metrics to rank models proposed by researchers for attack/anomaly detection problem. For SWaT (Table 1) and WADI (Table 2) datasets it was F-Score (also known as F-measure, more specifically F_1 score) and for BATADAL (Table 3) we have used S score defined by Aghashahi et al. [108] and listed S_{TTD} (Time Taken for Detection) as well. For each dataset, state of the art over the years has been marked with bold fonts on Tables 1, 2, 3.

Table 1 SWaT F1-Scores[a]

Authors	Model	F1-Score	Year
Ayas and Ayas [63]	**Modified DenseNet**	**0.9999**	**2020**
Alqurashi et al. [184]	MLP	0.9900	2021
Krithivasan et al. [217]	EPCA-HG-CNN	0.9805	2020
Xu et al. [218]	ATTAIN	0.9759	2021
Li et al. [219]	**MAD-GAN**	**0.9517**	**2019**
Kravchik and Shabtai [116]	**1D CNN**	**0.9200**	**2018**
Abdelaty et al. [220]	DAICS	0.8890	2021
Elnour et al. [221]	DIF	0.8820	2020
Kravchik and Shabtai [222]	AE Frequency	0.8730	2019
Kravchik and Shabtai [116]	1D CNN	0.8710	2018
Sapkota et al. [163]	CNN + LSTM w/ WT	0.8610	2020
Perales Gómez et al. [223]	MADICS	0.8510	2020
Lin et al. [94]	TABOR	0.8230	2018
Zizzo et al. [224]	LSTM	0.8170	2019
Shalyga et al. [225]	MLP	0.8120	2018
Li et al. [226]	GAN	0.8100	2019
Shalyga et al. [225]	CNN	0.8080	2018
Inoue et al. [106]	DNN	0.8030	2017
Faber et al. [227]	CNN 1D[b]	0.8000	2021
Inoue et al. [106]	**One-class SVM**	**0.7960**	**2017**
Shalyga et al. [225]	RNN	0.7960	2018
Inoue et al. [106]	SVM	0.7960	2017
Faber et al. [227]	USAD	0.7900	2021
Faber et al. [227]	CNN 1D	0.7800	2021
Goh et al. [104]	LSTM-CUSUM	0.7754	2017
Chakraborty et al. [228]	Random Forest	0.7700	2021
Li et al. [229]	GAN-AD	0.7500	2018
Toe et al. [70]	MARS	0.7480	2020
Faber et al. [227]	LSTM-VAE	0.7200	2021
Shalyga et al. [225]	RNN	0.6900	2018
Sapkota et al. [163]	CNN	0.6500	2020

[a] *Disclaimer*: These results are not validated as a part of this research

Table 2 WADI F1-Scores[a]

Authors	Model	F1-Score	Year
Xu et al. [218]	**ATTAIN**	**0.7444**	**2021**
Goh et al. [104]	**LSTM-CUSUM**	**0.6595**	**2017**
Li et al. [219]	MAD-GAN	0.5945	2019
Faber et al. [227]	CNN 1D	0.5400	2021
Faber et al. [227]	CNN 1D	0.5200	2021
Faber et al. [227]	USAD	0.4300	2021
Faber et al. [227]	LSTM-VAE	0.2800	2021

[a] *Disclaimer*: These results are not validated as a part of this research

Table 3 BATADAL S Scores[a]

Authors	Model	S Score	S_{TTD} Score	Year
Brentan et al. [230]	**Statistical analysis**	**0.9730**	**0.1900**	**2021**
Housh and Ohar [25]	**MILP**	**0.9700**	**0.9650**	**2018**
Abokifa et al. [160]	ANN and PCA	0.9660	0.9840	2019
Abokifa et al. [113]	**ANN**	**0.9490**	**0.9580**	**2017**
Ramotsoela et al. [231]	QDA	0.9400	0.9500	2019
Tsiami and Makropoulos [232]	TGCN	0.9310	0.9340	2021
Giacomoni et al. [111]	PCA	0.9270	0.9360	2017
Ramotsoela et al. [231]	MD	0.9100	0.9000	2019
Ramotsoela et al. [231]	iForest	0.9000	0.8600	2019
Brentan et al. [109]	RNN	0.8940	0.8570	2017
Ramotsoela et al. [231]	LOF	0.8700	0.8500	2019
Ramotsoela et al. [231]	SOD	0.8600	0.8300	2019
Mahmoud et al. [233]	SVM	0.8200	0.8400	2022
Mahmoud et al. [233]	3NN	0.8200	0.7500	2022
Mahmoud et al. [233]	RForest	0.8200	0.7800	2022
Mahmoud et al. [233]	XGBoost	0.8200	0.7500	2022
Mahmoud et al. [233]	BOSS	0.8200	0.7100	2022
Chandy et al. [110]	Convolutional variational auto-encoder	0.8000	0.8300	2017
Gjorgiev and Gievska [193]	VAE-D	0.8000	0.9750	2020
Gjorgiev and Gievska [193]	VAE-D-C	0.7780	0.9870	2020
Gjorgiev and Gievska [193]	LSTM-VAE-C	0.7780	0.9990	2020
Pasha et al. [114]	Statistical analysis	0.7730	0.8850	2017
Gjorgiev and Gievska [193]	LSTM-VAE-2E-C	0.7610	1.0000	2020

(continued)

Table 3 (continued)

Authors	Model	S Score	S_{TTD} Score	Year
Mahmoud et al. [233]	5NN	0.7600	0.6430	2022
Choi et al. [234]	SVM	0.7540	0.7220	2020
Gjorgiev and Gievska [193]	VAE-ReEncoder	0.7520	0.9350	2020
Mahmoud et al. [233]	7NN	0.7500	0.6345	2022
Ramotsoela et al. [231]	Naive Bayes	0.7500	1.0000	2019
Choi et al. [234]	ANN	0.7490	0.7590	2020
Gjorgiev and Gievska [193]	LSTM-VAE	0.7350	0.9790	2020
Mahmoud et al. [233]	1NN	0.7300	0.5720	2022
Gjorgiev and Gievska [193]	VAE-ReEncoder-C	0.7260	0.9400	2020
Gjorgiev and Gievska [193]	CNN-VAE-C	0.7130	0.9310	2020
Ramotsoela et al. [231]	OSVM	0.7100	0.6900	2019
Ramotsoela et al. [231]	LDA	0.6700	0.6500	2019
Gjorgiev and Gievska [193]	LSTM-VAE-2E	0.6640	0.8200	2020
Choi et al. [234]	ELM	0.5910	0.9410	2020
Aghashahi et al. [108]	RForest	0.5340	0.4290	2017
Gjorgiev and Gievska [193]	CNN-VAE	0.5230	0.5430	2020
Choi et al. [234]	5NN	0.4180	0.3230	2020

[a] *Disclaimer:* These results are not validated as a part of this research

Throughout the years efficiency of the neural network based models have drastically increased over numerous problems and attack/detection is one of them as well. Looking at highest ranked models on the SWaT F1-Scores Table 1, it can be seen that deep learning had a huge impact on the problem and following the success of Inoue et al. [106] with One-class SVM, in last 4 years breakthroughs were achieved using Deep Learning models Kravchik and Shabtai [116], Li et al. [219] and Ayas and Ayas [63]. This dominance can further be verified with the successful state of the art models developed by Goh et al. [104] and Xu et al. [218], once again using DNN models.

When it comes to the BATADAL dataset the picture slightly changes. Neural Network based models are still very effective on solving attack detection problem with BATADAL as well but they are not as dominant as they are with the other two datasets. Various types of approaches to the problem from many researchers provide a great understanding of the chaotic nature of data-driven problems on large physical systems. Dynamical essence of these systems requires researchers to approach the problem from many angles to ensure the models they would create to be trustworthy and secure. Some of the most successful researches to achieve these feats were, Abokifa et al. [113], Housh and Ohar [25] and Brentan et al. [230] as the state of the art holders.

6.2 Assessing the Cyberbiosecurity Literature

In this section, we discuss cyberbiosecurity further because it is a new discipline and there are different takes on exactly what it is. Unfortunately, most of the literature writes about cyberbiosecurity in a manner similar to cybersecurity for biological applications [8, 39, 81, 84, 90, 92, 95, 235–239].

This is not a fault, the focus of cyberbiosecurity is biology or related applications; however, most of the literature does not adequately define what sets cyberbiosecurity apart from IT or Computer Science in the life sciences. Gillum et al. [97] expressed a similar concern with the issues in the term "biosecurity," established fourteen years prior to their work. Multiple papers in the literature call for action or collaboration—"We call for analyses and publications to fully scope cyberbiosecurity and identify a comprehensive strategy to establish the discipline's goals and objectives" [2] and others, as called out by Drape et al. [29] and seen in Murch and DiEuliis [26]. This call from Richardson et al. [2] makes it seem like the field is still in the early planning stages, but this is not entirely true as there are papers that focus on concrete examples, lie case studies, surveys, and even one where the authors initiated an attack on a synthetic DNA supply chain that went undetected [29, 80, 86, 93, 97, 238].

Cyberbiosecurity systems are rooted in the physical sciences, but they can include pure information systems like databases for pathogens, genomics data, and land use data [4, 44, 83, 235]. We focus, however, on the physical supply chains and infrastructure, specifically water and food supply systems. Here, cyberbiosecurity secures supply through "the design of digital strategies, business models, technologies, standards and regulations" [240]. This does not exclude systems that rely on data, as even food systems depend on sharing and gathering insights from data. For example, in Duncan et al. [80] the authors discuss the need for sharing and protecting data to "design promising agricultural and food systems to better meet consumers' need." Data is just as much a part of physical systems.

Water systems are open to both natural anomalies and intentional attacks, something highlighted by Schmale III et al. [23], in their paper on a water supply system that is subject to harmful algal blooms, remote monitoring and control are incorporated to help ensure the water stays safe for drinking. However, this opens the system up to cyber-attacks, so cyberbiosecurity measures need to be taken to monitor and mitigate both sources of issues to ensure the safety of the water.

These systems are complex and multifaceted, which makes protections harder to implement and formalize, and this sentiment is highlighted in Duncan et al. [9] where the authors state current protections are not enough and "do not broadly exist across the food and agricultural system," and the "conversation on cyber security on the U.S. food and agricultural system (cyberbiosecurity) is incomplete and disjointed." There is a critical need to better incorporate cyberbiosecurity into the water and food supply chain infrastructures. Something easier said than done as these systems have multiple layers of weaknesses at the software level, the interface of cyber and physical, and the biological level. A sentiment that was expressed in Farbiash and Puzis [238] for the synthetic DNA supply chain, as those authors demonstrated an attack can bypass cybersecurity and biosecurity screenings to generate an attack based on gene editing in the synthetic data. In Bernal et al. [28], the work presented used bacteria in a DDoS style attack to demonstrate the unique risks to cyberbiosecurity that traditional cybersecurity measures cannot accommodate. These papers highlight the fact that there are biological exploits available to cyberbiosecurity systems an attacker can use without ever having physical access to a system. The multifaceted supply chains allow for multifaceted attacks that can slip through the cracks of traditional cybersecurity and biosecurity efforts.

6.3 Adoption of AI Assurance for Cyberbiosecurity

The goal of AI assurance is to mitigate any potential drawbacks or failures of AI in high-stakes applications. Assurance is a way of validating AI operates in a human-centered manner, and likewise the goals of cyberbiosecurity are to protect people from biological threats in many forms, they just happen to focus on cyber-systems and CPSs specifically. Despite this alignment of goals, we see little direct connections between cyberbiosecurity and AI in the surveyed papers (see the separation of cyberbiosecurity from the other papers in Fig. 4). There are, however, a handful of cyberbiosecurity papers we found that do overlap in topic with AI assurance, even if there is no connection via citations. Most of these papers deal with trustworthiness and safety [8, 28, 84, 241], and in fact these are also the most common assurances in the literature (see Fig. 3). Two of these papers also focus on fairness [84, 241], a little more surprising because fair AI was the least common assurance we found (again, see Fig. 3). There is one paper that focuses on explainability, specifically data and model transparency, in cyberbiosecurity [44], and how explainability ties more to security. The last paper focuses solely on trustworthiness in cyberbiosecurity [242].

Safety is a key AI assurance pillar (see Sect. 5.2), followed closely by trustworthiness (Sect. 5.5), that applies to cyberbiosecurity. The efforts of all the others are done in order to ensure the safety of the system or in the trust that the system operates in a safe manner. Ethical and fair AI (Sect. 5.1) ensures the AI system makes decisions that are correct and benefit everyone impacted equally, letting users trust that the AI makes safe decisions. Explainability (Sect. 5.3) gives us understanding of how the system operates and why it makes the decisions it does, letting users trust that the AI operates as it should to ensure the safety of those impacted. Secure AI (Sect. 5.4) ensures that if problems arise (anomalies or attacks) that the AI can handle them, either by correcting or mitigating negative effects, letting users trust that the AI system negates or limits possible harm to those impacted. Everything is done so we can trust the safety of the system.

Safety in cyberbiosecurity is mostly concerned with biosafety, or the protection from biological threats. We believe there should be more focus in the literature on food and water safety from a cyberbiosecurity perspective, especially as more technology is adopted in the water and agriculture sectors. However, there are some existing safety measures that can be adopted, like the Hazard Analysis and Critical Control Points (HACCP) for food safety and management which could be used as a starting point for safety assurances [9, 88].

Policy and regulations need to be part of the cyberbiosecurity solution, in part for the need of creating standard practices and metrics across the whole bioeconomy, and in part because cyberbiosecurity threats pose national and international security risks [243]. Cyberbiosecurity should be part of the national strategy for cybersecurity, part of the "Defend Forward" idealogy of national security [244]. This approach, however, requires the need for understanding the cyberbiosecurity field to create regulation and policy for federal agencies, something which is still lacking as "cyberbiosecurity roles, practices and metrics have not been defined and federal agencies appear uncertain regarding how to proceed" [93, 245].

The current state of the cyberbiosecurity literature focuses more on creating systems of awareness or best practices for mitigating security or safety threats, and there is little direct discussion on using explainable AI for cyberbiosecurity. Explainable AI lacks discourse in the cyberbiosecurity literature but is discussed frequently in the medical AI domain, where the goal is to create trust in AI in order to facilitate adoption by medical practitioners and to create transparency and traceability in the decisions made by the AI [246]. Explainable AI also allows for the combination of an interpretable, knowledge-based approach with that of an efficient neural based approach [247]. This means explainable AI is a way of augmenting human understanding of a problem when it uses models designed for human comprehension.

The augmentation of human intelligent via explainable AI feels like a particularly fitting application of AI for cyberbiosecurity. There is still more challenges to be addressed in the domain of explainable AI to show applicability in real-world deployments [246]; however, it does offer a lot of promise in applications where decisions are high-stakes, such as critical infrastructure including agricultural, food, and water supply chains. Richardson et al. [2] called for the implementation of "frameworks to facilitate responsible application of AI techniques to biology" and explainable AI is one way to do so.

This is particularly important to cyberbiosecurity and parts of the bioeconomy, where the sheer size and complexity of systems creates the potential for unintentional harm when trying to mitigate threats [22, 39]. Training and education of these systems (AI or otherwise) become a form of ensuring the continued safe operation of these complex systems. Training and education are also a form of creating awareness of threat mitigation to help ensure security. This is a common theme in the cyberbiosecurity literature [26, 29, 44, 45, 47, 80, 87, 88, 92, 95, 97].

All the pillars eventually boil down to ensuring trust that AI and cyberbiosecurity systems operate as intended. Section 5.5 discussed the connection of AI assurance to trustworthy AI. Society and the bioeconomy, in general, are built on trust, and if we do not trust them we will not use or participate in their activities. The same goes for AI in cyberbiosecurity, trust needs to be built so operators and all parties involved use them.

Developments in AI for cybersecurity and cyber-physical security could protect water, food, or other supply chains from intentional interference, while developments in AI for anomaly detection could protect the supply from natural phenomena [23, 25, 94, 102, 104, 106, 114, 119, 121, 225, 248–254]. Despite a clear alignment of incentives, there is not much direct overlap between these approaches in the cyberbiosecurity literature (see the separation of between cyberbiosecurity and attack/anomaly detection in Fig. 4). We conclude that although more of the cyberbiosecurity papers clearly make a call for action [2, 26, 29, 255], there is at best merely a brief attempt over existing solutions like the National Institute of Standards and Technology (NIST) cybersecurity framework [43, 47, 95, 256]. The safety and continuing function of any and all systems in the bioeconomy are important but "currently protections are minimal and do not broadly exist across the food and agricultural system" [9].

6.4 Merging the Water Security and Cyberbiosecurity Fields

Similar to AI assurance, there is not a large direct link in the literature between cyberbiosecurity and water systems. There is one series of links from cyberbiosecurity to water systems via Mueller [22], Schmale III et al. [23], Moyer et al. [24], and Housh and Ohar [25]. When we broadened our definition of cyberbiosecurity a little more from the literature we see a broader connection of papers that link the topic with water supply systems [6, 9, 23, 47, 48, 257]. What is also interesting to note is that none of these papers uses the open-source datasets we discussed in Sect. 4, instead these papers focus on broad topics of water within the food and agriculture sector [6, 9, 257] or the security of water sources [23, 47, 48]. Most of the water supply-related papers deal with security and attack/anomaly detection, aligning them more with AI assurance, but we feel they apply just as much to cyberbiosecurity as well.

There is not much existing cyber or cyber-physical security knowledge within the cyberbiosecurity field [2, 8, 29, 45, 86–88, 97]. This makes the openness of water supply testbeds and AI research critical, as these technologies can be developed and tested open-source in view of researchers focusing on cyberbiosecurity. More emphasis of the cyberbiosecurity research should be placed on using the open-source water testbeds from Sect. 4. This is the only way that water security (as a form of cyberbiosecurity) research can be performed using relevant data, and it also allows for training and hands-on experience, something a large portion of the literature called for [26, 29, 44, 45, 47, 80, 87, 88, 92, 95, 97]. This development of human understanding of cyberbiosecurity and water systems is a form of explainability and it significantly benefits from open-source data on how these systems operate.

6.5 Recommendations and Future Direction

Much of the work regarding AI assurance and cyberbiosecurity occurred in the last few years and developed separately. Figure 4 shows one link connecting cyberbiosecurity to water systems, which is then tied to the large web of anomaly and attack detection papers. Cyberbiosecurity research, however, still has a long way to meet its goal of wider adoption, and while we cannot speak for all possible sources of cross-collaboration, the expansion of cyberbiosecurity into the domains of water supply systems and AI assurance is wide open for future research.

Continuing the thread of expanding the research outside its immediate domain, cyberbiosecurity has a lot to gain from embracing open-source water supply testbeds. For one, the domain of water security is directly applicable to cyberbiosecurity, despite not making up much of the research. The literature mostly focuses on biology applications, but this feels narrow and collaborating with the established field of water security would be a great way to apply all those lessons learned to cyberbiosecurity. Many of the papers in the cyberbiosecurity literature call for more training, education, and hands-on experience. Open-source testbeds are ideal for developing resources for training and education, as well as developing new research into secure AI and other forms of AI assurance.

The goals of assurance are to validate AI aligns with the values of users impacted by an AI system, and likewise the goal of cyberbiosecurity is to protect users and citizens impacted by a biological system. AI has been instrumental in multiple agricultural applications [258–260] and offers many solutions to the threats of cyberbiosecurity but also includes several downsides; assurance nonetheless offers a way to apply AI to maximize its benefits while mitigating potential pitfalls. AI assurance should also be broadened to focus on the entirety of the system AI is deployed in, not just the assurance of the AI itself. For example, both applying AI to ensure the safety of drinking water via water quality monitoring and applying evaluation procedures to ensure the AI is operating properly are forms of assurance. In short, the cyberbiosecurity field should adopt AI measures to meet its goals and use AI assurance to validate both the AI employed is working properly and that the larger system the AI is used in is also operating properly.

7 Conclusions

In this survey, we investigated academic papers at the intersection of AI assurance, cyberbiosecurity, water and food supply systems. We assessed the application, both current and potential, of AI assurance to problems in cyberbiosecurity, specifically focusing on water and food supply systems. The survey focused on journal articles, conference proceedings, dissertations, books and book chapters, and industry white papers published from 2000 to April 2022 and at the intersection of two or more of the mentioned sectors.

A survey landscape (Sect. 2) was performed for an overview of the literature, showing most of the papers included were published since 2016, as researchers started applying AI more broadly and investigating AI assurance. Soon after in 2017, the field of cyberbiosecurity had traction and more water supply system papers were published. The increase in water supply papers since 2016 seems in part due to the start of open-source testbeds (SWaT in 2015, WADI in 2016, BATADAL in 2016, Smod in 2017, and DHALSIM in 2020), and because we specifically focused on papers that intersected with AI and cyberbiosecurity fields, both of which have seen sharp increases in the past few years. Although, looking at Fig. 4, we see there is little connection between the literature of cyberbiosecurity with the other sectors. We discussed how the papers covering these topics connected and how AI assurances apply in these fields, followed by our recommendations for future directions.

In the previous sections, we discussed the six pillars of AI assurance [1], the importance of each pillar, and the effects of the papers surveyed on water distribution systems and their applications. Figure 3, however, shows this distribution is not uniform. The pillars of Ethical AI and Fair AI were neglected, while the importance of these aspects kept growing over the last several years. This shows a great gap and opportunity for research in Ethical and Fair AI for agricultural and water systems.

We found less collaboration among the fields of AI assurance, cyberbiosecurity, and water or food supply systems than we initially expected. Figure 4 shows this disjoint well, and the literature for cyberbiosecurity does not directly discuss AI much, let alone AI assurance. The cyberbiosecurity definition should adapt a little more, as it feels too focused on cybersecurity for the life sciences. There is some acknowledgement that the current literature is not broad enough [9], especially when there are biological processes that can be exploited [28, 238].

Further research should emphasize collaboration across sectors and the use of open-source datasets and testbeds. The call for collaboration already exists with the cyberbiosecurity field, and one of our proposed solutions to that is publishing open-source datasets online. These open the field to broader research and hands-on training and experience, both of which have been expressed as needs for the cyberbiosecurity field. There are unique challenges, though these require expertise from biology, CPSs, and other domain specific knowledge for a desired application.

Lastly, we recommend that the cyberbiosecurity field adopts AI and AI assurances practices for better security while maintaining safe and trustworthy operations of these complex biological systems. There has been a lot of prior research applying AI for cybersecurity, and this would be a natural extension to incorporate into cyberbiosecurity. AI also offers more robust monitoring and an ability to make corrective actions, but this is not without issue as AI creates new vulnerabilities or failure modes. AI assurance can help mitigate these and help ensure the proper function of the overall cyberbiosecurity system.

Acknowledgments This work was supported in part by funding from Deloitte Touche Tohmatsu Limited.

We acknowledge the Center for Advanced Innovation in Agriculture (CAIA) at Virginia Tech and the Intelligent Systems Division (ISD) at The Hume Center for National Security and Technology, both for their support.

Additionally, a word of thanks to the members of Virginia Tech's A3 Research Lab (https://ai.bse.vt.edu/) for their inputs and feedback. Lastly, this work would of not been possible without the involvement of Dr. Susan Duncan (may she rest in peace) – to whom we dedicate this work.

References

1. F.A. Batarseh, L. Freeman, C.H. Huang, A survey on artificial intelligence assurance. J. Big Data **8**(1), 1–30 (2021)
2. L.C. Richardson, N.D. Connell, S.M. Lewis, E. Pauwels, R.S. Murch, Cyberbiosecurity: a call for cooperation in a new threat landscape. Front. Bioeng. Biotechnol. **7**, 99 (2019a)
3. J. Ayling, A. Chapman, Putting AI ethics to work: are the tools fit for purpose? AI Ethics, 1–25 (2021)
4. G.B. Frisvold, S.M. Moss, A. Hodgson, M.E. Maxon, Understanding the us bioeconomy: A new definition and landscape. Sustainability **13**(4), 1627 (2021)
5. The White House, National bioeconomy blueprint, April 2012. Industrial Biotechnology **8**(3), 97–102 (2012)
6. A. Aguilar, R. Wohlgemuth, T. Twardowski, Preface to the special issue bioeconomy (2018a)
7. Engineering National Academies of Sciences, Medicine, et al., *Safeguarding the Bioeconomy* (National Academies Press, 2020)
8. K.M. Berger, Addressing cyber threats in biology. IEEE Secur Privacy **18**(3), 58–61 (2020)
9. S.E. Duncan, R. Reinhard, R.C. Williams, F. Ramsey, W. Thomason, K. Lee, N. Dudek, S. Mostaghimi, E. Colbert, R. Murch, Cyberbiosecurity: A new perspective on protecting us food and agricultural system. Front. Bioeng. Biotechnol. **7**, 63 (2019)
10. R.A. Kemmerer, Cybersecurity, in *Proceedings of the 25th International Conference on Software Engineering, 2003* (IEEE, 2003), pp. 705–715
11. J.A. Lewis, Cybersecurity and critical infrastructure protection. Center Strategic Int. Stud. **1**, 12 (2006)
12. Department of Homeland Security, A glossary of common cybersecurity terminology. national initiative for cybersecurity careers and studies: Department of homeland security. http://niccs.us-cert.gov/glossary#letter_c (2022). Accessed: 2022-02-23
13. Z. Hu, J. Shi, Y. Huang, J. Xiong, X. Bu, Ganfuzz: a gan-based industrial network protocol fuzzing framework, in *Proceedings of the 15th ACM International Conference on Computing Frontiers* (2018), pp. 138–145
14. K. Lamshöft, T. Neubert, C. Krätzer, C. Vielhauer, J. Dittmann, Information hiding in cyber physical systems: Challenges for embedding, retrieval and detection using sensor data of the swat dataset, in *Proceedings of the 2021 ACM Workshop on Information Hiding and Multimedia Security* (2021), pp. 113–124
15. M. Dietz, M. Vielberth, G. Pernul, Integrating digital twin security simulations in the security operations center, in *Proceedings of the 15th International Conference on Availability, Reliability and Security* (2020), pp. 1–9
16. E.A. Lee, Cyber physical systems: Design challenges, in *2008 11th IEEE International Symposium on Object and Component-Oriented Real-Time Distributed Computing (ISORC)* (IEEE, 2008), pp. 363–369
17. N. Jazdi, Cyber physical systems in the context of industry 4.0, in *2014 IEEE International Conference on Automation, Quality and Testing, Robotics* (IEEE, 2014), pp. 1–4
18. J. Waage, J.D. Mumford, Agricultural biosecurity. Philos. Trans. R. Soc. B Biol. Sci. **363**(1492), 863–876 (2008)

19. FAO, Biosecurity in food and agriculture. https://www.fao.org/3/Y8453E/Y8453E.htm (2003). Accessed: 2022-02-26
20. S. Hinchliffe, J. Allen, S. Lavau, N. Bingham, S. Carter, Biosecurity and the topologies of infected life: from borderlines to borderlands. Trans. Inst. Brit. Geogr. **38**(4), 531–543 (2013)
21. J. Peiser, A hacker broke into a florida town's water supply and tried to poison it with lye, police said (2021). https://www.washingtonpost.com/nation/2021/02/09/oldsmar-water-supply-hack-florida/
22. S. Mueller, Facing the 2020 pandemic: What does cyberbiosecurity want us to know to safeguard the future? Biosafety Health **3**(01), 11–21 (2021)
23. D.G. Schmale III, A.P. Ault, W. Saad, D.T. Scott, J.A. Westrick, Perspectives on harmful algal blooms (habs) and the cyberbiosecurity of freshwater systems. Front. Bioeng. Biotechnol., 128 (2019)
24. J. Moyer, R. Dakin, R. Hewman, D. Groves, The case for cyber security in the water sector. J. Am. Water Works Assoc. **101**(12), 30–32 (2009)
25. M. Housh, Z. Ohar, Model-based approach for cyber-physical attack detection in water distribution systems. Water Research **139**, 132–143 (2018)
26. R. Murch, D. DiEuliis, Mapping the cyberbiosecurity enterprise. Front. Bioeng. Biotechnol., 235 (2019)
27. T. Dixon, The grey zone of cyber-biological security. International Affairs **97**(3), 685–702 (2021)
28. S.L. Bernal, D.P. Martins, A.H. Celdrán, Distributed denial of service cyberbioattack affecting bacteria-based biosensing systems, in *2020 17th International Conference on Electrical Engineering/Electronics, Computer, Telecommunications and Information Technology (ECTI-CON)* (IEEE, 2020), pp. 279–282
29. T. Drape, N. Magerkorth, A. Sen, J. Simpson, M. Seibel, R.S. Murch, S.E. Duncan, Assessing the role of cyberbiosecurity in agriculture: A case study. Front. Bioeng. Biotechnol., 742 (2021)
30. C. Perakslis, Cyberbiosecurity, ecopsychology, and beyond: Our formidable pit community [last word]. IEEE Technol. Soc. Mag. **39**(4), 84–84 (2020)
31. J. Goh, S. Adepu, K.N. Junejo, A. Mathur, A dataset to support research in the design of secure water treatment systems, in *International Conference on Critical Information Infrastructures Security* (Springer, 2016), pp. 88–99
32. T. Cruz, P. Simões, Down the rabbit hole: Fostering active learning through guided exploration of a scada cyber range. Applied Sciences **11**(20), 9509 (2021)
33. Q. Lin, S. Verwer, R. Kooij, A. Mathur, Using datasets from industrial control systems for cyber security research and education, in *International Conference on Critical Information Infrastructures Security* (Springer, 2019), pp. 122–133
34. C.M. Ahmed, V.R. Palleti, A.P. Mathur, Wadi: a water distribution testbed for research in the design of secure cyber physical systems, in *Proceedings of the 3rd International Workshop on Cyber-Physical Systems for Smart Water Networks* (2017), pp. 25–28
35. R. Taormina, S. Galelli, N.O. Tippenhauer, E. Salomons, A. Ostfeld, Characterizing cyber-physical attacks on water distribution systems. J. Water Resour. Plan. Manag. **143**(5), 04017009 (2017)
36. A. Ostfeld, E. Salomons, L. Ormsbee, J.G. Uber, C.M. Bros, P. Kalungi, R. Burd, B. Zazula-Coetzee, T. Belrain, D. Kang, et al., Battle of the water calibration networks. J. Water Resour. Plan. Manag. **138**(5), 523–532 (2012)
37. P.M. Laso, D. Brosset, J. Puentes, Dataset of anomalies and malicious acts in a cyber-physical subsystem. Data Brief **14**, 186–191 (2017)
38. A. Murillo, R. Taormina, N. Tippenhauer, S. Galelli, Co-simulating physical processes and network data for high-fidelity cyber-security experiments, in *Sixth Annual Industrial Control System Security (ICSS) Workshop* (2020), pp. 13–20
39. B.C. Wintle, C.R. Boehm, C. Rhodes, J.C. Molloy, P. Millett, L. Adam, R. Breitling, R. Carlson, R. Casagrande, M. Dando, et al., Point of view: A transatlantic perspective on 20 emerging issues in biological engineering. Elife **6**, e30247 (2017)

40. J.C. Reed, N. Dunaway, Cyberbiosecurity implications for the laboratory of the future. Front. Bioeng. Biotechnol., 182 (2019)
41. J.M. Bartoszewicz, A. Seidel, B.Y. Renard, Interpretable detection of novel human viruses from genome sequencing data. NAR Genomics Bioinforma. **3**(1), lqab004 (2021)
42. A. Salam, Internet of things for sustainability: perspectives in privacy, cybersecurity, and future trends, in *Internet of Things for Sustainable Community Development* (Springer, 2020), pp. 299–327
43. M. Walsh, W. Streilein, Security measures for safeguarding the bioeconomy. Health Security **18**(4), 313–317 (2020)
44. S.B. Jordan, S.L. Fenn, B.B. Shannon, Transparency as threat at the intersection of artificial intelligence and cyberbiosecurity. Computer **53**(10), 59–68 (2020)
45. F. Ramsey, H. Seyyedhasani, Cyber attacks in agriculture: protecting your farm and small business with cyberbiosecurity
46. L. Freeman, A. Rahman, F.A. Batarseh, Enabling artificial intelligence adoption through assurance. Social Sciences **10**(9), 322 (2021)
47. J. Germano, *Cybersecurity Risk & Responsibility in the Water Sector* (American Water Works Assn, 2018)
48. R.M. Clark, S. Panguluri, T.D. Nelson, R.P. Wyman, Protecting drinking water utilities from cyberthreats. J. Am. Water Works Assoc. **109**(INL/JOU-16-39302) (2017)
49. A. Aguilar, R. Wohlgemuth, T. Twardowski. Perspectives on bioeconomy (2018)
50. D. Wakabayashi, Self-driving uber car kills pedestrian in Arizona, where robots roam. The New York Times **19**(03) (2018)
51. A. Wilk, Teaching AI, ethics, law and policy (2019)
52. C. Rudin, Stop explaining black box machine learning models for high stakes decisions and use interpretable models instead. Nat. Mach. Intelli. **1**(5), 206–215 (2019)
53. C. Rudin, C. Wang, B. Coker, The age of secrecy and unfairness in recidivism prediction. Preprint (2018). arXiv:1811.00731
54. J. Angwin, J. Larson, S. Mattu, L. Kirchner, Machine bias, in *Ethics of Data and Analytics* (Auerbach Publications, 2016), pp. 254–264
55. L.K.J.A. J. Larson, S. Mattu, How we analyzed the compas recidivism algorithm. ProPublica (2016)
56. M. Arnold, R.K. Bellamy, M. Hind, S. Houde, S. Mehta, A. Mojsilović, R. Nair, K.N. Ramamurthy, A. Olteanu, D. Piorkowski, et al., Factsheets: Increasing trust in AI services through supplier's declarations of conformity. IBM J. Res. Dev. **63**(4/5), 6–1 (2019)
57. P. Laplante, D. Milojicic, S. Serebryakov, D. Bennett, Artificial intelligence and critical systems: from hype to reality. Computer **53**(11), 45–52 (2020)
58. R.V. Zicari, J. Brodersen, J. Brusseau, B. Düdder, T. Eichhorn, T. Ivanov, G. Kararigas, P. Kringen, M. McCullough, F. Möslein, et al., Z-inspection®: a process to assess trustworthy AI. IEEE Trans. Technol. Soc. **2**(2), 83–97 (2021)
59. C. Grady, S. Rajtmajer, L. Dennis, When smart systems fail: the ethics of cyber-physical critical infrastructure risk. IEEE Trans. Technol. Soc., 6–14 (2021)
60. R.A. Calvo, D. Peters, S. Cave, Advancing impact assessment for intelligent systems. Nature Mach. Intell. **2**(2), 89–91 (2020)
61. C.M. Hudson, N.D. Pattengale, R.K. Iyer, Z.T. Kalbarczyk, N. Alli, Genomic and synthetic biology digital biosecurity, in *Pacific Symposium On Biocomputing 2022* (World Scientific, 2021), pp. 402–406
62. M. Gardezi, R. Stock, Growing algorithmic governmentality: Interrogating the social construction of trust in precision agriculture. J. Rural Stud. **84**, 1–11 (2021)
63. S. Ayas, M.S. Ayas, A modified densenet approach with nearmiss for anomaly detection in industrial control systems. Multimedia Tools Appl., 1–14 (2021)
64. C. Rodríguez Martínez, M. Quiñones-Grueiro, C. Verde, O. Llanes-Santiago, A novel approach for detection and location of cyber-attacks in water distribution networks, in *International Workshop on Artificial Intelligence and Pattern Recognition* (Springer, 2021), pp. 79–90

65. Y. Wu, S. Liu, A review of data-driven approaches for burst detection in water distribution systems. Urban Water J. **14**(9), 972–983 (2017)
66. H.H. Addeen, Y. Xiao, J. Li, M. Guizani, A survey of cyber-physical attacks and detection methods in smart water distribution systems. IEEE Access **9**, 99905–99921 (2021)
67. N. Tuptuk, P. Hazell, J. Watson, S. Hailes, A systematic review of the state of cyber-security in water systems. Water **13**(1), 81 (2021)
68. S. Athalye, C.M. Ahmed, J. Zhou, A tale of two testbeds: a comparative study of attack detection techniques in cps, in *International Conference on Critical Information Infrastructures Security* (Springer, 2020), pp. 17–30
69. M. Abdelaty, R. Doriguzzi-Corin, D. Siracusa, Aads: A noise-robust anomaly detection framework for industrial control systems, in *International Conference on Information and Communications Security* (Springer, 2019), pp. 53–70
70. T.T. Toe, L.H. Yi, E.F.M. Josephlal, Advanced predictive techniques for detection of cyber-attacks in water infrastructures, in *2020 IEEE International Conference on Signal Processing, Communications and Computing (ICSPCC)* (IEEE, 2020), pp. 1–6
71. S. Abba, V. Nourani, G. Elkiran, Multi-parametric modeling of water treatment plant using ai-based non-linear ensemble. J. Water Supply Re. Technol. Aqua **68**(7), 547–561 (2019)
72. M. Al-Yaari, T.H. Aldhyani, S. Rushd, Prediction of arsenic removal from contaminated water using artificial neural network model. Applied Sciences **12**(3), 999 (2022)
73. A. Jain, L.E. Ormsbee, Short-term water demand forecast modeling techniques–conventional methods versus AI. J. Am. Water Works Assoc. **94**(7), 64–72 (2002)
74. L. Karamoutsou, A. Psilovikos, Deep learning in water resources management: The case study of kastoria lake in greece. Water **13**(23), 3364 (2021)
75. L. Nishi, M. Baesso, R. Santana, P. Fregadolli, D. Falavigna, A. Falavigna-Guilherme, Investigation of cryptosporidium spp. and giardia spp. in a public water-treatment system. Zoonoses Public Health **56**(5), 221–228 (2009)
76. M. Florjanič, J. Kristl, Microbiological quality assurance of purified water by ozonization of storage and distribution system. Drug Dev. Ind. Pharm. **32**(10), 1113–1121 (2006)
77. U. Gentile, S. Marrone, F. De Paola, R. Nardone, N. Mazzocca, M. Giugni, Model-based water quality assurance in ground and surface provisioning systems, in *2015 10th International Conference on P2P, Parallel, Grid, Cloud and Internet Computing (3PGCIC)* (IEEE, 2015), pp. 527–532
78. D. Ghernaout, B. Ghernaout, On the concept of the future drinking water treatment plant: algae harvesting from the algal biomass for biodiesel production—a review. Desalin. Water Treat. **49**(1-3), 1–18 (2012)
79. I. Montalvo, J. Izquierdo, R. Pérez, M.M. Tung, Particle swarm optimization applied to the design of water supply systems. Comput. Math. Appl. **56**(3), 769–776 (2008)
80. S.E. Duncan, B. Zhang, W. Thomason, M. Ellis, N. Meng, M. Stamper, R. Carneiro, T. Drape, Securing data in life sciences—a plant food (edamame) systems case study. Front. Sustain., 10 (2020)
81. A. Adler, J. Beal, M. Lancaster, D. Wyschogrod, Cyberbiosecurity and public health in the age of covid-19, in *Emerging Threats of Synthetic Biology and Biotechnology* (Springer, Dordrecht, 2021), pp. 103–115
82. D. Greenbaum, Cyberbiosecurity: An emerging field that has ethical implications for clinical neuroscience. Camb. Q. Healthc. Ethics **30**(4), 662–668 (2021)
83. J. Caswell, J.D. Gans, N. Generous, C.M. Hudson, E. Merkley, C. Johnson, C. Oehmen, K. Omberg, E. Purvine, K. Taylor, et al., Defending our public biological databases as a global critical infrastructure. Front. Bioeng. Biotechnol. **7**, 58 (2019)
84. J. Li, H. Zhao, L. Zheng, W. An, Advances in synthetic biology and biosafety governance. Front. Bioeng. Biotechnol. **9**, 173 (2021)
85. P.M. Ney, Securing the future of biotechnology: A study of emerging bio-cyber security threats to dna-information systems. Ph.D. thesis (2019)
86. K. Millett, E. Dos Santos, P.D. Millett, Cyber-biosecurity risk perceptions in the biotech sector. Front. Bioeng. Biotechnol. **7**, 136 (2019)

87. L.C. Richardson, S.M. Lewis, R.N. Burnette, Building capacity for cyberbiosecurity training. Front. Bioeng. Biotechnol. **7**, 112 (2019b)
88. S. Duncan, R. Carneiro, J. Braley, M. Hersh, F. Ramsey, R. Murch, Beyond ransomware: Securing the digital food chain (2021)
89. X.L. Palmer, E. Powell, L. Potter, Biocyberwarfare and crime: A juncture of rethought, in *European Conference on Cyber Warfare and Security* (Academic Conferences International Limited, 2021), pp. 517–XIV
90. R.J. Hester, Bioveillance: A techno-security infrastructure to preempt the dangers of informationalised biology. Sci. Culture **29**(1), 153–176 (2020)
91. K.M. Berger, P.A. Schneck, National and transnational security implications of asymmetric access to and use of biological data. Front. Bioeng. Biotechnol. **7**, 21 (2019)
92. J. Peccoud, J.E. Gallegos, R. Murch, W.G. Buchholz, S. Raman, Cyberbiosecurity: from naive trust to risk awareness. Trends Biotechnol. **36**(1), 4–7 (2018)
93. G. Turner, The growing need for cyberbiosecurity, in *InSITE 2019: Informing Science+ IT Education Conferences: Jerusalem* (2019), pp. 207–215
94. Q. Lin, S. Adepu, S. Verwer, A. Mathur, Tabor: A graphical model-based approach for anomaly detection in industrial control systems, in *Proceedings of the 2018 on Asia Conference on Computer and Communications Security* (2018), pp. 525–536
95. J.L. Mantle, J. Rammohan, E.F. Romantseva, J.T. Welch, L.R. Kauffman, J. McCarthy, J. Schiel, J.C. Baker, E.A. Strychalski, K.C. Rogers, et al., Cyberbiosecurity for biopharmaceutical products. Front. Bioeng. Biotechnol. **7**, 116 (2019)
96. C.O. Adetunji, O.T. Olugbemi, O.A. Anani, D.I. Hefft, N. Wilson, A.S. Olayinka, K.E. Ukhurebor, Cyberespionage: Socioeconomic implications on sustainable food security, in *AI, Edge and IoT-based Smart Agriculture* (Elsevier, 2022), pp. 477–486
97. D. Gillum, L.A.O. Carrera, I.A. Mendoza, P. Bates, D. Bowens, Z. Jetson, J. Maldonado, C. Mancini, M. Miraldi, R. Moritz, et al., The 2017 arizona biosecurity workshop: an open dialogue about biosecurity. Applied Biosafety **23**(4), 233–241 (2018)
98. L. Potter, X.L. Palmer, Human factors in biocybersecurity wargames, in *Future of Information and Communication Conference* (Springer, 2021), pp. 666–673
99. S. Adepu, A. Mathur, Introducing cyber security at the design stage of public infrastructures: A procedure and case study, in *Complex Systems Design & Management Asia* (Springer, 2016a), pp. 75–94
100. A. Ilyas, L. Engstrom, A. Athalye, J. Lin, Black-box adversarial attacks with limited queries and information, in *Proceedings of the 35th International Conference on Machine Learning, ICML 2018* (2018). https://arxiv.org/abs/1804.08598
101. A. Hassanzadeh, A. Rasekh, S. Galelli, M. Aghashahi, R. Taormina, A. Ostfeld, M.K. Banks, A review of cybersecurity incidents in the water sector. J. Environ. Eng. **146**(5), 03120003 (2020)
102. F. Pasqualetti, F. Dörfler, F. Bullo, Attack detection and identification in cyber-physical systems. IEEE Trans. Automatic Control **58**(11), 2715–2729 (2013)
103. S. Hochreiter, J. Schmidhuber, Long short-term memory. Neural Computation **9**(8), 1735–1780 (1997)
104. J. Goh, S. Adepu, M. Tan, Z.S. Lee, Anomaly detection in cyber physical systems using recurrent neural networks, in *2017 IEEE 18th International Symposium on High Assurance Systems Engineering (HASE)* (IEEE, 2017), pp. 140–145
105. A.P. Mathur, N.O. Tippenhauer, Swat: A water treatment testbed for research and training on ics security, in *2016 International Workshop on Cyber-Physical Systems for Smart Water Networks (CySWater)* (IEEE, 2016), pp. 31–36
106. J. Inoue, Y. Yamagata, Y. Chen, C.M. Poskitt, J. Sun, Anomaly detection for a water treatment system using unsupervised machine learning, in *2017 IEEE International Conference on Data Mining Workshops (ICDMW)* (IEEE, 2017), pp. 1058–1065
107. R. Taormina, S. Galelli, N.O. Tippenhauer, E. Salomons, A. Ostfeld, D.G. Eliades, M. Aghashahi, R. Sundararajan, M. Pourahmadi, M.K. Banks, et al., Battle of the attack detection algorithms: Disclosing cyber attacks on water distribution networks. J. Water Res. Plann. Manag. **144**(8), 04018048 (2018)

108. M. Aghashahi, R. Sundararajan, M. Pourahmadi, M.K. Banks, Water distribution systems analysis symposium–battle of the attack detection algorithms (batadal), in *World Environmental and Water Resources Congress 2017* (2017), pp. 101–108
109. B.M. Brentan, E. Campbell, G. Lima, D. Manzi, D. Ayala-Cabrera, M. Herrera, I. Montalvo, J. Izquierdo, E. Luvizotto Jr, On-line cyber attack detection in water networks through state forecasting and control by pattern recognition. in *World Environmental and Water Resources Congress 2017* (2017), pp. 583–592
110. S.E. Chandy, A. Rasekh, Z.A. Barker, B. Campbell, M.E. Shafiee, Detection of cyber-attacks to water systems through machine-learning-based anomaly detection in scada data, in *World Environmental and Water Resources Congress 2017* (2017), pp. 611–616
111. M. Giacomoni, N. Gatsis, A. Taha, Identification of cyber attacks on water distribution systems by unveiling low-dimensionality in the sensory data, in *World Environmental and Water Resources Congress 2017* (2017), pp. 660–675
112. M. Mardani, G. Mateos, G.B. Giannakis, Recovery of low-rank plus compressed sparse matrices with application to unveiling traffic anomalies. IEEE Trans. Inf. Theory **59**(8), 5186–5205 (2013)
113. A.A. Abokifa, K. Haddad, C.S. Lo, P. Biswas, Detection of cyber physical attacks on water distribution systems via principal component analysis and artificial neural networks, in *World Environmental and Water Resources Congress 2017* (2017), pp. 676–691
114. M.F.K. Pasha, B. Kc, S.L. Somasundaram, An approach to detect the cyber-physical attack on water distribution system, in *World Environmental and Water Resources Congress 2017* (2017), pp. 703–711
115. M. Housh, Z. Ohar, Integrating physically based simulators with event detection systems: Multi-site detection approach. Water Research **110**, 180–191 (2017)
116. M. Kravchik, A. Shabtai, Detecting cyber attacks in industrial control systems using convolutional neural networks, in *Proceedings of the 2018 Workshop on Cyber-Physical Systems Security and PrivaCy* (2018), pp. 72–83
117. M.A. Umer, A. Mathur, K.N. Junejo, S. Adepu, Generating invariants using design and data-centric approaches for distributed attack detection. Int. J. Crit. Infrastruct. Prot. **28**, 100341 (2020)
118. K.N. Junejo, J. Goh, Behaviour-based attack detection and classification in cyber physical systems using machine learning, in *Proceedings of the 2nd ACM International Workshop on Cyber-Physical System Security* (2016), pp. 34–43
119. S. Adepu, A. Mathur, Distributed detection of single-stage multipoint cyber attacks in a water treatment plant, in *Proceedings of the 11th ACM on Asia Conference on Computer and Communications Security* (2016), pp. 449–460
120. S. Adepu, A. Mathur, An investigation into the response of a water treatment system to cyber attacks, in *2016 IEEE 17th International Symposium on High Assurance Systems Engineering (HASE)* (IEEE, 2016), pp. 141–148
121. S. Adepu, A. Mathur, Distributed attack detection in a water treatment plant: Method and case study. IEEE Trans. Dependable Secure Comput. **18**(1), 86–99 (2018)
122. A. Al-Abassi, H. Karimipour, A. Dehghantanha, R.M. Parizi, An ensemble deep learning-based cyber-attack detection in industrial control system. IEEE Access **8**, 83965–83973 (2020)
123. M. Sermesant, H. Delingette, H. Cochet, P. Jaïs, N. Ayache, Applications of artificial intelligence in cardiovascular imaging. Nat. Rev. Cardiol. **18**(8), 600–609 (2021)
124. P. Sinčak, J. Ondo, D. Kaposztasova, M. Virčikova, Z. Vranayova, J. Sabol, Artificial intelligence in public health prevention of legionelosis in drinking water systems. Int. J. Environ. Res. Public Health **11**(8), 8597–8611 (2014)
125. J.M. Wing, Trustworthy AI. Commun. ACM **64**(10), 64–71 (2021)
126. S. Thiebes, S. Lins, A. Sunyaev, Trustworthy artificial intelligence. Electronic Markets **31**(2), 447–464 (2021)
127. V. Morckel, K. Terzano, Legacy city residents' lack of trust in their governments: An examination of flint, michigan residents' trust at the height of the water crisis. J. Urban Aff. **41**(5), 585–601 (2019)

128. O. Inderwildi, C. Zhang, X. Wang, M. Kraft, The impact of intelligent cyber-physical systems on the decarbonization of energy. Energy Environ. Sci. **13**(3), 744–771 (2020)
129. C.S. Wickramasinghe, D.L. Marino, J. Grandio, M. Manic, Trustworthy AI development guidelines for human system interaction, in *2020 13th International Conference on Human System Interaction (HSI)* (IEEE, 2020), pp. 130–136
130. R. Kaasschieter. The "why" in building trust in AI (2020). https://www.capgemini.com/2020/09/the-why-in-building-trust-in-ai/#:~:text=Accountability2C%20transparency%2C%20fairness%2C%20etc,they%20will%20not%20buy%20it
131. N.A. Smuha, The eu approach to ethics guidelines for trustworthy artificial intelligence. Comput. Law Rev. Int. **20**(4), 97–106 (2019)
132. H. Liu, Y. Wang, W. Fan, X. Liu, Y. Li, S. Jain, Y. Liu, A.K. Jain, J. Tang, Trustworthy AI: A computational perspective. Preprint (2021). arXiv:2107.06641
133. E. Toreini, M. Aitken, K.P. Coopamootoo, K. Elliott, V.G. Zelaya, P. Missier, M. Ng, A. van Moorsel, Technologies for trustworthy machine learning: A survey in a socio-technical context. Preprint (2020). arXiv:2007.08911
134. B.W. Israelsen, N.R. Ahmed, "dave... i can assure you... that it's going to be all right..." a definition, case for, and survey of algorithmic assurances in human-autonomy trust relationships. ACM Comput. Surv. (CSUR) **51**(6), 1–37 (2019)
135. G. Bernieri, M. Conti, F. Turrin, Evaluation of machine learning algorithms for anomaly detection in industrial networks, in *2019 IEEE International Symposium on Measurements & Networking (M&N)* (IEEE, 2019), pp. 1–6
136. S.D. Anton, S. Kanoor, D. Fraunholz, H.D. Schotten, Evaluation of machine learning-based anomaly detection algorithms on an industrial modbus/tcp data set, in *Proceedings of the 13th International Conference on Availability, Reliability and Security* (2018), pp. 1–9
137. H. Wiemer, A. Dementyev, S. Ihlenfeldt, A holistic quality assurance approach for machine learning applications in cyber-physical production systems. Applied Sciences **11**(20), 9590 (2021)
138. B. Li, P. Qi, B. Liu, S. Di, J. Liu, J. Pei, J. Yi, B. Zhou, Trustworthy AI: From principles to practices. Preprint (2021b). arXiv:2110.01167
139. J. Mökander, L. Floridi, Ethics-based auditing to develop trustworthy AI. Minds Mach. **31**(2), 323–327 (2021)
140. E. Daglarli, Explainable artificial intelligence (xai) approaches and deep meta-learning models for cyber-physical systems, in *Artificial Intelligence Paradigms for Smart Cyber-Physical Systems* (IGI Global, 2021), pp. 42–67
141. D. Kaur, S. Uslu, A. Durresi, Requirements for trustworthy artificial intelligence—a review, in *International Conference on Network-Based Information Systems* (Springer, 2020), pp. 105–115
142. C. Louisell, K. Heaslip, Securing the digitally managed water supply, in *World Environmental and Water Resources Congress 2020: Emerging and Innovative Technologies and International Perspectives* (American Society of Civil Engineers Reston, VA, 2020), pp. 1–11
143. J. Bergal, Florida hack exposes danger to water systems (2021). https://www.pewtrusts.org/en/research-and-analysis/blogs/stateline/2021/03/10/florida-hack-exposes-danger-to-water-systems
144. B. Kerstein, Israel thwarts major coordinated cyber-attack on its water infrastructure command and control systems (2020). https://www.algemeiner.com/2020/04/26/israel-thwarts-major-coordinated-cyber-attack-on-its-water-infrastructure-command-and-control-systems/
145. M. Taddeo, T. McCutcheon, L. Floridi, Trusting artificial intelligence in cybersecurity is a double-edged sword. Nat. Mach. Intell. **1**(12), 557–560 (2019)
146. N. Nicolaou, D.G. Eliades, C. Panayiotou, M.M. Polycarpou, Reducing vulnerability to cyber-physical attacks in water distribution networks, in *2018 international workshop on cyber-physical systems for smart water networks (CySWater)* (IEEE, 2018), pp. 16–19
147. A. Khaled, S. Ouchani, Z. Tari, K. Drira, Assessing the severity of smart attacks in industrial cyber-physical systems. ACM Trans. Cyber Phys. Syst. **5**(1), 1–28 (2020)

148. F. Pasqualetti, F. Dörfler, F. Bullo, Cyber-physical security via geometric control: Distributed monitoring and malicious attacks, in *2012 IEEE 51st IEEE Conference on Decision and Control (CDC)* (IEEE, 2012), pp. 3418–3425
149. Y. Wu, H.N. Dai, H. Tang, Graph neural networks for anomaly detection in industrial internet of things. IEEE Internet Things J. (2021)
150. B. Siegel, Industrial anomaly detection: A comparison of unsupervised neural network architectures. IEEE Sens. Lett. **4**(8), 1–4 (2020)
151. L. Rosa, T. Cruz, M.B. de Freitas, P. Quitério, J. Henriques, F. Caldeira, E. Monteiro, P. Simões, Intrusion and anomaly detection for the next-generation of industrial automation and control systems. Future Gener. Comput. Syst. **119**, 50–67 (2021)
152. L.A. Maglaras, J. Jiang, Intrusion detection in scada systems using machine learning techniques, in *2014 Science and Information Conference* (IEEE, 2014), pp. 626–631
153. C.M. Ahmed, G.R. MR, A.P. Mathur, Challenges in machine learning based approaches for real-time anomaly detection in industrial control systems, in *Proceedings of the 6th ACM on Cyber-Physical System Security Workshop* (2020), pp. 23–29
154. J. Zhang, L. Pan, Q.L. Han, C. Chen, S. Wen, Y. Xiang, Deep learning based attack detection for cyber-physical system cybersecurity: A survey. IEEE/CAA J. Automat. Sin. **9**(3), 377–391 (2021)
155. Y. Luo, Y. Xiao, L. Cheng, G. Peng, D. Yao, Deep learning-based anomaly detection in cyber-physical systems: Progress and opportunities. ACM Comput. Surv. (CSUR) **54**(5), 1–36 (2021)
156. N. Kadosh, A. Frid, M. Housh, Detecting cyber-physical attacks in water distribution systems: One-class classifier approach. J. Water Resour. Plann. Manag. **146**(8), 04020060 (2020)
157. D.C.L. Sung, G.R. MR, A.P. Mathur, Design-knowledge in learning plant dynamics for detecting process anomalies in water treatment plants. Comput. Secur. **113**, 102532 (2022)
158. D. Garcia, V. Puig, J. Quevedo, Prognosis of water quality sensors using advanced data analytics: Application to the barcelona drinking water network. Sensors **20**(5), 1342 (2020)
159. R. Taormina, S. Galelli, Real-time detection of cyber-physical attacks on water distribution systems using deep learning, in *World Environmental and Water Resources Congress 2017* (2017), pp. 469–479
160. A.A. Abokifa, K. Haddad, C. Lo, P. Biswas, Real-time identification of cyber-physical attacks on water distribution systems via machine learning–based anomaly detection techniques. J. Water Resour. Plann. Manag. **145**(1), 04018089 (2019)
161. N. Neha, S. Priyanga, S. Seshan, R. Senthilnathan, V. Shankar Sriram, Sco-rnn: A behavioral-based intrusion detection approach for cyber physical attacks in scada systems, in *Inventive Communication and Computational Technologies* (Springer, 2020), pp. 911–919
162. J. Kim, J.H. Yun, H.C. Kim, Anomaly detection for industrial control systems using sequence-to-sequence neural networks, in *Computer Security* (Springer, 2019), pp. 3–18
163. S. Sapkota, A. Mehdy, S. Reese, H. Mehrpouyan, Falcon: Framework for anomaly detection in industrial control systems. Electronics **9**(8), 1192 (2020)
164. C.H. Yoong, J. Heng, Framework for continuous system security protection in swat, in *Proceedings of the 2019 3rd International Symposium on Computer Science and Intelligent Control* (2019), pp. 1–6
165. L.H.A. Reis, A. Murillo Piedrahita, S. Rueda, N.C. Fernandes, D.S. Medeiros, M.D. de Amorim, D.M. Mattos, Unsupervised and incremental learning orchestration for cyber-physical security. Trans. Emerg. Telecommun. Technol. **31**(7), e4011 (2020)
166. M. Gauthama Raman, N. Somu, A.P. Mathur, Anomaly detection in critical infrastructure using probabilistic neural network, in *International Conference on Applications and Techniques in Information Security* (Springer, 2019), pp. 129–141
167. S. Kim, W. Jo, T. Shon, Apad: autoencoder-based payload anomaly detection for industrial ioe. Appl. Soft Comput. **88**, 106017 (2020)
168. S.K. Alabugin, A.N. Sokolov, Applying of generative adversarial networks for anomaly detection in industrial control systems, in *2020 Global Smart Industry Conference (GloSIC)* (IEEE, 2020), pp. 199–203

169. D.D. Tiwari, S. Naskar, A.S. Sai, V.R. Palleti, Attack detection using unsupervised learning algorithms in cyber-physical systems, in *Computer Aided Chemical Engineering*, vol. 50 (Elsevier, 2021), pp. 1259–1264
170. W. Zhou, X.-m. Kong, K.-l. Li, X.-m. Li, L.-l. Ren, Y. Yan, Y. Sha, X.-y. Cao, X.-j. Liu, Attack sample generation algorithm based on data association group by gan in industrial control dataset. Computer Communications **173**, 206–213 (2021)
171. M.G. Raman, W. Dong, A. Mathur, Deep autoencoders as anomaly detectors: Method and case study in a distributed water treatment plant. Comput. Secur. **99**, 102055 (2020)
172. R. Taormina, S. Galelli, Deep-learning approach to the detection and localization of cyber-physical attacks on water distribution systems. J. Water Resour. Plann. Manag. **144**(10), 04018065 (2018)
173. H. Wijaya, M. Aniche, A. Mathur, Domain-based fuzzing for supervised learning of anomaly detection in cyber-physical systems, in *Proceedings of the IEEE/ACM 42nd International Conference on Software Engineering Workshops* (2020), pp. 237–244
174. P. Schneider, K. Böttinger, High-performance unsupervised anomaly detection for cyber-physical system networks, in *Proceedings of the 2018 Workshop on Cyber-Physical Systems Security and Privacy* (2018), pp. 1–12
175. M. Elnour, N. Meskin, K.M. Khan, Hybrid attack detection framework for industrial control systems using 1d-convolutional neural network and isolation forest, in *2020 IEEE Conference on Control Technology and Applications (CCTA)* (IEEE, 2020), pp. 877–884
176. R. Alguliyev, Y. Imamverdiyev, L. Sukhostat, Hybrid deepgcl model for cyber-attacks detection on cyber-physical systems. Neural Comput. Appl. **33**(16), 10211–10226 (2021)
177. Z. Chen, D. Chen, X. Zhang, Z. Yuan, X. Cheng, Learning graph structures with transformer for multivariate time series anomaly detection in iot. IEEE Internet Things J. (2021)
178. Y. Chen, C.M. Poskitt, J. Sun, S. Adepu, F. Zhang, Learning-guided network fuzzing for testing cyber-physical system defences, in *2019 34th IEEE/ACM International Conference on Automated Software Engineering (ASE)* (IEEE, 2019), pp. 962–973
179. A. Meleshko, V. Desnitsky, I. Kotenko, Machine learning based approach to detection of anomalous data from sensors in cyber-physical water supply systems, in *IOP Conference Series: Materials Science and Engineering*, vol. 709 (IOP Publishing, 2020), p. 033034
180. P. Perrone, F. Flammini, R. Setola, Machine learning for threat recognition in critical cyber-physical systems, in *2021 IEEE International Conference on Cyber Security and Resilience (CSR)* (IEEE, 2021), pp. 298–303
181. S. Athalye, C. Mujeeb Ahmed, J. Zhou, Model-based cps attack detection techniques: Strengths and limitations, in *Security in Cyber-Physical Systems* (Springer, 2021), pp. 155–187
182. A. Robles-Durazno, N. Moradpoor, J. McWhinnie, G. Russell, Z. Tan, Newly engineered energy-based features for supervised anomaly detection in a physical model of a water supply system. Ad Hoc Networks **120**, 102590 (2021)
183. J. Sun, Z. Yang, Objsim: efficient testing of cyber-physical systems, in *Proceedings of the 4th ACM SIGSOFT International Workshop on Testing, Analysis, and Verification of Cyber-Physical Systems and Internet of Things* (2020), pp. 1–2
184. S. Alqurashi, H. Shirazi, I. Ray, On the performance of isolation forest and multi layer perceptron for anomaly detection in industrial control systems networks, in *2021 8th International Conference on Internet of Things: Systems, Management and Security (IOTSMS)* (IEEE, 2021), pp. 1–6
185. M. Balaji, S. Shrivastava, S. Adepu, A. Mathur, Super detector: An ensemble approach for anomaly detection in industrial control systems, in *International Conference on Critical Information Infrastructures Security* (Springer, 2021), pp. 24–43
186. A.N. Jahromi, H. Karimipour, A. Dehghantanha, K.K.R. Choo, Toward detection and attribution of cyber-attacks in iot-enabled cyber–physical systems. IEEE Internet Things J. **8**(17), 13712–13722 (2021)
187. M. Baptiste, F. Julien, S. Franck, Systematic and efficient anomaly detection framework using machine learning on public ics datasets, in *2021 IEEE International Conference on Cyber Security and Resilience (CSR)* (IEEE, 2021), pp. 292–297

188. T. Chalongvorachai, K. Woraratpanya, A data generation framework for extremely rare case signals. Heliyon **7**(8), e07687 (2021)
189. G.R. MR, N. Somu, A.P. Mathur, A multilayer perceptron model for anomaly detection in water treatment plants. Int. J. Crit. Infrastruct. Prot. **31**, 100393 (2020)
190. P.F. de Araujo-Filho, G. Kaddoum, D.R. Campelo, A.G. Santos, D. Macêdo, C. Zanchettin, Intrusion detection for cyber–physical systems using generative adversarial networks in fog environment. IEEE Internet Things J. **8**(8), 6247–6256 (2020)
191. F. Turrin, A. Erba, N.O. Tippenhauer, M. Conti, A statistical analysis framework for ics process datasets, in *Proceedings of the 2020 Joint Workshop on CPS&IoT Security and Privacy* (2020), pp. 25–30
192. G. Sebestyen, A. Hangan, Z. Czako, Anomaly detection in water supply infrastructure systems, in *2021 23rd International Conference on Control Systems and Computer Science (CSCS)* (IEEE, 2021), pp. 349–355
193. L. Gjorgiev, S. Gievska, Time series anomaly detection with variational autoencoder using mahalanobis distance, in *International Conference on ICT Innovations* (Springer, 2020), pp. 42–55
194. S. Chockalingam, W. Pieters, A. Teixeira, P. van Gelder, Bayesian network model to distinguish between intentional attacks and accidental technical failures: a case study of floodgates. Cybersecurity **4**(1), 1–19 (2021)
195. R. Qadeer, C. Murguia, C.M. Ahmed, J. Ruths, Multistage downstream attack detection in a cyber physical system, in *Computer Security* (Springer, 2017), pp. 177–185
196. C.M. Ahmed, S. Adepu, A. Mathur, Limitations of state estimation based cyber attack detection schemes in industrial control systems, in *2016 Smart City Security and Privacy Workshop (SCSP-W)* (IEEE, 2016), pp. 1–5
197. C.M. Ahmed, M. Ochoa, J. Zhou, A.P. Mathur, R. Qadeer, C. Murguia, J. Ruths, Noiseprint: Attack detection using sensor and process noise fingerprint in cyber physical systems, in *Proceedings of the 2018 on Asia Conference on Computer and Communications Security* (2018), pp. 483–497
198. T.K. Das, S. Adepu, J. Zhou, Anomaly detection in industrial control systems using logical analysis of data. Comput. Secur. **96**, 101935 (2020)
199. S. Adepu, J. Prakash, A. Mathur, Waterjam: An experimental case study of jamming attacks on a water treatment system, in *2017 IEEE International Conference on Software Quality, Reliability and Security Companion (QRS-C)* (IEEE, 2017), pp. 341–347
200. S. Liyakkathali, F. Furtado, G. Sugumar, A. Mathur, A mechanism to assess the effectiveness anomaly detectors in industrial control systems. J. Integr. Des. Process Sci. (Preprint), 1–26 (2022)
201. G. Sugumar, A. Mathur, Testing the effectiveness of attack detection mechanisms in industrial control systems, in *2017 IEEE International Conference on Software Quality, Reliability and Security Companion (QRS-C)* (IEEE, 2017), pp. 138–145
202. A. Mathur, Secwater: A multi-layer security framework for water treatment plants, in *Proceedings of the 3rd International Workshop on Cyber-Physical Systems for Smart Water Networks* (2017), pp. 29–32
203. D. Dovžan, V. Logar, I. Škrjanc, Implementation of an evolving fuzzy model (efumo) in a monitoring system for a waste-water treatment process. IEEE Trans. Fuzzy Syst. **23**(5), 1761–1776 (2014)
204. S. Adepu, S. Shrivastava, A. Mathur, Argus: An orthogonal defense framework to protect public infrastructure against cyber-physical attacks. IEEE Internet Comput. **20**(5), 38–45 (2016)
205. S. Adepu, A. Mathur, Assessing the effectiveness of attack detection at a hackfest on industrial control systems. IEEE Trans. Sustain. Comput. **6**(2), 231–244 (2018b)
206. D. Urbina, J. Giraldo, N.O. Tippenhauer, A. Cardenas, Attacking fieldbus communications in ics: Applications to the swat testbed, in *Proceedings of the Singapore Cyber-Security Conference (SG-CRC) 2016* (IOS Press, 2016), pp. 75–89

207. K. Pal, S. Adepu, J. Goh, Effectiveness of association rules mining for invariants generation in cyber-physical systems, in *2017 IEEE 18th International Symposium on High Assurance Systems Engineering (HASE)* (IEEE, 2017), pp. 124–127
208. M.A. Umer, A. Mathur, K.N. Junejo, S. Adepu, Integrating design and data centric approaches to generate invariants for distributed attack detection, in *Proceedings of the 2017 Workshop on Cyber-Physical Systems Security and Privacy* (2017), pp. 131–136
209. E. Kang, S. Adepu, D. Jackson, A.P. Mathur, Model-based security analysis of a water treatment system, in *2016 IEEE/ACM 2nd International Workshop on Software Engineering for Smart Cyber-Physical Systems (SEsCPS)* (IEEE, 2016), pp. 22–28
210. S. Shrivastava, G.R. MR, A. Mathur, Pcat: Plc command analysis tool for automatic incidence response in water treatment plants, in *2021 IEEE International Conference on Big Data (Big Data)* (IEEE, 2021), pp. 2151–2159
211. A. Robles-Durazno, N. Moradpoor, J. McWhinnie, G. Russell, I. Maneru-Marin, Plc memory attack detection and response in a clean water supply system. Int. J. Crit. Infrastruct. Prot. **26**, 100300 (2019)
212. A. Agrawal, C.M. Ahmed, E.C. Chang, Poster: Physics-based attack detection for an insider threat model in a cyber-physical system, in *Proceedings of the 2018 on Asia Conference on Computer and Communications Security* (2018), pp. 821–823
213. N. Chikhalia, Y. Dhawan, Security of industrial cyberspace: Fair clustering with linear time approximation, in *Handbook of Big Data Analytics and Forensics* (Springer, 2022), pp. 75–88
214. J. Wang, J. Sun, Y. Jia, S. Qin, Z. Xu, Towards 'verifying' a water treatment system, in *International Symposium on Formal Methods* (Springer, 2018), pp. 73–92
215. G.R. MR, A.P. Mathur, Aicrit: A unified framework for real-time anomaly detection in water treatment plants. J. Inf. Secur. Appl. **64**, 103046 (2022)
216. S. Uslu, D. Kaur, S.J. Rivera, A. Durresi, M. Babbar-Sebens, J.H. Tilt, A trustworthy human–machine framework for collective decision making in food–energy–water management: The role of trust sensitivity. Knowl. Based Syst. **213**, 106683 (2021)
217. K. Krithivasan, S. Pravinraj, V.S. Shankar Sriram, et al., Detection of cyberattacks in industrial control systems using enhanced principal component analysis and hypergraph-based convolution neural network (epca-hg-cnn). IEEE Trans. Ind. Appl. **56**(4), 4394–4404 (2020)
218. Q. Xu, S. Ali, T. Yue, Digital twin-based anomaly detection in cyber-physical systems, in *2021 14th IEEE Conference on Software Testing, Verification and Validation (ICST)* (IEEE, 2021), pp. 205–216
219. Z. Li, J. Li, Y. Wang, K. Wang, A deep learning approach for anomaly detection based on sae and lstm in mechanical equipment. Int. J. Adv. Manuf. Technol. **103**(1), 499–510 (2019)
220. M.F. Abdelaty, R.D. Corin, D. Siracusa, Daics: A deep learning solution for anomaly detection in industrial control systems. IEEE Trans. Emerg. Top. Comput. (2021)
221. M. Elnour, N. Meskin, K. Khan, R. Jain, A dual-isolation-forests-based attack detection framework for industrial control systems. IEEE Access **8**, 36639–36651 (2020)
222. M. Kravchik, A. Shabtai, Efficient cyber attack detection in industrial control systems using lightweight neural networks and pca. IEEE Trans. Dependable Secure Comput. (2021)
223. Á.L. Perales Gómez, L. Fernández Maimó, A. Huertas Celdrán, F.J. García Clemente, Madics: A methodology for anomaly detection in industrial control systems. Symmetry **12**(10), 1583 (2020)
224. G. Zizzo, C. Hankin, S. Maffeis, K. Jones, Intrusion detection for industrial control systems: Evaluation analysis and adversarial attacks. Preprint (2019). arXiv:1911.04278
225. D. Shalyga, P. Filonov, A. Lavrentyev, Anomaly detection for water treatment system based on neural network with automatic architecture optimization. Preprint (2018). arXiv:1807.07282
226. D. Li, D. Chen, B. Jin, L. Shi, J. Goh, S.K. Ng, Mad-gan: Multivariate anomaly detection for time series data with generative adversarial networks, in *International Conference on Artificial Neural Networks* (Springer, 2019), pp. 703–716

227. K. Faber, M. Pietron, D. Zurek, Ensemble neuroevolution-based approach for multivariate time series anomaly detection. Entropy **23**(11), 1466 (2021)
228. S. Chakraborty, A. Onuchowska, S. Samtani, W. Jank, B. Wolfram, Machine learning for automated industrial iot attack detection: an efficiency-complexity trade-off. ACM Trans. Manag. Inf. Syst. (TMIS) **12**(4), 1–28 (2021)
229. D. Li, D. Chen, J. Goh, S.k. Ng, Anomaly detection with generative adversarial networks for multivariate time series. Preprint (2018). arXiv:1809.04758
230. B. Brentan, P. Rezende, D. Barros, G. Meirelles, E. Luvizotto, J. Izquierdo, Cyber-attack detection in water distribution systems based on blind sources separation technique. Water **13**(6), 795 (2021)
231. D.T. Ramotsoela, G.P. Hancke, A.M. Abu-Mahfouz, Attack detection in water distribution systems using machine learning. HCIS **9**(1), 1–22 (2019)
232. L. Tsiami, C. Makropoulos, Cyber-physical attack detection in water distribution systems with temporal graph convolutional neural networks. Water **13**(9), 1247 (2021)
233. H. Mahmoud, W. Wu, M.M. Gaber, A time-series self-supervised learning approach to detection of cyber-physical attacks in water distribution systems. Energies **15**(3), 914 (2022)
234. Y.H. Choi, A. Sadollah, J.H. Kim, Improvement of cyber-attack detection accuracy from urban water systems using extreme learning machine. Applied Sciences **10**(22), 8179 (2020)
235. B.A. Vinatzer, L.S. Heath, H.M. Almohri, M.J. Stulberg, C. Lowe, S. Li, Cyberbiosecurity challenges of pathogen genome databases. Front. Bioeng. Biotechnol. **7**, 106 (2019)
236. J. Diggans, E. Leproust, Next steps for access to safe, secure dna synthesis. Front. Bioeng. Biotechnol. **7**, 86 (2019)
237. R. Puzis, D. Farbiash, O. Brodt, Y. Elovici, D. Greenbaum, Increased cyber-biosecurity for DNA synthesis. Nature Biotechnology **38**(12), 1379–1381 (2020)
238. D. Farbiash, R. Puzis, Cyberbiosecurity: Dna injection attack in synthetic biology. Preprint (2020). arXiv:2011.14224
239. S. Mueller, On DNA signatures, their dual-use potential for gmo counterfeiting, and a cyber-based security solution. Front. Bioeng. Biotechnol. **7**, 189 (2019)
240. D. Guttieres, S. Stewart, J. Wolfrum, S.L. Springs, Cyberbiosecurity in advanced manufacturing models. Front. Bioeng. Biotechnol., 210 (2019)
241. Z. Li, H. Zhao, J. Shi, Y. Huang, J. Xiong, An intelligent fuzzing data generation method based on deep adversarial learning. IEEE Access **7**, 49327–49340 (2019)
242. P. Rana, L.R. Varshney, Trustworthy predictive algorithms for complex forest system decision-making. Front. Forests Global Change, 153 (2021)
243. A.M. George, The national security implications of cyberbiosecurity. Front. Bioeng. Biotechnol. **7**, 51 (2019)
244. X.L. Palmer, S. Karahan, Defending forward: an exploration through the lens of biocybersecurity, in *ICCWS 2020 15th International Conference on Cyber Warfare and Security* (Academic Conferences and Publishing Limited, 2020), p. 373
245. X.L. Palmer, L. Potter, S. Karahan, On the emerging area of biocybersecurity and relevant considerations, in *Future of Information and Communication Conference* (Springer, 2020), pp. 873–881
246. A.F. Markus, J.A. Kors, P.R. Rijnbeek, The role of explainability in creating trustworthy artificial intelligence for health care: a comprehensive survey of the terminology, design choices, and evaluation strategies. J. Biomed. Inf. **113**, 103655 (2021)
247. A. Holzinger, C. Biemann, C.S. Pattichis, D.B. Kell, What do we need to build explainable AI systems for the medical domain? Preprint (2017). arXiv:1712.09923
248. M. Quiñones-Grueiro, A. Prieto-Moreno, C. Verde, O. Llanes-Santiago, Decision support system for cyber attack diagnosis in smart water networks. IFAC-PapersOnLine **51**(34), 329–334 (2019)
249. S. Adepu, A. Mathur, Using process invariants to detect cyber attacks on a water treatment system, in *IFIP International Conference on ICT Systems Security and Privacy Protection* (Springer, 2016), pp. 91–104

250. M. Macas, C. Wu, An unsupervised framework for anomaly detection in a water treatment system, in *2019 18th IEEE International Conference On Machine Learning And Applications (ICMLA)* (IEEE, 2019), pp. 1298–1305
251. A. Deng, B. Hooi, Graph neural network-based anomaly detection in multivariate time series, in *Proceedings of the AAAI Conference on Artificial Intelligence*, vol. 35 (2021), pp. 4027–4035
252. C. Gehrmann, M. Gunnarsson, A digital twin based industrial automation and control system security architecture. IEEE Trans. Ind. Inf. **16**(1), 669 (2019)
253. Y. Jia, J. Wang, C.M. Poskitt, S. Chattopadhyay, J. Sun, Y. Chen, Adversarial attacks and mitigation for anomaly detectors of cyber-physical systems. Int. J. Crit. Infrastruct. Prot. **34**, 100452 (2021)
254. J.H. Moon, J.H. Yu, K.A. Sohn, An ensemble approach to anomaly detection using high-and low-variance principal components. Comput. Electr. Eng. **99**, 107773 (2022)
255. R.S. Murch, W.K. So, W.G. Buchholz, S. Raman, J. Peccoud, Cyberbiosecurity: an emerging new discipline to help safeguard the bioeconomy. Front. Bioeng. Biotechnol., 39 (2018)
256. D.S. Schabacker, L.A. Levy, N.J. Evans, J.M. Fowler, E.A. Dickey, Assessing cyberbiosecurity vulnerabilities and infrastructure resilience. Front. Bioeng. Biotechnol. **7**, 61 (2019)
257. K. Demestichas, N. Peppes, T. Alexakis, Survey on security threats in agricultural iot and smart farming. Sensors **20**(22), 6458 (2020)
258. S. Gurrapu, F.A. Batarseh, P. Wang, M.N.K. Sikder, N. Gorentala, M. Gopinath, Deepag: Deep learning approach for measuring the effects of outlier events on agricultural production and policy. in *2021 IEEE Symposium Series on Computational Intelligence (SSCI)* (IEEE, 2021), pp. 1–8
259. M. Gopinath, F.A. Batarseh, J. Beckman, Machine learning in gravity models: An application to agricultural trade. Tech. rep., National Bureau of Economic Research (2020)
260. A. Monken, F. Haberkorn, M. Gopinath, L. Freeman, F.A. Batarseh, Graph neural networks for modeling causality in international trade, in *The International FLAIRS Conference Proceedings*, vol. 34 (2021)

23. M. M. [illegible], C. W. [illegible], An unsupervised transfer learning for anomaly detection in a water treatment system. In: [illegible] IEEE International Conference on Industrial Informatics And Applications (ICIIA) [illegible] pp. [illegible]
24. [illegible] Deng, B. Hooi, Graph neural network-based anomaly detection in multivariate time series. In: Proceedings of the AAAI Conference on Artificial Intelligence, vol. 35 (2021), pp. 48092–[illegible]
25. [illegible] J. C. [illegible], A digital twin-based anomaly detection [illegible] industrial control systems. [illegible]
[illegible]

Artificial Intelligence and the Weaponization of Genetic Data

Sterling Sawaya, Erin Kenneally, Demetrius Nelson, and Garrett Schumacher

Abstract Advancements in genetics have the ability to rapidly improve medicine, with a number of factors converging to push the integration of genomics into mainstream healthcare. As technologies that use genetic data begin to expand, so does exposure to risks. The cost and harms from the misuse of genetic data can be latent. The immutability, uniqueness, and information-rich nature of DNA renders it a high-value target. Knowledge and control asymmetries exist between individuals, industry, and governments. Genetic data's value as an asset is mirrored in the potential degree of harm if abused. This article highlights the critical threats and vulnerabilities associated with genetic data risk, from identification and profiling, to exposure of health and medical conditions and susceptibility, as well as to the broader social welfare risk associated with biowarfare. All of these threats are rapidly actualizing from the advancement of artificial intelligence (AI). Here we outline the ways in which data science is improving genetics and how that can ultimately lead to its weaponization.

Keywords AI · Genetics · Bioweapon · Privacy · Biosecurity

List of Abbreviations

AI	Artificial intelligence
CCPA	California Consumer Privacy Act
GDPA	General Data Protection Regulation
HIPAA	Health Insurance Portability and Accountability Act
NIH	National Institute of Health
PHI	Protected health information
PII	Personally identifiable information
SNP	Single-nucleotide polymorphisms

S. Sawaya (✉) · E. Kenneally · D. Nelson · G. Schumacher
GeneInfoSec Inc., Boulder, CO, USA
e-mail: s@geneinfosec.com; geneinfosec.com

D. Greenbaum (ed.), *Cyberbiosecurity*, https://doi.org/10.1007/978-3-031-26034-6_14

1 Introduction and Background

New risks are emerging as the collection and use of genetic data becomes commonplace. These emerging risks are partially due to the rapid drop in the cost of DNA sequencing as well as advances in synthetic biology. Whole genome sequencing has reached $100 per genome,[1] down from nearly $1K over the past few years and around $1B just over twenty years ago[2] [61]. Furthermore, the use of microarray technology allows millions of genetic variants, called single-nucleotide polymorphisms (SNPs), to be genotyped rapidly and inexpensively. Consequently, genetic data has become increasingly available from multiple sources. Private industry is collecting genetic information, from direct-to-consumer (DTC) genetic and genealogy services to pharmaceutical companies collecting consumer and healthcare data. Data is also being generated through research funded by the United States government, much of which is made available to the public alongside other information about study participants, for example, from the 1000 Genomes project and data on dgGap [42, 66]. Currently, two large programs seek to expand these databases: the National Institute of Health's (NIH) All of Us program obtaining genetic data from one million civilian participants, and the Veteran Affairs' Million Veterans Program obtaining genetic data from one million former military service members. These research programs have great potential to advance science and medicine. Through a better understanding of the way genetics can influence and cause disease, we can improve our ability to predict disease susceptibility and provide genetic-specific treatments. These advancements require large genetic databases that include a variety of personal data, ranging from health data to lifestyle choices.

Such databases are typically "anonymized," or de-identified, by simply replacing participants' names with a numeric identifier. Unfortunately, as we detail in this article, this data can be easily re-identified. As far back as the early 2000s, the identifiability of poorly-anonymized data was recognized, and the threat landscape has become appreciably richer since then [11]. Furthermore, security around the generation and storage of genetic data is typically weak, and a number of potential attack vectors exist [14, 18, 50]. A myriad of techniques have been disclosed that adversaries can use to compromise genetic data [64], such as exploiting API-enabled features that are created for third-party data uses [14, 50]. This insecurity is compounded because third-party access to genetic information does not have adequate oversight. In the U.S., DTC companies are outside the purview of the Health Insurance Portability and Accountability Act (HIPAA). Similarly, the Genetic Information Non-discrimination Act does not limit data access; it only prohibits employers and health insurers from using that data for discrimination in certain contexts. Researchers obtaining federal funding are bound by the Common

[1] https://www.technologyreview.com/s/615289/china-bgi-100-dollar-genome/.

[2] https://www.genome.gov/about-genomics/fact-sheets/DNA-Sequencing-Costs-Data.

Rule's[3] requirement to obtain informed consent from data subjects, but this protection is rife with substantive and procedural deficiencies[4] Furthermore, Certificates of Confidentiality[5] issued by NIH to safeguard research subjects' privacy do not prevent the discretionary release of data by principal researchers and their institutions. NIH's Genomic Data Sharing Policy[6] provides soft law guidelines for protecting subject privacy, but not all data is deemed controlled access.

When it comes to advances in biometric-data-specific privacy laws or amendments to existing laws, the EU under the General Data Protection Regulation (GDPR) and about half of states in the U.S. have hard laws on the books governing genomic privacy and penalize the illegitimate use of genetic data (e.g. the California Consumer Privacy Act, CCPA). However, there are inconsistencies and a lack of uniformity regarding triggering provisions such as the scope of what is unauthorized (collection, analysis and/or disclosure), the conditions for consent requirements, and enforcement provisions. As well, not all include genetic data or DNA in their definitions of biometrics, limiting the scope to physiological and behavioral data. While the GDPR and CCPA do cover biological data collection, including genetic data and sale for commercial purposes, GDPR does not apply to non-EU individuals, and CCPA is not germane to the large corpus of research data that comprise the bulk of publicly available genetic databases. Lastly, there are no protections in place for the genetic privacy of relatives of individuals who choose to disclose their own data; any rights afforded only apply to the person who is the data source. Absent a legal or market-based force to drive the protection of these data, entities stewarding genetic data lack incentives to bear the cost of security protections, especially if it impedes their time-to-market or competitive advantage.

The threats surrounding these insecure databases are compounded with advancements in AI, especially in machine learning, the subset of AI that learns from data. The application of machine learning to genetic data is rapidly advancing the ability of medical science and clinical practice to generate genetic-based diagnostics and therapeutics. At the same time, these advancements open up dangerous opportunities for adversaries to exploit genetic data to cause harm (Fig. 1). Machine learning allows adversaries to amplify threats by improving their ability to identify targets within genetic databases, uncover sensitive information from the data, and then weaponize that data for exploitation. Genetic data has become the latest target in the cat-and-mouse game that is information security, and the risks that are emerging are amplified by developments in data science. The immutability of DNA leads us to consider threats that not only exist today or in the near future; threats that are over twenty years away must also be considered, especially when handling the genetic information of children. A recent article discusses the threats outlined here, highlighting the challenges that arise when advances in AI and genetic data are made

[3] http://www.hhs.gov/ohrp/humansubjects/commonrule/index.html.

[4] http://irb.ufl.edu/irb02/informed-consent-instructions-procedures/ifcprob.html.

[5] https://grants.nih.gov/policy/humansubjects/coc.htm.

[6] https://www.genome.gov/about-nhgri/Policies-Guidance/Genomic-Data-Sharing.

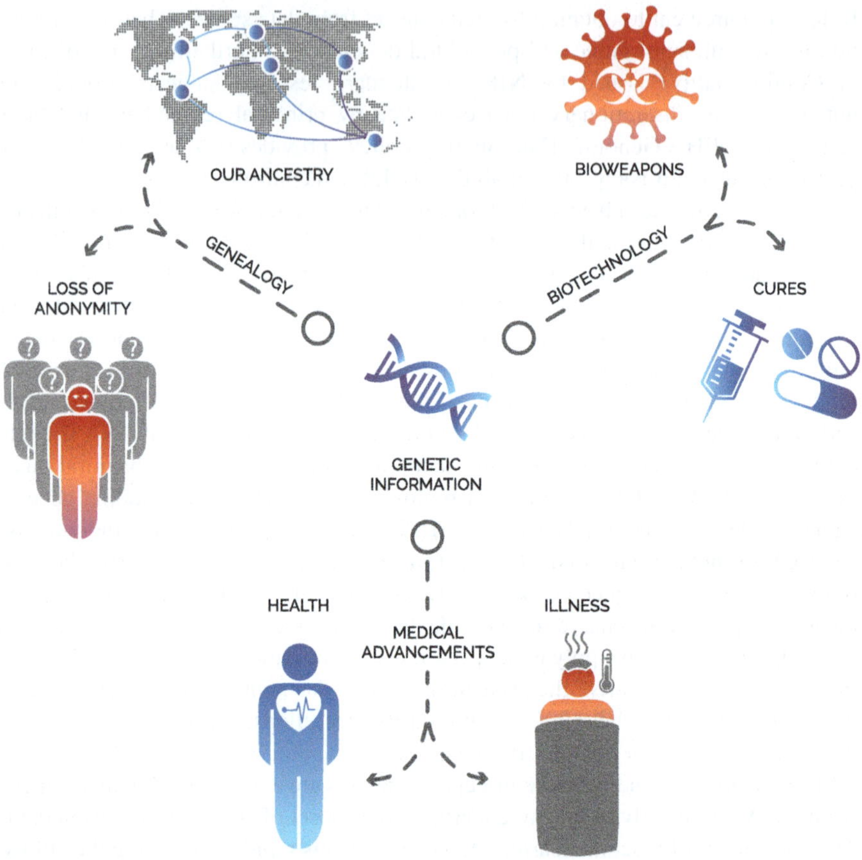

Fig. 1 The application of AI to genetic information will lead to great advancements that can also be weaponized. Advances in genetic medicine will lead to healthier lives, but these advances can also be used to make us ill. Advances in biotechnology will lead to revolutionary cures, but some of these advances can also be utilized for the development of bioweapons. Advances in genealogy are giving us an unprecedented understanding of our history but can lead to the loss of anonymity

open source [34]. Here, we expand on their research, discussing specific examples of current and future dangers surrounding genetic information.

2 Identifiability

Genetic information along with other biometrics, such as face, fingerprints, retina-scans, and voiceprints, are natural constructs—they are inherent and unique to individuals. Individuals are also identifiable from socioeconomic constructs, such as income, education, employment, and age. All of these data, known as personally identifiable information (PII), comprise a mosaic of our respective identities. Consequently, databases that contain PII can be traced back to individuals. Database structures that link PII to information in another data repository can allow adversaries to obtain sensitive information. Genetic data is distinct from other types of personal information. It is inherently PII and also contains protected health information (PHI), information about relations between individuals, and other potentially sensitive information about individuals or populations. Genetic data is also immutable and remains with an individual for his or her entire life, tying it intrinsically to our identity. Due to this immutability, genetic data breaches can have long-lasting consequences and must be considered distinct from other types of data breaches.

Genetic data is shared between family members, so a genetic data breach for one individual can be transitive to his/her relatives [17]. This has a significant effect on identifiability. If a distant relative has his/her genetic data released or breached, then many related individuals become easier to identify via familial matching of genetic data [17]. Consequently, genetic data insecurity will have lasting effects for generations. A number of open and semi-open genetic databases are already available publicly [2, 22, 25], and other databases have been found to be vulnerable to unauthorized access [14, 50]. Using this data, many individuals can already be identified by their DNA alone. For example, the Golden State Killer was recently caught because DNA he left at a crime scene could be traced to a relative that was in a genetic database [26].

Even if one's relatives are not present in an open database, genetic data can still be identifiable. These databases often contain other potentially identifying information, such as age or other demographic information, so re-identification can occur using rather trivial methods [15, 16, 62]. Here, identity can be uncovered by matching metadata in the genetic database (e.g. demographics or health conditions) with publicly available demographic identifiers (physical location, race, age or date of birth, gender, etc.) from voter registries, public record search engines, and/or social media [16, 66]. The likelihood of re-identification increases if a genetic database contains phenotypic information or other health information [15, 28, 32, 62]. Moreover, genetic data can also be re-identified by estimating an individual's surname [27], so individuals with unique surnames become uniquely susceptible. Similarly, ethnicity and ancestry can be estimated with genetic data, so individuals with a unique ethnicity become easily identifiable with genetic data, particularly within certain geographies. Consequently, even unsophisticated analyses can already uncover the identities of at least a portion of most genetic databases.

Importantly, genetic data contains information about an individual's traits, such as height or hair color. Simply estimating these traits can facilitate re-identification for individuals with unique traits. Furthermore, advances in trait prediction through machine learning are rapidly improving the ability for individuals to be identified from genetic information. For example, using machine learning methods height can be estimated to within a few centimeters [39]. Complex traits can also be predicted with machine learning, and facial features can even be reconstructed [43, 45]. If a database of faces attached to names is available, such as a driver's license photo database, then genetic data can be directly re-identified with machine learning[65]. As we increasingly place our photographs in the public domain,[7] we improve the likelihood that our genetic data can be re-identified. Individuals with unique traits or with a wealth of publicly available photographs are more susceptible to this type of re-identification. This methodology is being utilized by law enforcement to identify suspects [73]. Other advances in re-identification will inevitably be developed in this area, again enabled by machine learning and AI. Due to its inherent identifiability, a genetic data breach can have widespread and long-lasting consequences.

3 Exposure of Medical Information

The use of machine learning on genetic data is rapidly improving our ability to predict disease predisposition, as well as other traits of interest in humans [5, 7, 9, 13, 19, 31, 33, 39, 40, 48, 54–56, 63, 77] and agriculture [3, 23, 24, 29, 55, 69]. The highly accurate predictions made by machine learning are perhaps surprising, considering that the genetic architecture of many genetic diseases remains limited, often referred to as the "missing heritability problem" [46]. The ability for machine learning to pick up information about culture, epigenetics, and ethnicity can lead to models that make accurate predictions, even for traits that do not have high genetic heritability. The ability to predict one's susceptibility to disease will become essential to the practice of medicine, allowing preventative medicine to supplant reactive disease treatment [56]. For example, one's susceptibility to cancer can be strongly influenced by genes such as BRCA1 and BRCA2 [10]. Furthermore, as environmental and behavioral cofactors are added into the predictive model, the power of these predictions increases [40]. Analysis of these cofactors can help medical practitioners make lifestyle suggestions to prevent and/or treat diseases. For example, the genetic predisposition to Crohn's disease can be predicted with near-clinical significance using machine learning on only SNP data [7, 63], and such predictions may eventually allow disease symptoms to be avoided through early-life intervention. Genetic models can also predict susceptibility to osteoporotic fracture, which could be used to prevent dangerous bone fractures in elderly patients [19]. These capabilities highlight the genetic advancements that can be expected in the near future. As full genome sequence data replaces SNP data, these predictors will rapidly improve (e.g.,[45]).

Although some studies have shown that clinical genetic predictions will be soon available [19, 48], many of these machine learning predictions are far from clinical relevance. However, the capability to make even weak disease predictions en masse can allow adversaries to target weaknesses within groups or populations. Genetic prediction can be used on a population to uncover a range of potential targets.

[7] https://www.nytimes.com/2020/01/18/technology/clearview-privacy-facial-recognition.html.

Even if the genetic predictor lacks clinical accuracy, the efficacy of the attack could increase by targeting individuals with the predicted genetic weakness. Thus, for many genetic predictions the overall risk of harm to any given individual may be low, but the risk of damage at a population level may remain significant. Causing disease in even a small percentage of a population can induce both economic and societal disruption, especially if the affected group is a high-value political, military, or industrial target. Perhaps for these and other reasons, the Department of Defense recently released a memo advising service members to avoid genetic testing because of their potential to "create unintended security consequences and increased risk to the joint force and mission."[8]

The prediction of behavioral traits is particularly troubling in this regard. Mental diseases, such as bi-polar disorder and schizophrenia, have a strong genetic component, and genetic data can be used to predict susceptibility to and severity of these diseases [9, 52, 77]. At an individual level, for most people, these relatively weak predictors are unlikely to be a threat. At the population level, however, these predictions could allow an adversary to generate an enriched list of targets or find vulnerable sub-populations. Other behavioral diseases, such as post-traumatic stress disorder, can have a measurable genetic influence [21], and these diseases may be exploited, especially in vulnerable populations like military veterans. A wide range of behavioral traits must be considered when assessing this threat, not simply known behavioral diseases. The accuracy of these predictions is certain to improve as technology advances and more data is generated. These predictions, even when imperfect, allow an adversary to increase their success rate by targeting a small number of potentially susceptible individuals instead of the population at large.

Diseases can be influenced by environmental factors that have a greater effect on individuals if they are genetically susceptible. A concerning example is found in the effect of metals on a wide range of diseases (reviewed in [67]). Metal accumulation and toxicity has a genetic component and is influenced by a range of several different genes [51, 71]. Furthermore, some metals are a dietary necessity, such as zinc, and optimal intake can depend on an individual's genetic variants [12]. However, adversarial knowledge about an individual's genetic metal tolerance could be used to induce disease. Machine learning applied to protein-metal interaction has improved our understanding of how different genetic variations influence the binding of metals to proteins [38]. Large genetic studies are therefore not necessarily required to uncover novel genetic variants related to metal toxicity. Introducing toxic metals into a population could induce a range of diseases [67], and susceptible individuals could be targeted. For example, Alzheimer's disease may be influenced by dietary aluminum [47, 70]. Similarly, some individuals are genetically susceptible to mercury poisoning, causing an increased vulnerability to a range of neurological disorders [1]. Hence, genetic susceptibility to metals, or other potential

[8] https://www.military.com/daily-news/2019/12/27/pentagon-leaders-tell-troops-stop-using-mail-genealogy-dna-kits.html.

environmental toxins, can be determined from genetic data and can be used to cause harm.

An understanding of how our DNA can influence disease will be necessary for the advancement of medicine. However, the potential misuse of these scientific advances is widely being ignored. If we allow this genetic information to be obtained by adversaries, they will have deep knowledge about the weaknesses of individuals and populations. When this knowledge is clear, such as with Huntington's disease [58], exploiting it may be as simple as blackmailing an individual with the threat of publicly releasing the weakness. However, strong genetic prediction is not required for a population-level attack. By neglecting genetic information security, our health information can be exploited, directly harming susceptible individuals and putting entire populations at risk.

4 Bioweapons Utilizing Genetic Data

The threat from biological weapons has been long standing, with international treaties signed banning them from 1925 and 1972 [20]. Most of the biological weapons that have been stockpiled are non-contagious organisms, such as anthrax, which can kill individuals without the possibility of spreading the infections. These non-contagious organisms avoid the threat of a blowback, in which a contagious weapon damages the group or country that uses it. Furthermore, should they accidentally escape from their containment in the lab, pathogens could directly harm the groups that are attempting to develop them.

The risk that we may face dangerous and highly contagious pathogens is rapidly increasing, driven by the advancements in synthetic biology [8, 30, 34, 49] and access to human genetic information [6]. Advancements in CRISPR-Cas9 techniques allow scientists to genetically modify organisms with ease and precision. Organisms can now be designed to have specific genes, and there is widespread fear that dangerous diseases could be engineered by genetically modifying a similar yet benign organism [68]. This genetic modification requires DNA to be synthesized and implanted in a microbe. A proposed method to prevent the development of these dangerous organisms is monitoring of DNA synthesis to prevent the DNA found in dangerous genes from being generated [37, 74]. However, this preventative measure will not be easy to implement when DNA synthesis becomes widely available on-site [57]. Therefore, in the very near future, small groups will have the capability to secretly generate dangerous pathogens with only limited resources. Recently, the COVID-19 pandemic is instructive, demonstrating the wide-scale health and economic damage that can be caused by a pathogen. Designer pathogens could potentially be more difficult to detect and treat than COVID-19 and could therefore cause significantly more damage.

Adversarial groups have various methods by which they can avoid blowback from designer pathogens. The development of vaccines for designer pathogens is one possibility, but this would require significant resources and also run the risk of vaccine failure. Another possibility is for the group to remain isolated after they release their pathogens, but indefinite isolation becomes a challenge for larger groups and is nearly impossible for nation states. Alternatively, designer pathogens can be engineered so that they only threaten individuals with specific genetics that are not shared by the adversary. Hence, a group could design a pathogen so that they are naturally resistant, while other groups or ethnicities are naturally vulnerable. Note that designing a bioweapon for which an adversarial group is not susceptible is easier for small groups than large groups or nations.

Designing targeted pathogens requires knowledge about how proteins fold to understand how they interact with our immune system and infect our cells. Protein folding is complicated and can be difficult to estimate [75]. However, as more information about protein evolution and protein sequence is made available, the accuracy of these predictions will improve [44]. Utilizing these advances in machine learning, novel drugs or disease agents can be generated with the ability to bind to specific proteins. Accurate prediction of protein folding remains limited, but recent advancements by AlphaFold demonstrate rapid development of protein prediction [35]. The source code for AlphaFold has recently been released [36], and researchers were able to reproduce these capabilities using only a description of the algorithm [4]. For at least some types of proteins, this method will soon approach accuracy found in traditional, labor intensive methods of protein folding prediction. As this technology advances further, scientists will be able to design organisms with a range of properties, including novel enzymatic activity and the ability to infect a range of different cell types.

Recent advances in machine learning also allow the design of metabolic pathways [60, 76]. This can facilitate the design of organisms that produce highly toxic chemicals. While these poison producing microbes are unlikely to become highly contagious, they may nevertheless be used in targeted attacks, potentially targeting politicians and other high profile individuals [59]. Designing these organisms to become dangerous weapons remains a challenge, but motivated groups could leverage advancements in AI to compensate for their lack of domain expertise. These capabilities could expand to non-state-actors, making them difficult to regulate or control, introducing an emerging challenge that needs to be addressed [41, 72].

5 Conclusion

The threats outlined here either presently exist or have a strong probability of manifesting just over the horizon. As biotechnology advances, we may also face more dangerous threats that are not currently contemplated or defy current imagination, the proverbial unknown-unknowns. While a range of measures can be applied to avoid or mitigate these threats,[9] including design trade-offs that limit AI model

[9] https://councilonstrategicrisks.org/2020/03/09/release-can-the-u-s-make-bioweapons-obsolete/.

deployments [34], risk prevention should be considered concomitant with biotech innovation, and a framework needs to be developed that includes a multidisciplinary effort [53]. Critically, genetic information must be protected from adversarial exposure, and the sharing of genetic information must be limited to organizations and persons with legitimate and authorized use purposes. Advancements in AI will usher in medical breakthroughs as well as the weaponization of genetic information. These advancements must be considered as we seek to balance the need to access genetic data with the need to keep it secure. Genetic data sharing and technology advancement cannot and rightfully should not be stopped. However, they should be responsibly controlled and governed. The individuals, populations, and nations for whom this data and technology are not appropriately governed will be the most vulnerable and exploited.

Once an individual's genetic data is breached it can no longer be protected, so we cannot for harm to manifest before we take systemic measures to protect our genetic information. The uncontrolled public sharing of genetic databases needs to be assessed and corrected proportional to the risk it poses, and controlled sharing of private genetic databases must be appropriately secured. Ultimately this problem stems from a lack of awareness. Individuals and institutions need to be educated about these threats, but awareness of the problem is just the first step. Methods to secure genetic data need to be thoroughly evaluated, implemented, continuously reviewed, and meaningfully enforced. This security paradigm for genetic data will require a multi-stakeholder effort, appropriate resources, and widespread acknowledgement that genetic information is critical infrastructure that demands security considerations paid to other essential assets. Standards and regulations will be necessary to ensure that institutions that handle genetic data are taking appropriate precautions. Any groups or institutions that collect, use, or share genetic information must work together to generate and implement these standards. We hope the threats outlined in this article will advance the dialog about the nature of the risk, raise awareness about the gaps in our current capabilities, and encourage action to be taken before it is too late.

Declarations

Availability of data and materials—Not applicable

Competing interests—The authors work for GeneInfoSec Inc. and are generating technology to protect genetic data.

Funding—GeneInfoSec Inc. fully funded the reported research and as the authors are employees, GeneInfoSec Inc therefore had a role in the design of the study.

Authors' contributions—All authors contributed to the conception of the research, its implementation and writing of the manuscript.

Acknowledgements—Not applicable

References

1. V. Andreoli, F. Sprovieri, Genetic aspects of susceptibility to mercury toxicity: an overview. Int. J. Environ. Res. Public Health **14**(1), 93 (2017)
2. M. Angrist, Eyes wide open: the personal genome project, citizen science and veracity in informed consent. Personalized Medicine **6**(6), 691–699 (2009)
3. C.B. Azodi, A. McCarren, M. Roantree, G. de los Campos, S.-H. Shiu, Benchmarking algorithms for genomic prediction of complex traits. bioRxiv, 614479 (2019)
4. M. Baek, F. DiMaio, I. Anishchenko, J. Dauparas, S. Ovchinnikov, G.R. Lee, J. Wang, Q. Cong, L.N. Kinch, R. Dustin Schaeffer, et al., Accurate prediction of protein structures and interactions using a three-track neural network. Science **373**(6557), 871–876 (2021)
5. P. Bellot, G. de los Campos, M. Pérez-Enciso, Can deep learning improve genomic prediction of complex human traits? Genetics **210**(3), 809–819 (2018)
6. J.L. Black III, Genome projects and gene therapy: gateways to next generation biological weapons. Military Medicine **168**(11), 864–871 (2003)
7. V. Botta, G. Louppe, P. Geurts, L. Wehenkel, Exploiting SNP correlations within random forest for genome-wide association studies. PloS One **9**(4), e93379 (2014)
8. R. Breitling, E. Takano, T.S. Gardner, Judging Synthetic Biology Risks (2015)
9. L.-C. Chuang, P.-H. Kuo, Building a genetic risk model for bipolar disorder from genome-wide association data with random forest algorithm. Scientific Reports **7**, 39943 (2017)
10. F.J. Couch, K.L. Nathanson, K. Offit, Two decades after BRCA: setting paradigms in personalized cancer care and prevention. Science **343**(6178), 1466–1470 (2014)
11. T. Data, G.T. Duncan, S.E. Fienberg, R. Krishnan, Confidentiality, disclosure and data access: Theory and practical applications for statistical agencies (2001)
12. K.J. Day, M.M. Adamski, A.L. Dordevic, C. Murgia, Genetic variations as modifying factors to dietary zinc requirements: A systematic review. Nutrients **9**(2), 148 (2017)
13. J.A. Diao, I.S. Kohane, A.K. Manrai, Biomedical informatics and machine learning for clinical genomics. Hum. Mol. Genet. **27**(R1), R29–R34 (2018)
14. M.D. Edge, G. Coop, Attacks on genetic privacy via uploads to genealogical databases. Elife 9 (2020)
15. Y. Erlich, A. Narayanan, Routes for breaching and protecting genetic privacy. Nat. Rev. Genet. **15**(6), 409–421 (2014)
16. Y. Erlich, Major flaws in "identification of individuals by trait prediction using whole-genome sequencing data". bioRxiv (2017)
17. Y. Erlich, T. Shor, I. Peter, S. Carmi, Identity inference of genomic data using long-range familial searches. Science **362**(6415), 690–694 (2018)
18. I. Fayans, Y. Motro, L. Rokach, Y. Oren, J. Moran-Gilad, Cyber security threats in the microbial genomics era: implications for public health. Eurosurveillance **25**(6), 1900574 (2020)
19. V. Forgetta, J. Keller-Baruch, M. Forest, A. Durand, S. Bhatnagar, J. Kemp, J.A. Morris, J.A. Kanis, D.P. Kiel, E.V. McCloskey, et al., Machine learning to predict osteoporotic fracture risk from genotypes. bioRxiv, 413716 (2018)
20. F. Frischknecht, The history of biological warfare. EMBO Reports **4**(S1), S47–S52 (2003)
21. J. Gelernter, N. Sun, R. Polimanti, R. Pietrzak, D.F. Levey, J. Bryois, Q. Lu, Y. Hu, B. Li, K. Radhakrishnan, et al., Genome-wide association study of post-traumatic stress disorder reexperiencing symptoms in > 165,000 us veterans. Nature Neuroscience **22**(9), 1394–1401 (2019)
22. Genomes Project Consortium et al., A global reference for human genetic variation. Nature **526**(7571), 68–74 (2015)
23. J.M. González-Camacho, J. Crossa, P. Pérez-Rodríguez, L. Ornella, D. Gianola, Genome-enabled prediction using probabilistic neural network classifiers. BMC Genomics **17**(1), 208 (2016)
24. O. González-Recio, S. Forni, Genome-wide prediction of discrete traits using Bayesian regressions and machine learning. Genet. Sel. Evol. **43**(1), 7 (2011)

25. B. Greshake, P.E. Bayer, H. Rausch, J. Reda, OpenSNP—a crowdsourced web resource for personal genomics. PLoS One **9**(3), e89204 (2014)
26. C.J. Guerrini, J.O. Robinson, D. Petersen, A.L. McGuire, Should police have access to genetic genealogy databases? capturing the golden state killer and other criminals using a controversial new forensic technique. PLoS Biology **16**(10), e2006906 (2018)
27. M. Gymrek, A.L. McGuire, D. Golan, E. Halperin, Y. Erlich, Identifying personal genomes by surname inference. Science **339**(6117), 321–324 (2013)
28. A. Harmanci, M. Gerstein, Quantification of private information leakage from phenotype-genotype data: linking attacks. Nature Methods **13**(3), 251 (2016)
29. N. Heslot, H.-P. Yang, M.E. Sorrells, J.-L. Jannink, Genomic selection in plant breeding: a comparison of models. Crop Science **52**(1), 146–160 (2012)
30. A. Hessel, M. Goodman, S. Kotler, Hacking the president's DNA. The Atlantic **310**(4), 83 (2012)
31. D.S.W. Ho, W. Schierding, M. Wake, R. Saffery, J. O'Sullivan, Machine learning SNP based prediction for precision medicine. Front. Genet. 10 (2019)
32. M. Humbert, K. Huguenin, J. Hugonot, E. Ayday, J.-P. Hubaux, De-anonymizing genomic databases using phenotypic traits. Proc. Priv. Enhanc. Technol. **2015**(2), 99–114 (2015)
33. W.-Y. Hwang, Biological feature selection and disease gene identification using new stepwise random forests. Ind. Eng. Manag. Syst. **16**(1), 64–79 (2017)
34. S.B. Jordan, S.L. Fenn, B.B. Shannon, Transparency as threat at the intersection of artificial intelligence and cyberbiosecurity. Computer **53**(10), 59–68 (2020)
35. J. Jumper, R. Evans, A. Pritzel, T. Green, M. Figurnov, K. Tunyasuvunakool, O. Ronneberger, R. Bates, A. Zidek, A. Bridgland, et al., High accuracy protein structure prediction using deep learning, in *Fourteenth Critical Assessment of Techniques for Protein Structure Prediction (Abstract Book),* vol. 22, p. 24 (2020)
36. J. Jumper, R. Evans, A. Pritzel, T. Green, M. Figurnov, O. Ronneberger, K. Tunyasuvunakool, R. Bates, A. Žídek, A. Potapenko, et al., Highly accurate protein structure prediction with alphafold. Nature, 1 (2021)
37. A. Kobokovich, R. West, M. Montague, T. Inglesby, G.K. Gronvall, Strengthening security for gene synthesis: Recommendations for governance. Health Security **17**(6), 419–429 (2019)
38. M. Koohi-Moghadam, H. Wang, Y. Wang, X. Yang, H. Li, J. Wang, H. Sun, Predicting disease-associated mutation of metal-binding sites in proteins using a deep learning approach. Nat. Mach. Intell. **1**(12), 561–567 (2019)
39. L. Lello, S.G. Avery, L. Tellier, A.I. Vazquez, G. de los Campos, S.D.H. Hsu, Accurate genomic prediction of human height. Genetics **210**(2), 477–497 (2018)
40. L. Lello, T.G. Raben, S.Y. Yong, L.C.A.M. Tellier, S.D.H. Hsu, Genomic prediction of 16 complex disease risks including heart attack, diabetes, breast and prostate cancer. Scientific Reports **9**(1), 1–16 (2019)
41. F. Lentzos, How to protect the world from ultra-targeted biological weapons. Bull. Atomic Sci. **76**(6), 302–308 (2020)
42. J. Li, T.B. Conzalez Zarzar, J. White, K. Indencleef, H. Hoskens, A. Ortega Castrillon, N. Nauwelaers, A. Zaidi, R. Eller, T. Gunther, et al., Robust genome-wide ancestry inference for heterogeneous datasets and ancestry facial imaging based on the 1000 genomes project. biorxiv (2019)
43. J. Li, T.B. Conzalez Zarzar, J. White, K. Indencleef, H. Hoskens, A.O. Castrillon, N. Nauwelaers, A. Zaidi, R. Eller, T. Gunther, et al., Robust genome-wide ancestry inference for heterogeneous datasets and ancestry facial imaging based on the 1000 genomes project. biorxiv (2019)
44. B. Li, M. Fooksa, S. Heinze, J. Meiler, Finding the needle in the haystack: towards solving the protein-folding problem computationally. Crit. Rev. Biochem. Mol. Biol. **53**(1), 1–28 (2018)
45. C. Lippert, R. Sabatini, M.C. Maher, E.Y. Kang, S. Lee, O. Arikan, A. Harley, A. Bernal, P. Garst, V. Lavrenko, et al., Identification of individuals by trait prediction using whole-genome sequencing data. Proc. Natl. Acad. Sci. **114**(38), 10166–10171 (2017)

46. T.A. Manolio, F.S. Collins, N.J. Cox, D.B. Goldstein, L.A. Hindorff, D.J. Hunter, M.I. McCarthy, E.M. Ramos, L.R. Cardon, A. Chakravarti, et al., Finding the missing heritability of complex diseases. Nature **461**(7265), 747–753 (2009)
47. M. Mold, C. Linhart, J. Gómez-Ramírez, A. Villegas-Lanau, C. Exley, Aluminum and amyloid-β in familial Alzheimer's disease. J. Alzheimer's Disease (Preprint), 1–9 (2019)
48. C.A.C. Montaez, P. Fergus, A.C. Montaez, A. Hussain, D. Al-Jumeily, C. Chalmers, Deep learning classification of polygenic obesity using genome wide association study SNPs, in *2018 International Joint Conference on Neural Networks (IJCNN)* (IEEE, 2018), pp.1–8
49. National Academies of Sciences Engineering and Medicine, *Biodefense in the Age of Synthetic Biology* (National Academies Press, 2018)
50. P. Ney, L. Ceze, T. Kohno, Genotype extraction and false relative attacks: security risks to third-party genetic genealogy services beyond identity inference. Preprint Posted **10**(29), 19 (2020)
51. E. Ng, P.M. Lind, C. Lindgren, E. Ingelsson, A. Mahajan, A. Morris, L. Lind, Genome-wide association study of toxic metals and trace elements reveals novel associations. Hum. Mol. Genet. **24**(16), 4739–4745 (2015)
52. M. Nieuwenhuis, *The Ghost in the Machine: Machine learning models of the brain and genome in patients with schizophrenia and bipolar disorder*. PhD thesis, Utrecht University, 2016
53. J.T. O'Brien, C. Nelson, Assessing the risks posed by the convergence of artificial intelligence and biotechnology. Health Security **18**(3), 219–227 (2020)
54. J.H. Oh, S. Kerns, H. Ostrer, S.N. Powell, B. Rosenstein, J.O. Deasy, Computational methods using genome-wide association studies to predict radiotherapy complications and to identify correlative molecular processes. Scientific Reports **7**, 43381 (2017)
55. S. Okser, T. Pahikkala, A. Airola, T. Salakoski, S. Ripatti, T. Aittokallio, Regularized machine learning in the genetic prediction of complex traits. PLoS Genetics **10**(11), e1004754 (2014)
56. A.L. Oliveira, Biotechnology, big data and artificial intelligence. Biotechnology J. **14**(8), 1800613 (2019)
57. S. Palluk, D.H. Arlow, T. De Rond, S. Barthel, J.S. Kang, R. Bector, H.M. Baghdassarian, A.N. Truong, P.W. Kim, A.K. Singh, et al., De novo DNA synthesis using polymerase-nucleotide conjugates. Nature Biotechnology **36**(7), 645 (2018)
58. J.S. Paulsen, D.R. Langbehn, J.C. Stout, E. Aylward, C.A. Ross, M. Nance, M. Guttman, S. Johnson, M. MacDonald, L.J. Beglinger, et al., Detection of Huntington's disease decades before diagnosis: the predict-HD study. J. Neurol. Neurosurgery Psychiatr. **79**(8), 874–880 (2008)
59. E. Pauwels, *The New Geopolitics of Converging Risks: The UN and Prevention in the Era of AI* (United Nations University Centre for Policy Research, New York, 2019)
60. T. Radivojević, Z. Costello, K. Workman, H.G. Martin, A machine learning automated recommendation tool for synthetic biology. Nature Communications **11**(1), 1–14 (2020)
61. J.A. Reuter, D.V. Spacek, M.P. Snyder, High-throughput sequencing technologies. Molecular Cell **58**(4), 586–597 (2015)
62. L. Rocher, J.M. Hendrickx, Y.-A. De Montjoye, Estimating the success of re-identifications in incomplete datasets using generative models. Nature Communications **10**(1), 1–9 (2019)
63. A. Romagnoni, S. Jégou, K. Van Steen, G. Wainrib, J.-P. Hugot, Comparative performances of machine learning methods for classifying Crohn disease patients using genome-wide genotyping data. Scientific Reports **9**(1), 1–18 (2019)
64. G.J. Schumacher, S. Sawaya, D. Nelson, A.J. Hansen, Genetic information insecurity as state of the art. bioRxiv (2020)
65. D. Sero, A. Zaidi, J. Li, J.D. White, T.B.G. Zarzar, M.L. Marazita, S.M. Weinberg, P. Suetens, D. Vandermeulen, J.K. Wagner, et al., Facial recognition from DNA using face-to-DNA classifiers. Nature Communications **10**(1), 2557 (2019)
66. X. Shi, X. Wu, An overview of human genetic privacy. Ann. N. Y. Acad. Sci. **1387**(1), 61 (2017)
67. M. Umair, M. Alfadhel, Genetic disorders associated with metal metabolism. Cells **8**(12), 1598 (2019)

68. J. Van Aken, E. Hammond, Genetic engineering and biological weapons. EMBO Reports **4**(S1), S57–S60 (2003)
69. P. Waldmann, Genome-wide prediction using Bayesian additive regression trees. Genet. Sel. Evol. **48**(1), 42 (2016)
70. Z. Wang, X. Wei, J. Yang, J. Suo, J. Chen, X. Liu, X. Zhao, Chronic exposure to aluminum and risk of Alzheimer's disease: A meta-analysis. Neuroscience Letters **610**, 200–206 (2016)
71. J.B. Whitfield, V. Dy, R. McQuilty, G. Zhu, A.C. Heath, G.W. Montgomery, N.G. Martin, Genetic effects on toxic and essential elements in humans: arsenic, cadmium, copper, lead, mercury, selenium, and zinc in erythrocytes. Environ. Health Perspect. **118**(6), 776–782 (2010)
72. J.K. Wickiser, K.J. O'Donovan, M. Washington, S. Hummel, F.J. Burpo, Engineered pathogens and unnatural biological weapons: The future threat of synthetic biology. CTC Sentinel **13**, 8 (2020)
73. M. Wienroth, Socio-technical disagreements as ethical fora: Parabon NanoLab's forensic DNA snapshot service at the intersection of discourses around robust science, technology validation, and commerce. BioSocieties, 1–18 (2018)
74. World Economic Forum; Nuclear Threat Initiative, Biosecurity innovation and risk reduction: A global framework for accessible, safe and secure DNA synthesis, January 2019
75. Y. Zhang, Progress and challenges in protein structure prediction. Curr. Opin. Struct. Biol. **18**(3), 342–348 (2008)
76. J. Zhang, S.D. Petersen, T. Radivojevic, A. Ramirez, A. Pérez-Manríquez, E. Abeliuk, B.J. Sánchez, Z. Costello, Y. Chen, M.J. Fero, et al., Combining mechanistic and machine learning models for predictive engineering and optimization of tryptophan metabolism. Nature Communications **11**(1), 1–13 (2020)
77. A.B. Zheutlin, A.M. Chekroud, R. Polimanti, J. Gelernter, F.W. Sabb, R.M. Bilder, N. Freimer, E.D. London, C.M. Hultman, T.D. Cannon, Multivariate pattern analysis of genotype–phenotype relationships in schizophrenia. Schizophrenia Bulletin **44**(5), 1045–1052 (2018)

The Attack Surface of Wet Lab Automation

Naor Dalal, Yossi Oren, Yuval Dorfan, Jonathan Giron, and Rami Puzis

Abstract Robotic liquid handlers save human effort and are, in many cases, faster and more precise than a human operator. They can be operated and controlled remotely and do not require technical programming skills from their operators. Unfortunately, like many other high-tech products, robotic wet lab automation may have exploitable vulnerabilities and design weaknesses that allow subversion by an adversary. The distributed nature and remote control capabilities of wet lab automation expand its attack surface increasing the opportunities for an attack to interfere with the executed biological protocols, affect medical products, and alter test results. Perimeter defenses are known to be insufficient for proper protection of systems. Security needs to be considered throughout the entire pipeline of wet lab operations, including machinery, local- and cloud-based software, and even biological protocols. In this chapter, we review the most prominent types of robots in a biological laboratory through the lens of cyber-biosecurity and map the general attack surface of wet lab automation.

Keywords Cybersecurity · Laboratory automation · Liquid handlers · Cyberbiosecurity

N. Dalal
Software and Information Systems Engineering, Ben-Gurion University of the Negev, Beer-Sheva, Israel

Y. Oren · R. Puzis (✉)
Software and Information Systems Engineering, Ben-Gurion University of the Negev, Beer-Sheva, Israel

Cyber@BGU, Ben-Gurion University of the Negev, Beer-Sheva, Israel

Y. Dorfan · J. Giron
Innovation center, Reichman University, Herzliya, Israel

D. Greenbaum (ed.), *Cyberbiosecurity*, https://doi.org/10.1007/978-3-031-26034-6_15

1 Introduction

Pipetting, preparing, and transferring liquids require considerable time and effort in a traditional wet lab environment. Robotic liquid handlers save human effort and are, in many cases, faster and more precise than a human operator. Wet lab automation goes further, allowing a biologist to automate experiments or production via robotic control. It does not require technical programming skills and saves time and effort allowing the biologist to focus on the experimental design and data analysis. Wet lab automation frameworks can be operated and controlled remotely via a local network [1, 2] or even through a cloud [3, 4].

The core component of wet lab automation solutions is the lab robot. These robots have different capabilities, such as precise work with a pipette, liquid temperature control module, and precise liquid distribution, which can replace and scale up the work of a human lab technician. These robots carry out multiple steps in a biological protocol pipeline, starting with external biological inputs and ending with biological products, scientific data, or even clinical recommendations.

Unfortunately, like many other high-tech products, wet lab automation may have exploitable vulnerabilities and design weaknesses that allow subversion by an adversary. Regardless of the financial, ideological, or political motivation of the attackers, control over the production or experimental pipeline may result in serious adverse impacts ranging from disruption of the production to unintended and unanticipated dangerous biological byproducts.

The more distributed a wet lab automation control system is, and the more it is exposed to the Internet, the higher is the risk of an attack. Attacks can interfere with biological processes, affect medical products, and alter test results. Perimeter defenses, such as password-protected access and encrypted communication, are known to be insufficient for proper protection of systems. Security needs to be considered throughout the entire pipeline of wet lab operations, including machinery, local- and cloud-based software, and even biological protocols. Cross-site scripting, insecure applications, and insecure Internet-of-Things (IoT) controllers wired to the robots are just a few examples of potential attack vectors.

While there are many articles on cyberbiosecurity [5], biosafety and biosecurity [6, 7], cyberbiosecurity for DNA synthesis [8], assessing cyberbiosecurity vulnerabilities [9], protecting US food and agricultural system [10], harmful algal blooms (HABs) and the cyberbiosecurity of freshwater systems [11], and risk perceptions in the biotech sector [12], nevertheless, no previous work has discussed the particular security context of wet lab automation throughout the multiple steps of running the protocol pipeline.

In this research, we try to bridge this gap and try to shed light on the dangers and possible impacts of intervening with the running of a biological protocol in wet lab automation and the need to secure its proper execution.

Our contributions are as follows: First, we build a wet-lab automation ecosystem taxonomy and expand on each variable in the taxonomy. We also review a number of diverse robots in the field of biological laboratory automation and their capabilities. Next, we build and examine the pipeline of a running protocol, mapping the relevant parts for each step in the pipeline, and we describe what its role in the pipeline. For each step in the pipeline, we examine if it may be vulnerable and describe the required permission and access conditions which enable an adversary to attack this step. We create the connection between wet lab automation capabilities and the attack vectors, which attack vector can affect which capability. Finally, we perform a case study on several important lab automation protocols and show how an attacker can adversely intervene with them and what are the possible impacts of such attacks.

Wet lab automation is becoming more widespread supporting increased number of applications and deployment possibilities. Thus, it is important to consider the security aspects of wet lab automation as early as possible. By doing so, the community can prevent security-related configuration blunders with possibly fatal consequences.

2 The Wet Lab Automation Ecosystem

In this section, we analyze and present the taxonomy of the wet lab automation ecosystem as demonstrated in Fig. 1. The taxonomy shows the ecosystem of wet lab automation in general. Each leaf in the graph is variable of robot's criteria. Each wet lab automation robot can omit or add the variable in the taxonomy and implement him in his way. We present the different implementations and the generic way to implement it for each node in the taxonomy tree. The taxonomy breaks down the robot into logical parts, the ecosystem part that contains the hardware and software of the robot, and its commercial part.

2.1 Hardware

This section describes the hardware and physical (nonprogrammable) capabilities and component specification of the robot. This section is divided into several subsections; each subsection describes the hardware, ability, or physical feature of the robot.

2.1.1 Deployment

Deployment of robots involves placing the robots and their resources in specific location where they can perform their intended tasks. When you are setting up the robot laboratory infrastructure, you will be faced with multiple decisions, convenience, cost, and quality. There are two main options standing for you: on-premise robots and cloud robots (lab as a service). In this section, we introduce these two options and compare them.

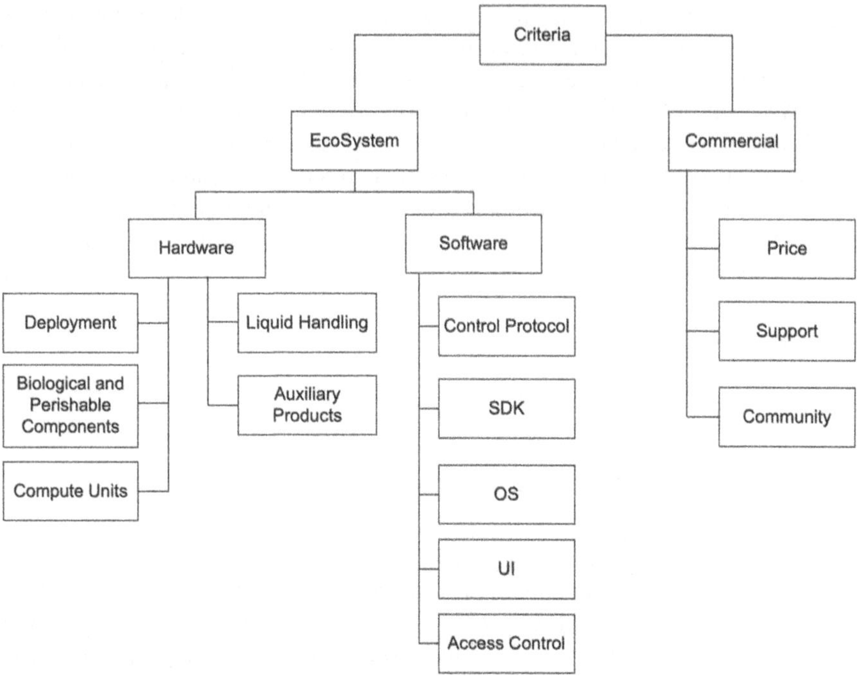

Fig. 1 Criteria hierarchy for wet lab automation ecosystem

On-Premise Most of the robots are 3D robots that provide open-source 3D models for do it yourself. You need to build the robots and store it in your lab. This requires you to access a 3D printer. You'll need to reserve an area in your lab for the robot and make sure you have basic knowledge of hardware assembly. You may need to purchase IoT devices (e.g., Arduino) for robot control and other connectable modules, for example, tip racks, well plates, and a syringe reactor. These robots are more dynamic and can be modified more easily and adapted to the needs of the laboratory, but they are less quality and simpler.

Cloud Lab Automation as-a-Service (CLAaS) Another type of robots are the CLAaS. These robots contain work cells that are woven together by an integrated stack of control software. A robotic cloud lab is a deeply integrated technology stack of biology, hardware, and software made available to its users via the cloud. Unlike traditional on-premise robots, a robotic cloud lab flexibly supports multiple assay types and is built from the ground up to be controlled remotely. These robots are more complex and have more capabilities and are usually also of better quality, but sometimes it is more difficult to adapt them to the needs of the laboratory.

On-premise robots in contrast to CLAaS are more available because the robot is located in your lab and can easily adapt to your purpose. But on the other side, CLAaS is more maintained and the quality is higher. On-premise lab prices are according to the level of equipment of the robot and quantity of the pluggable modules you buy. There are robots that you can buy from the company instead of assembling it yourself (i.e., OT-2).

2.1.2 Biological and Perishable Components

Here we list the physical components required to operate biological protocols.

Single and Multichannel Pipette A pipette is a laboratory instrument used to measure out or transfer small quantities of liquid. Multichannel pipettes generally come with either 8 or 12 pipette heads, easily allowing for a single device to fill multiple wells at a single time.

Pipette Tip Pipette tips are disposable attachments for the uptake and dispensing of liquids using a pipette.

Tip Racks Holders and replacement trays for disposable pipette tips are designed and packaged to facilitate the reuse of pipette tip boxes to reduce the overall amount of plastic waste.

Well Plates The well plate is a flat plate that looks like a tray with multiple wells that are used as small test tubes.

Tube Rack Test tube racks are laboratory equipment used to hold upright multiple test tubes at the same time. They are most commonly used when various different solutions are needed to work with simultaneously, for safety reasons, for safe storage of test tubes, and to ease the transport of multiple tubes.

2.1.3 Compute Units

Next is a list of common hardware compute units that are responsible for communication, processing, and control of the robot actuators.

Stepper Motor Driver Carrier (i.e., DRV8825) Stepper motor drivers are specifically designed to drive stepper motors, which are capable of continuous rotation with precise position control, even without a feedback system. Stepper motors are used for moving the robots in multiple axes (2 and 3 axes) separately and simultaneously.

Arduino Arduino is an open-source electronic platform based on easy-to-use hardware and software. It's intended for anyone making interactive projects. Arduino can be used for two purposes: as an endpoint that can be connected to robots via Wi-Fi and an actuator that communicates with the robots via a proprietary protocol.

Raspberry Pi Raspberry Pi is a tiny computer about the size of a deck of cards. It uses what is called a system on a chip, which integrates the CPU and GPU in a single integrated circuit, with the RAM, universal serial bus (USB) ports, and other components soldered onto the board for an all-in-one package. Raspberry is used for communicating withrobots, that is, EvoBot Raspberry sends the G-code commands to the robot through a USB connection.

2.1.4 Liquid Handling

Liquid handling is the act of transferring liquid from one location to another in a laboratory, usually for testing purposes. The robots have varied types of liquid handling capabilities:

Shake the Tube A hardware module controlled by firmware that is designed to mix liquids in different frequencies.

Vacuum Aspiration A hardware module for pulling liquid up into the pipette tip.

Blow Out A hardware module for pushing an extra amount of air through the pipette tip, so as to make sure that any remaining droplets are expelled.

Dispense Liquids A hardware module for pushing out liquid from the pipette tip into plate or another implement.

2.1.5 Auxiliary Products

Some of the robots came with connectable modules that optimize and help with the experiment. We mention a short list of these products:

Camera Module Some robots have the ability to put a camera on top of the robot that will record all the experiments; this helps in exploring and understanding the experiment.

Microscope Module It is a pluggable module that helps biologists better observe the liquid during the experiment. It is an instrument used to examine objects that are too small to be seen by the naked eye. The camera and microscope can combine together by recording the experiment through the microscope.

Temperature Module It is a pluggable module that can control accurately the temperature of the liquids. Temperature module is a hot and cold plate module.

Magnetic Module The magnetic module is a magnetic bead-based chemistry block for extraction and purification. It automatically engages and disengages high-strength magnetic bars to seated well plates for magnetic bead-based purification protocols.

Thermocycler Module Thermocyclers are instruments used to amplify DNA and RNA samples by the polymerase chain reaction.

High-Efficiency Particulate Air (HEPA) Module HEPA is an efficiency standard of air filter.

Sensors In addition to the auxiliary components listed above, some robotic frameworks for the wet lab also include various sensors, such as: motion sensor, ultrasonic sensor, sound sensor, and light sensor.

2.2 Software

This section describes the programmable parts in the robot, according to the taxonomy tree in Fig. 1. Programmable parts could be software, firmware, or even the protocol between the components of the robot. Each subsection describes these programmable parts.

2.2.1 Control Protocol

The robots use various control protocols, some of which are proprietary and some are known standards.

G-code G-code is a software programming language used to control a computer numerical control (CNC) machine. It is used mainly in computer-aided manufacturing to control automated machine tools and has many variants. Raspberry Pi sends the G-code commands to the robot through a USB connection.

uArm Swift Pro Protocol The uArm Swift Pro is an open-source Arduino-based robot arm designed for desktop use. Based on the standard G-code protocol, they add a new protocol head in front of the G-code so that it can be more easily used and debugged. What is more, it is designed to be compatible with the standard G-code.

Proprietary Protocols Some robot designers created their own simple control protocols suitable for their robot. They programmed Arduino using the analog write and read pin functions.

2.2.2 Software Development Kit (SDK)

SDK is a collection of software development tools in one installable package. The robots provide an SDK for controlling the robots; most of the robots provide a python SDK. The SDK contains functionality for full control of the robots. Usually the SDK simply sends a hypertext transfer protocol (HTTP) request to a server that actually controls the robots, but some run on the computer that controls the robots. In some devices, the SDK command translates to Extensible Markup Language (XML)-Remote Procedure Call (RPC) (XML-RPC), a protocol that uses XML to encode its calls and HTTP as a transport mechanism. You can automate the robot action and create protocols by python script and API the robots reveals to the user.

On-premise robots can be modified, and you can automate it yourself because you have the firmware of the IoT devices.

2.2.3 Operating System (OS)

Arduino lacks a full operating system, usually writing code that is interpreted by its firmware. However, Raspberry Pi has all the features of a computer; it needs an operating system to run and comes with a fully functional operating system called Raspberry Pi OS. In addition, sometimes there is a personal computer (PC) that controls robots or runs the HTTP server; its operating system can be any operating system that runs python (especially Windows or Linux).

2.2.4 User Interface (UI)

Several robots have interactive webpage graphical user interface (GUI) to control the robot, and some have smartphone applications. The GUI displays the entire protocol and robot control process and can be changed in any time. Behind the scenes, the beautiful GUI is converted to either code running on the IoT device or to Application Programming Interface (API) commands. OpenLH, for example, builds their GUI with Google's Blockly interface [13] which is converted to python code running on the computer which controls the arm of the robot. Another type of UI is the command-line interface (CLI); some robots provide commands that you can run from the CLI and automate the robots with it. Another type of robot does not provide GUI or CLI; the programmer needs to write the protocol using integrated development environment (IDE).

2.2.5 Access Control

For most of on-premise robots, there are no security aspects in the software. Some of them [1] created an open Wi-Fi by one of the IoT devices, and everyone in the same local area network (LAN) can control the robot. Others just need to connect to HTTP server through specific port, and you are free to go and run every protocol you want. However, CLAaS place more emphasis on security and use well-known security models such as hypertext transfer protocol secure (HTTPS) and two-factor authentication (2FA); 2FA is a security method that adds an additional layer of protection on top of just your username and password. It is a method of verifying that the person who is trying to access your account is who they say they are.

2.3 Commercial Aspects

This section describes the commercial aspect of the robots. This section describes the price ranges of wet lab automation robot types, the support that the developers of the robot give, and the community and the distribution of the robots.

2.3.1 Price

The robot prices range from $400 for open source and do-it-yourself robots to $9000 for robots that you got full assembly with multiple hardware components as described above. There are open-source robots that offer you full assembly instead of do-it-yourself. CLAaS robots are for subscription.

2.3.2 Support

Commercial robots run by companies are including contact support, return policy, warranty, and documentation. In contrast, open-source robots are less maintained; this is reflected in the lack of good documentation, contact support, and quality.

2.3.3 Community

Commercial robots are widely distributed; there are many companies that collaborate with the company that builds the robot; either it's CLAaS or on-premise. But some robots (do-it-yourself) are less distributed, and there are not too many sources on the community of these robots.

2.4 Summary

Detailing and mapping the taxonomy of wet lab automation ecosystem help us to understand better how the robots are built and what their capabilities are. It sheds light on where the security failures may be found and where a potential attacker could intervene in the system. We analyzed each property in the taxonomy considering whether this property may have security failures and whether an adversary can utilize it to his advantage.

3 Biological Laboratory Robots

In this section, we present some examples of wet lab automation robots. For each of them, we detail about its capabilities, hardware, and software. We showcase the uniqueness of each robot and how it differs from the other robots shown in this section. Furthermore, we attached figures of the robots and links for their open source and cited academic articles if exist.

Review Methodology In order to obtain the information about the robots, we read datasheets and published articles describing the robots [1, 14, 15]. To understand the developer's perspective on creating and running custom protocols, we examine the robots' APIs [16, 17] and attempt using the API ourselves. Further, in order to understand how a reviewed framework operates behind the scenes, we inspect the open-source code of the robots [18–22]. Such inspection often reveals issues not listed in the datasheets and API specifications. This review methodology is limited in a sense that we did not have access to the source code of all robots. The close source robots were examined in a less profound way. In Sect. 6, we elaborate the limitations in more detail.

3.1 Fully Integrable Noncommercial Dispensing Utility System (FINDUS)

FINDUS [1] is an on-premise open-source [20] 3D Printable Liquid-Handling Workstation for Laboratory Automation in Life Sciences. FINDUS hardware contains: (i) 3D-printed parts with an Anycubic 4Max printer; (ii) four stepper motors, for XY drives, Z drive, and pipet drive; (iii) DRV8825 controller boards and controlled stepper motors using a motor library provided by Laurentiu Badea; and (iv) two Arduino NodeMCU 1.0 (ESP-12E Module).

FINDUS software builds from python package for controlling the robot from PC through Wi-Fi and Arduino code that implements API server for commands from PC and controls the movements and shakers.

FINDUS is able to (i) start/stop shake the tube; (ii) start/stop vacuum aspiration; (iii) move in X, Y, and Z axes; (iv) move pipet; (v) move in X and Y axes simultaneously; and (vi) set position for X, Y, and Z axis pipette.

We can see in Fig. 2 the FINDUS workstation and its components. There are three-axis motion motors, syringe reactor, tip rack, and shaker motor, and more.

3.2 EvoBot

EvoBot [14] is an open-source [19], modular, liquid-handling robot for scientific experiments. Figure 3 shows a schematic view of the electronics of EvoBot and

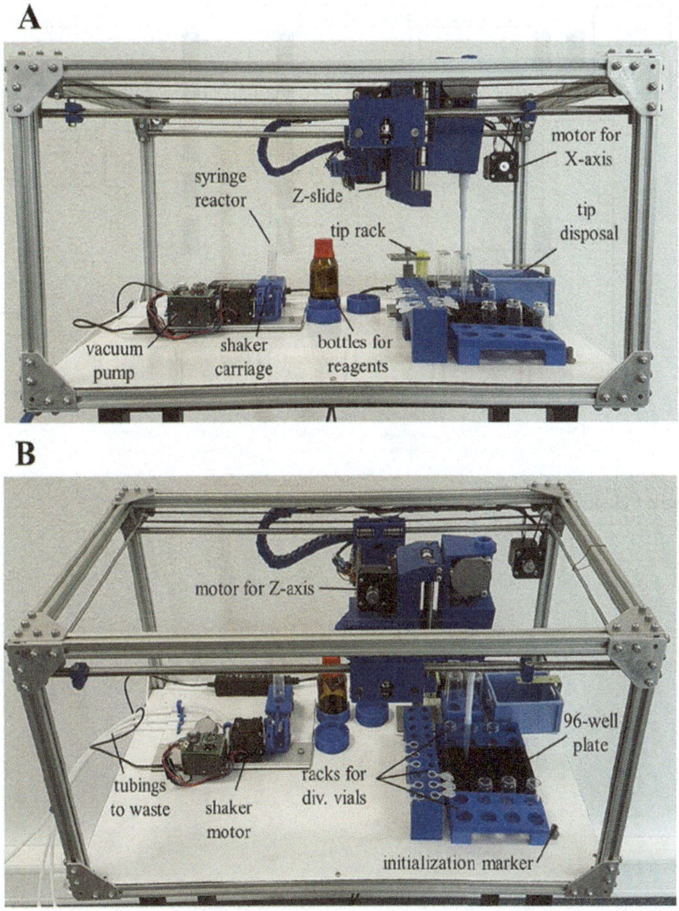

Fig. 2 FINDUS workstation, from FINDUS [1]

its different printed circuit boards (PCBs). The core of the electronics is based on electronics used in the open-source 3D printer community. EvoBot is built from the following, as shown in Fig. 4:

(i) Three layers: an actuation, an experimental, and an observation layers.
(ii) Actuation layer holds modules and can be moved in the horizontal plane by using two stepper motors.
(iii) Experimental layer supports the objects of the experiment such as petri dishes, microscope slides, or tubes.
(iv) Observation layer is optional, and most modules plugged into this layer are used to sense or observe the ongoing experiments, for example, camera and microscope.

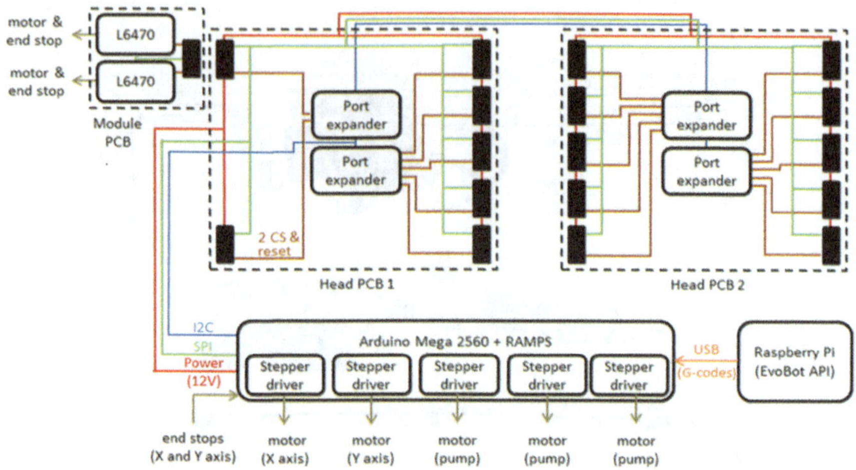

Fig. 3 EvoBot electronic schematic view, from EvoBot [14]

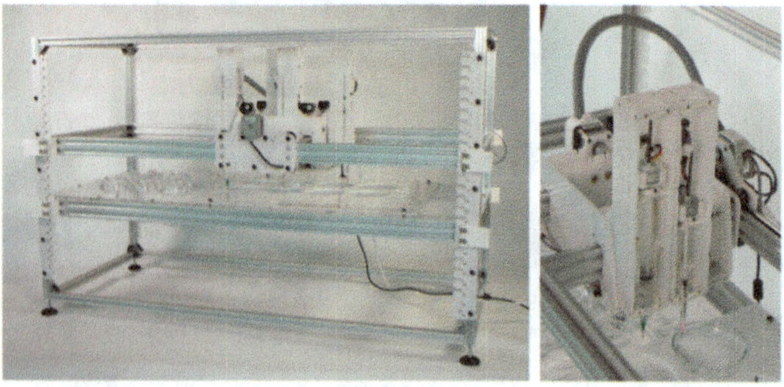

Fig. 4 EvoBot liquid-handling robot, from EvoBot OpenLH [14]

(v) Three different kinds of modules: a syringe module, a pump-based dispensing module, and a heavy payload module (microscope, three-dimensional scanner).
(vi) Arduino and Raspberry Pi 3.
(vii) Stepper motors.

EvoBot includes a software part that contains the following:

(i) Arduino runs a modified version of the Marlin firmware, which is widely used to control 3D printers using G-code.
(ii) Raspberry Pi sends the G-code commands to the robot through a USB connection.
(iii) Python API gives users access to control the robot.

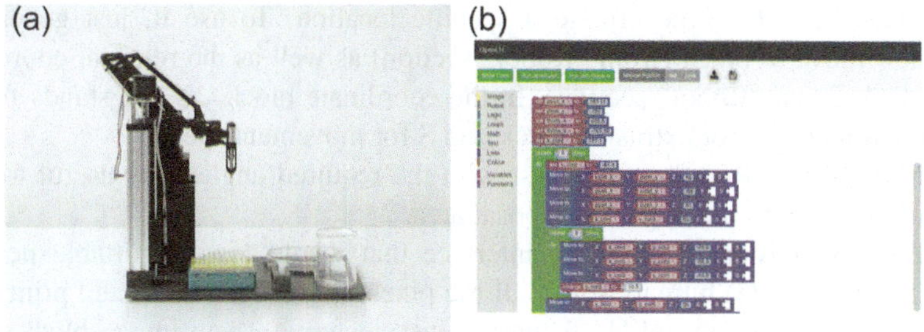

Fig. 5 OpenLH, from Ref. [21]. (**a**) uArm Swift Pro. (**b**) OpenLH blockly interface

(iv) Users can interact directly with the robot using a GUI, or they can run programs directly on the Raspberry Pi.

EvoBot is able to perform the following:

(i) The syringe module moves liquids with precision, it can move the syringe up and down in addition to the movement of the plunger.
(ii) The syringes can be easily replaced by just loosening and tightening one screw.
(iii) The dispensing module can pump up to four liquids and is used to wash Petri dishes or dispense pure reagents into vessels' start/stop vacuum aspiration.
(iv) Heavy payload module to hold a 3D scanner.

3.3 OpenLH

The OpenLH [15] is an open-source [21] liquid-handling system based on an available robotic arm platform (uArm Swift Pro) which allows for creative exploration by biologists and bio-enthusiasts. OpenLH is built from three main parts: (i) an open-source robotic arm, uArm Swift Pro [18]; (ii) a linear actuator-operated syringe pump; and (iii) the custom-made liquid-handling attachment, as can be seen in Fig. 5a.

The uArm runs on top of an Arduino Mega 2560 with a custom version of Marlin firmware (available under GPL license). The robot operates using G-code definitions sent through universal asynchronous receiver transmitter (UART) protocol. OpenLH software is built from several parts as the following:

(i) The user may generate different programs manipulating the arm using Google's Blockly interface [13] as can be seen in Fig. 5b.
(ii) The generated program is then compiled to python code, using the Swift API (which compiles to G-code commands).
(iii) It is possible to save programs for later use and upload images for the Bitmap to bioprint feature.

OpenLH has the following main features:

(i) Move To: Move the arm to a specific location. To use it, just generate a new move to block (from "Robot" section) as well as the relevant coordinate block (from "Robot" section). In the coordinate block, X Y Z stands for the coordinates, E for extrusion level, and S for movement speed.
(ii) Move Wrist: Rotate arm's wrist with the required angle. It is useful to drop used tips from the arm to a disposal area.
(iii) Bitmap to Bioprint: It is an interface that would load a portable network graphics (PNG) bitmap, select all the pixels of a single color, and print these pixels with the OpenLH. To use it, just generate a new image block (from "Image" section) as well as the relevant coordinate blocks (from "Robot" section).
(iv) Manual Position: Puts the arm in disjoint mode, allowing the user to move it around manually and sample coordinates. After reaching a desired location, a tip to pick up, for example, hit set button to generate the location's coordinates as a new usable block.

3.4 Opentrons OT-2

Opentrons [2] OT-2 is an open-source [22] liquid-handling robot. Opentrons OT-2 is built from following three sections:

(i) Labware – You must tell the protocol context about what should be present on the deck (well plate, tube rack), Labware Library.
(ii) Pipettes – You define the instruments required for your protocol. You tell the protocol context about which pipettes should be attached and which slot they should be attached to (11 slots on the deck).
(iii) Commands define the commands that make up the protocol. The most common commands are aspirate, dispense, pickup tip, and drop tip. Opentrons OT-2 pipette configurations: Single- and eight-channel pipetting, two-pipette mounts, for a configuration of one or two single- or eight-channel pipettes. Pipettes are easily interchangeable. Opentrons OT-2 contains 11 deck slots that enable countless configurations; deck slots are compatible with standard SBS dimensions. Deck also includes a removable trash bin. Its connectivity is through Wi-Fi 2.4 GHz IEEE 802.11b/g/n, USB 2.0. It can be applied to connectable pluggable hardware modules; modules are peripherals that attach to the OT-2 to extend its capabilities: (i) temperature, (ii) magnetic, and (iii) thermocycler modules, as shown in Fig. 6b.

Opentrons OT-2 is an open-source do-it-yourself but can be bought from Opentrons starting at $5000. The OT-2 Python Protocol API is a simple python framework designed to make writing automated biology lab protocols easy. The python script is running by Opentrons App.

Opentrons OT-2 has advanced control: sometimes, you may write a protocol that is not suitable for execution through the Opentrons App. Perhaps it requires user

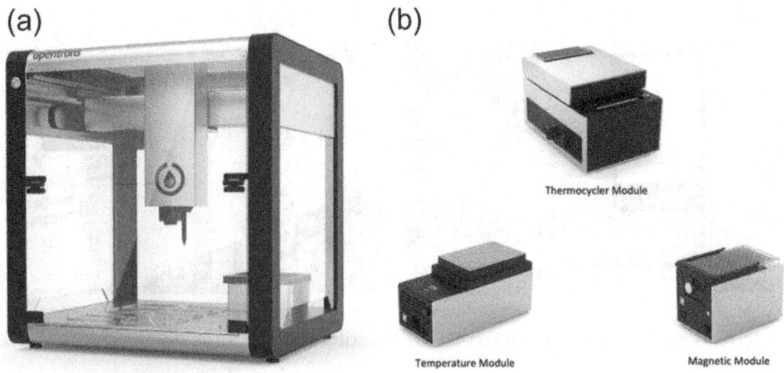

Fig. 6 Opentrons, from Ref. [2]. (**a**) OT-2 (**b**) Pluggable Modules

input; perhaps it needs to do a lot of things it cannot do when being simulated. There are two ways to run a protocol on the OT-2 without using the Opentrons App: Jupyter Notebook and CLI.

Opentrons OT-2 can be used for many purposes, for example, the following articles [23–25] show how it was used for COVID-19 polymerase chain reaction (PCR) testing automation.

3.5 Strateos

Strateos [3, 26] is a CLAaS [27] solution provider. The company's platform enables scientists to design, run, and analyze experiments remotely utilizing Strateos' robotic cloud labs. In addition, Strateos designs, builds, and implements modular cloud labs in their clients' facilities. Clients have the option of toggling between their own on-site facilities and Strateos' remote-controlled cloud labs for small molecule drug discovery, biologics, and synthetic biology workflows, advancing the digitization of laboratories via a hybrid lab solution.

Strateos' core technology is their lab control software that integrates and controls various instrument types, ranging from liquid handlers, bioreactors, and high content imagers to an array of ultrahigh-throughput screening instruments and devices. This modular, cloud-based software addresses common challenges in research operations and scientific experiment execution, even in labs with no automation currently. Strateos' software also focuses on solving operational challenges found in labs, such as managing experimental requests, asset and workflow calendaring, and executing between teams and individual users and enabling automatic data capture and centralization of scientific workflows to accelerate the design-make-test-analyze cycle and generate AI-enabled data that will aid in the discovery of new scientific insights. The software is modular and scalable from control of work cells to multiple client facilities as can be seen in Fig. 7b.

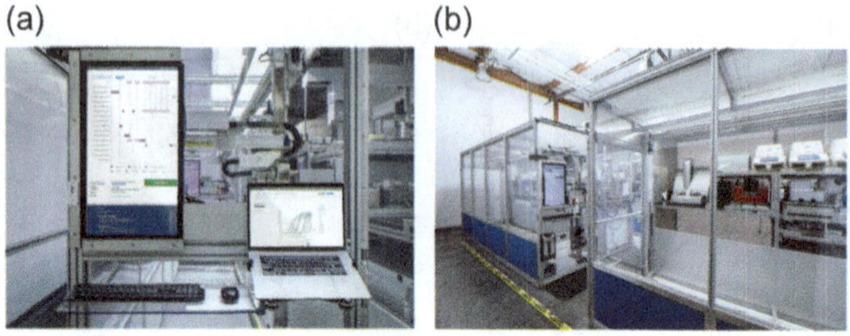

Fig. 7 Strateos, from Ref. [3]. (**a**) Strateos CLAaS solution. (**b**) Strateos work cell

Strateos developed and maintained Autoprotocol [28], as the open-source standard helping define experiments that are run over the Internet on remote robotic automation, moving research into the cloud. Open-source software packages are used to organize a collection of protocols and allow customers to build protocols using Python, or alternatively clients can access Strateos GUI to build and automate protocols. Autoprotocol is a JavaScript Object Notation (JSON) formatted data structure that provides a precise way of describing and automating biological and chemical protocols in the lab. A run can be submitted by posting properly formatted Autoprotocol to the server via the Strateos API.

3.6 Summary

In Table 1, we summarize the main components of each robot. Most solutions are deployed on-premise with a great deal of customization and lack of standardized security controls. This naturally increases the attack surface of open-source wet lab automation frameworks. Access control in majority of the solution is based on plain HTTP allowing man-in-the-middle attacks. We also observed high similarity among the biological protocol processing, server components, and control in different solutions. Vulnerabilities in the implementation and processing of biological protocols as well as components responsible for the protocol's execution may have the most severe impacts and thus deserve the most attention of security researchers.

4 Attack Surface

This section describes the potential attack vectors among the wet lab automation ecosystem. In this section, we analyze each entry point of the system and the all-pipeline of running protocol from his design and planning until it is running. We examine and describe each step in the pipeline and analyze whether it is possible

Table 1 Comparison of wet lab automation solutions

Component	Robot				
	FINDUS	EvoBot	OpenLH	Opentrons OT-2	Strateos
Deployment	On-premise	On-premise	On-premise	On-premise	CLAaS
Auxiliary products (short)	None	Camera	Sensors, temperature, electromagnet, and camera	Magnetic, thermocycler, temperature, and HEPA	Magnetic, thermocycler, temperature, and Illumina sequence
Control proto-Col	Proprietary protocol	G-code	G-code	G-code	Secretive
SDK	Python SDK	Python SDK and API	Python SDK and API	Python SDK and API	Python SDK and API
OS	Arduino	Marlin firmware and raspberry pi OS	Marlin firmware	Proprietary embedded hardware	Secretive
UI	IDE	GUI	GUI	GUI and CLI	GUI and CLI
Access control	Open Wi-fi	XML-RPC	HTTP server	HTTP server	HTTPS server and 2FA

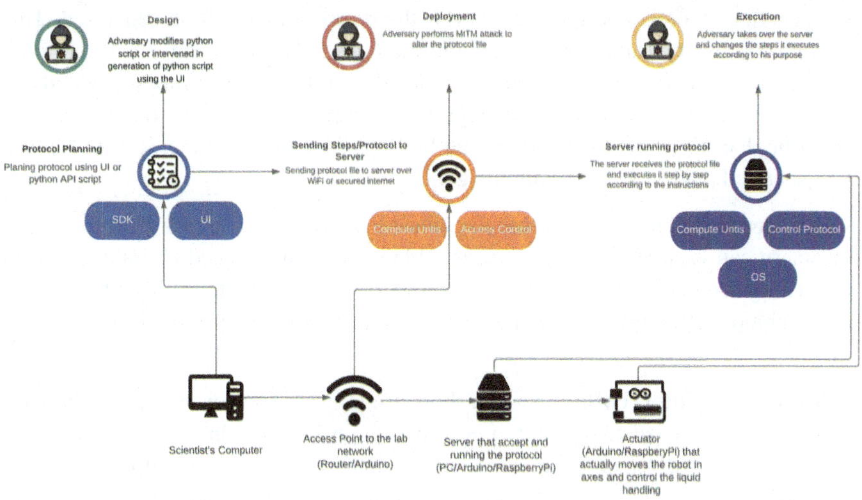

Fig. 8 Attack surfaces on wet lab automation ecosystem

to intervene at this step in the final protocol, and if possible, we describe how can adversary do this.

We can see in Fig. 8 the pipeline of running protocol and the potential intervention of adversary. The bottom items describe the physical components and their connection to the proper step in the pipeline of running protocol. Each step is accompanied by a description and variable in the taxonomy tree in Fig. 1. The upper arrows coming out of each step express the possible intervention in the protocol or in the laboratory spec at this step. The variables that accompanied each step were designed to explain which parts are relevant to the intervention at this step. There are several articles that describe similar general structures such as on cyber-physical system (CPS) networks [29] but not on wet lab automation ecosystem.

4.1 Design

When the biologists start planning their protocol, mostly they have two main options to design the protocol as described in taxonomy Sect. 2: SDK or UI. Adversary that aims to change the protocol can do it with a grip on the biologist's computer by the following techniques:

Script Injection Adversary can inject himself to protocol script and inject his steps to protocol flow and prevent some steps in the protocol. This technique assumes that the adversary knows the protocol and knows how to replace specific steps to gain his goal.

Script Generator Intervention Adversary can inject himself to the UI application that generates protocol script and controls the protocol that will be generated. This technique doesn't assume anything about the knowledge of the adversary with the generated protocol; if the adversary knows the protocol the biologist intends to create, he can replace only specific steps with minimal intervention to gain his goal. But if he doesn't know the protocol the biologist intends to create, he can replace the generated protocol file to gain his goal.

Script Replacement If the adversary doesn't know the biologist's protocol, the adversary can treat the protocol file as a black box, and instead of intervening in an existing protocol, he can replace the protocol file with another file as he wishes. This technique assumes that the adversary doesn't know the protocol the biologist intends to create.

Labware Spec Intervention Some robots use labware spec or manifest. The spec uses to provide metadata and parameters and describes both labware's dimensions and properties. Adversary can intervene in this manifest and manipulate it in a malicious way to influence or damage the results of the protocol.

Affected Capabilities Design state script manipulation can influence among other things the following capabilities:

(i) Temperature module. Adversary can change the temperature of the temperature module and affect the proper procedure of the protocol.
(ii) Magnetic Module. Adversary can raise the magnets to induce a magnetic field in the labware.

(iii) Thermocycler module. Adversary can change the temperature of the block in which samples are located and temperature of the lid heating pad.
(iv) Vacuum aspiration. Adversary can change the amount of liquid that pulls into the pipette tip.
(v) Dispense liquids. Adversary can change the amount of liquid that push out from pipette tip into plate.
(vi) Blow Out. Adversary can prevent from blowing out the remaining droplets.

4.2 Deployment

After the biologist created the protocol script, either by SDK or UI, he needs to send the protocol (the script itself or JSON file that represents the protocol (Autoprotocol [28])). Adversary that aims to change the protocol can do it with a grip on the biologist's lab network/LAN by the following techniques:

Man in the Middle (MITM) Attack MITM is a known approach in many cases including CPS networks [30, 31]. Most of the open-source on-premise robots come with unsecured Wi-Fi and HTTP server and not HTTPS which reveals the biologist to MITM attacks. In such robots, adversary can perform MITM between the biologist's computer to HTTP server and modify the command the protocol that is sent to the server without the knowledge of the biologist. In addition, the attacker can change the labware spec to manipulate the lab environment for malicious purpose and damage or manipulate experiment results.

Impersonate Biologist Due to insecure control on on-premise robots, adversary can impersonate biologist's computer and control the robot and run any protocol he wants. This technique assumes that adversary has grip on some device in biologist's lab, that is, another computer in the network, the access point (Arduino).

Wi-Fi Sniffing Adversary can sniff the traffic on an open Wi-Fi using grip if a malware is present on some computer near the robot. Operating sniffing tools in a monitoring mode to collect traffic in open Wi-Fi networks does not require authentication. Thus, unencrypted Wi-Fi channel opened by one of the robot's components facilitates leakage of information about the running protocols in the lab and their results.

4.3 Execution

Finally, the protocol (Python script or JSON file) arrived to server that controls the robot and runs the protocol. Because in some robots the protocol represents using python script that runs as is in the server, adversary could run an arbitrary code (remote code execution (RCE)) in the server without the knowledge of the biologist. Running on the server that controls the robot could lead to dire consequences on the lab and protocol results. Execution is too late for intervention in the labware spec. Adversary with a grip on the server can make the following actions:

File System Manipulation Adversary that changed the protocol using one of the ways we mentioned above can manipulate the file system by inserting python commands into the protocol that accesses the file system. Adversary can do this because script execution is performed in an unsafe environment and not using restricted python [32]. Lack of restricted python and trust in the script itself without integrity validation of the script can lead to unwanted results of the protocol and leakage of results of previous protocols running on the robot. This attack vector assumes that adversary has grip on the server that runs the protocols and the protocols are sent to the server, which is a Python script that the server is running.

G-Code Command Intervention Most of the robots are controlled by G-code commands as mentioned in the taxonomy of wet lab automation. The server that runs the protocol actually sends G-code commands to the robot according to the command in the script. Adversary with knowledge on the G-code commands that are sent to the robot, which is not an unfounded requirement because all the robots we mention are open-source, with grip in the server can create its own G-code commands and send them to the robot to manipulate the running of the protocol or run preliminary steps to control the results of the protocol that the server intends to run.

Actuator Hijacking Actuator is the controller (Arduino/Raspberry Pi) that actually moves the robot in axes and controls the liquid handling. Those actuators are usually unsecured and written in the simplest way; in most of the types, it is Arduino that gets the G-code commands through USB or UART. If the adversary could hijack actuator through a vulnerability he exploits via USB or UART, he can run arbitrary code on the actuator and actually do whatever he wants, run commands as desired, skip commands from the biologist, and even leak the protocol using Wi-Fi that is sometimes found on these actuators. Adversary can control the Arduino using one of the vulnerabilities it exposed as detailed in security analysis and exploitation of Arduino devices in the Internet of Things [33].

5 Misuse Cases

In this section, we describe several automation processes in the biological field; we explain the process and how automation fits into them and its importance. In each of the processes, we explain how an attacker can intervene in it and what the possible damages are to such an attack.

5.1 Personalized Medicine

As genetic technology is improving, personalized medicine will replace conventional treatment [34, 35]. Together with the great hopes of tailor-made medicine, we identify a greater potential of damage due to automation failures.

For example, tissue testing is applied in cases of cancer to choose the best biological chemotherapy or its combined treatment using automated screening methods to comply each specific case.

DNA or RNA aptamer use for general and personalized therapeutics has shown great promise [36, 37]; production of template RNA/DNA aptamers is commonly done using automated DNA synthesis methods [38] that produce the required strands in a sequential way.

Attack on the DNA synthesis is presented in [39], who show acoustic side-channel attack methodology which can be used on DNA synthesizers to breach their confidentiality and steal valuable oligonucleotide sequences. The potential attack could result in null effect of the treatment or damage to the tissue.

High-throughput screening of dedicated medicine and factors including repurposing of generic drugs to measure hit conformation with a robust effect on the tested tissue.

Intervention in personalized medicine process can be performed by the following:

(i) Damage of the tissue while using nondrug-related influence and temperature can cause mismatches of hit confirmation and treatment of patients using noneffective drugs.
(ii) Damage the process of the drug administration for the screening.
(iii) Its weak point is its flexibility because it supports many types of treatments, so it can be disrupted relatively easily and substances can be omitted or added to the drug if adversary has grip on the robot.

5.2 COVID-19 PCR Tests

Automation of PCR testing [40, 41] is already used everywhere. PCR is the duplication and amplification of short and long DNA oligos. This process allows for the detection of minuscule samples of DNA [42] and is used to detect the presence of the COVID-19 virus in human samples. The impact of mistakes here could be dramatic on public health.

Intervention in automation of PCR testing can cause several damages as the following:

(i) Integrity. Adversary can damage the integrity of the test through swapping the samples of people and cause impairment of the test's integrity. Moreover, adversary can damage the integrity using malware on the PCR reader that alters the screening report and swapping people results. Adversary using a malware on the robot can manipulate the liquid's temperature and cause false-negative or positive results.

(ii) Confidentiality. Adversary with grip on the automation ecosystem can release the PCR test results to the public.

5.3 Sportsmen Doping Test Control

Doping is an old well-known issue in professional sport [43]. Throughout the process of competitive competition, many drug tests are conducted. The process includes sampling the sportsmen and women and identifying illegal substances in their blood or urine. The preparation and measurements of the samples using automated measures could provide a solution for many cases where the results are needed in a short time. While automation provides great advantages, it also poses risks to the integrity and coding of the samples and possibilities of cyber-attacks that can meddle in the analysis and reporting process of the results.

Similar to PCR tests, adversary can swap the samples of sportsmen and cause impairment of the test's integrity, swapping test of sportsmen that took drugs with test of clean sportsmen to evade punishment. In addition, adversary can contaminate the test with the standard that the test compares to and cause to a clean athlete to be considered to have taken drugs.

5.4 On-site Drug Production and Dispensing

In many cases, on-site production is necessary [44, 45]. It can reduce shipping costs and improve the quality of multiple products, especially medicine, such as vaccines that need special preserving conditions. The on-site robots will prepare the drugs using liquid and powder handling automation. Furthermore, they will provide necessary dispensing of ready-made tablets and liquids per client.

While providing great economic advantages, on-site production also poses threats through cyber-attacks that interfere or meddle in the preparation or tagging process of the prepared drugs.

Because of the automation of the process, adversary can cause robots to return the wrong type of medicine without the knowledge of the patient. Adversary also can damage the production of the medicine by omitting important substances of the medicine. Moreover, adversary could change the labels of allergen on the medicine package and cause people to have allergic reaction that could endanger their lives.

6 Summary and Conclusions

Key Takeaways In this chapter, we have analyzed the potential attack surface in the wet lab automation ecosystem. The most important finding from the analysis is that the biological protocol implemented as a Python script is dangerous. Vulnerabilities in the protocol editing tools, components that interpret and execute the protocols, and components operated by the protocols can lead to severe adverse consequences as described in Sect. 5. Some consequences can be prevented digitally by signing the protocol and executing it in a secured environment such as restricted python [32]. In general, a user-provided script should never be considered trusted.

Furthermore, many components in the robot environment are distributed and wireless. This allows the attacker to intervene with the robot operation in several stages. Therefore, secure operation of the robot requires hardened communication between the components.

Limitations Our collection data methodology for each device has some limitations because there are robots we haven't their source code and only API and datasheet documentations. It could be that some of the data we had on the rest of the robots does not quite fit these robots. Moreover, we didn't actually implement a real attack on these robots; it could be that the vulnerabilities we mentioned are not existing in some robots also because of a possible mismatch between the design of the system and its actual implementation. On the one hand, the attacks can be less deadly and dangerous than we have described, but on the other hand, the opposite is also true and attacks can be even more dangerous than we described.

Related Work Cyberbiosecurity is proposed as a new discipline at the interface of cybersecurity, CPS, and biosecurity to help safeguard the bioeconomy [5]. The first paper on cyberbiosecurity was focused on biotechnology and its security concerns [6]; it explained that biotechnology workflows are cyber-physical processes and illustrated with biomanufacturing process [7]. map the cyberbiosecurity landscape, in biotechnology and digitization of traditional technology. They explained about biosecurity on automation processes similar to us and cyberbiosecurity on artificial intelligence (AI) techniques across the biology sector [8]. illustrate malicious DNA injection performed by a remote cyber-criminal in DNA synthesis process and offer some mitigations. Protecting US food and agricultural system is reviewed [10], the cyberbiosecurity concepts from food production to the end user are explored, challenges are described, and solutions to integrate cyberbiosecurity in food and agricultural sectors are recommended. According to [12], cyberbiosecurity risks are difficult to characterize due to their diversity in types of threats, targets, and potential impacts.

Future Work The prevalence of attack vectors in wet lab automation frameworks suggests that we cannot rely only on perimeter protection and standard security controls. In order to continue providing the flexibility and power of custom design of biological protocols, future wet lab automation frameworks should be robust against subversion of their components.

Future research is required to illustrate the dangers of cyber-attacks on biological protocols and offer better protections of the wet lab automation systems: (i) identifying vulnerable protocol whose results can be manipulated without alerting the biologist, (ii) investigating process signing approaches for biological protocols, and (iii) designing adversary resilient distributed wet lab automation systems where every component ensures the correct operation of other components.

Acknowledgments This study was partially supported by the Cyber Security Research Center at the Ben-Gurion University of the Negev. All images in this chapter, except Fig. 7, are used under open-source licenses of their respective owners. Figure 7 is reproduced with permission of its owner.

References

1. F. Barthels, U. Barthels, M. Schwickert, T. Schirmeister, Findus: An open-source 3d printable liquid-handling workstation for laboratory automation in life sciences. SLAS TECHNOLOGY: Translating Life Sciences Innovation **25**(2), 190–199 (2020). https://doi.org/10.1177/2472630319877374. PMID: 31540570
2. Opentrons ot-2 – opentrons open source lab automation. https://opentrons.com/ot-2
3. Strateos – cloud-base lab automation solution. https://strateos.com/strateos-control-our-lab/
4. Automata labs – cloud-base lab automation for life sciences. https://automata.tech/products/automata-labs/
5. R.S. Murch, W.K. So, W.G. Buchholz, S. Raman, J. Peccoud, Cyberbiosecurity: An emerging new discipline to help safeguard the bioeconomy. Front. Bioeng. Biotechnol. **39** (2018)
6. J. Peccoud, J.E. Gallegos, R. Murch, W.G. Buchholz, S. Raman, Cyberbiosecurity: From naive trust to risk awareness. Trends Biotechnol. **36**(1), 4–7 (2018)
7. L.C. Richardson, N.D. Connell, S.M. Lewis, E. Pauwels, R.S. Murch, Cyberbiosecurity: A call for cooperation in a new threat landscape. Front. Bioeng. Biotechnol. **7**, 99 (2019)
8. R. Puzis, D. Farbiash, O. Brodt, Y. Elovici, D. Greenbaum, Increased cyber-biosecurity for DNA synthesis. Nat. Biotechnol. **38**(12), 1379–1381 (2020)
9. D.S. Schabacker, L.-A. Levy, N.J. Evans, J.M. Fowler, E.A. Dickey, Assessing cyberbiosecurity vulnerabilities and infrastructure resilience. Front. Bioeng. Biotechnol. **7**, 61 (2019)
10. S.E. Duncan, R. Reinhard, R.C. Williams, F. Ramsey, W. Thomason, K. Lee, N. Dudek, S. Mostaghimi, E. Colbert, R. Murch, Cyberbiosecurity: A new perspective on protecting us food and agricultural system. Front. Bioeng. Biotechnol. **7**, 63 (2019)
11. G. David, I.I.I. Schmale, A.P. Ault, W. Saad, D.T. Scott, J.A. Westrick, Perspectives on harmful algal blooms (habs) and the cyberbiosecurity of freshwater systems. Front. Bioeng. Biotechnol. **128** (2019)
12. K. Millett, E. Dos Santos, P.D. Millett, Cyberbiosecurity risk perceptions in the biotech sector. Front. Bioeng. Biotechnol. **7**, 136 (2019)
13. Blockly – client-side library for the programming language javascript for creating block-based visual programming languages and editors. https://developers.google.com/blockly
14. A. Faiña, B. Nejati, K. Stoy, Evobot: An open-source, modular, liquid handling robot for scientific experiments. Appl. Sci. **10**(3) (2020). https://doi.org/10.3390/app10030814. ISSN 2076-3417. https://www.mdpi.com/2076-3417/10/3/814

15. G. Gome, J. Waksberg, A. Grishko, I.Y. Wald, O. Zuckerman, Openlh: Open liquid-handling system for creative experimentation with biology, in *Proceedings of the Thirteenth International Conference on Tangible, Embedded, and Embodied Interaction*, (2019), pp. 55–64. https://doi.org/10.1145/3294109.3295619
16. Ot-2 python protocol api version 2. https://docs.opentrons.com/v2/
17. Strateos developer center. https://developers.strateos.com/docs
18. uarm developer – python library for uarm software. https://github.com/uArm-Developer/pyuarm
19. Evobot developer – software for the evobot robot. https://bitbucket.org/afaina/evobliss-software/src/master/
20. Findus developer- an open-source 3d printable liquid-handling workstation for laboratory automation in life sciences. https://github.com/FBarthels/FINDUS
21. Idc milab openlh – open source liquid handling system. https://github.com/idc-milab/openlh
22. Opentrons open source – source code for the opentrons api and ot app. https://github.com/Opentrons
23. J. Rader, K. Watson, Affordable covid-19 testing automation with the opentrons ot-2
24. Fernando L'azaro-Perona, Carlos Rodriguez-Antol'ın, Marina AlguacilGuill'en, Almudena Guti'errez-Arroyo, Jesu's Mingorance, Julio Garc'ıaRodriguez, and SARS-CoV-2 Working Group, Evaluation of two automated low-cost rna extraction protocols for sars-cov-2 detection. PLoS One **16**(2), e0246302 (2021)
25. Jos'e Luis Villanueva-Can˜as, Eva Gonzalez-Roca, Aitor Gastaminza Unanue, Esther Titos, Miguel Juli'an Mart'ınez Yoldi, Andrea Vergara G'omez, and Joan Anton Puig-Butill'e, Implementation of an open-source robotic platform for sars-cov-2 testing by real-time rt-pcr. PLoS One **16**(7), e0252509 (2021)
26. Robert P Goldman, Puja Trivedi, Daniel Bryce, Matthew DeHaven, Alex Plotnick, Peter L Lee, Joshua Nowak, Vanessa M Biggers, Trissha R Higa, and Jeremy P Hunt. A bayesian model for experiment choice in synthetic biology
27. J. Hayes, Technology laboratory automation: Labs go auto. Engineering & Technology **16**(7), 58–60 (2021)
28. Autoprotocol – language for specifying experimental protocols. https://autoprotocol.org/
29. A. Chattopadhyay, A. Prakash, M. Shafique, Secure cyber-physical systems: Current trends, tools and open research problems, in *Design, Automation & Test in Europe Conference & Exhibition (DATE), 2017*, (IEEE, 2017), pp. 1104–1109
30. C. Cheh, A. Fawaz, M.A. Noureddine, B. Chen, W.G. Temple, W.H. Sanders, Determining tolerable attack surfaces that preserves safety of cyber-physical systems, in *2018 IEEE 23rd Pacific Rim International Symposium on Dependable Computing (PRDC)*, (IEEE, 2018), pp. 125–134
31. D. Antonioli, N.O. Tippenhauer, Minicps: A toolkit for security research on cps networks, in *Proceedings of the First ACM Workshop on Cyber-Physical Systems-Security and/or Privacy*, (2015), pp. 91–100
32. Restrictedpython – provide a program input into a trusted environment. https://pypi.org/project/RestrictedPython/
33. C. Alberca, S. Pastrana, G. Suarez-Tangil, P. Palmieri, Security analysis and exploitation of arduino devices in the internet of things, in *Proceedings of the ACM International Conference on Computing Frontiers*, (2016), pp. 437–442. https://doi.org/10.1145/2903150.2911708
34. M.J. Joyner, N. Paneth, Seven questions for personalized medicine. JAMA **314**(10), 999–1000 (2015)
35. K. Gorshkov, C.Z. Chen, R.E. Marshall, N. Mihatov, Y. Choi, D.-T. Nguyen, N. Southall, K.G. Chen, J.K. Park, W. Zheng, Advancing precision medicine with personalized drug screening. Drug Discov. Today **24**(1), 272–278 (2019)
36. M.M. Soldevilla, D. Meraviglia-Crivelli, A.P. de Caso, Menon, and Fernando Pastor., Aptamer-irnas as therapeutics for cancer treatment. Pharmaceuticals **11(4):108** (2018)

37. M.S. Nabavinia, A. Gholoobi, F. Charbgoo, M. Nabavinia, M. Ramezani, K. Abnous, Anti-muc1 aptamer: A potential opportunity for cancer treatment. Med. Res. Rev. **37**(6), 1518–1539 (2017)
38. S. Kosuri, G.M. Church, Large-scale de novo dna synthesis: Technologies and applications. Nat. Methods **11**(5), 499–507 (2014)
39. S. Faezi, S.R. Chhetri, A.V. Malawade, J.C. Chaput, W. Grover, P. Brisk, M.A. Al Faruque, Oligo-snoop: A non-invasive side channel attack against DNA synthesis machines, in *Network and Distributed Systems Security (NDSS) Symposium 2019*, vol. 2019,
40. F. de Jesus, D.G. Cortez, D. Tandel, P.V. Robinson, D. Seftel, D.M. Wilson, D.M. Maahs, B.A. Buckingham, K.W.P. Miller, C.-t. Tsai, Automation of a multiplex agglutination-pcr (adap) type 1 diabetes (t1d) assay for the rapid analysis of islet autoantibodies. SLAS Technology (2021)
41. The opentrons covid-19 testing system. https://blog.opentrons.com/how-to-use-opentrons-to-test-for-covid-19/
42. J. Bartlett, D. Stirling, A short history of the polymerase chain reaction, in *PCR protocols*, (Springer, 2003), pp. 3–6
43. J. Salm, O. Sefiha, Restorative justice in sports: Does restorative justice have a place in anti-doping governance? Sport in Society, 1–16 (2021)
44. Emelie Ohnstedt, Hava Lofton Tomenius, Peter Frank, Stefan Roos,̈ E. V°agesjö, and Mia Phillipson. Accelerated wound healing in minipigs by on-site production and delivery of cxcl12 by transformed lactic acid bacteria. Pharmaceutics, 14(2):229, 2022
45. Chloe Kent, Drug dispensing goes digital. https://www.pharmaceutical-technology.com/features/robotic-drug-dispensing-digital-pharmacy/

Index

D. Greenbaum (ed.), *Cyberbiosecurity*, https://doi.org/10.1007/978-3-031-26034-6

SPRINGER NATURE

GPSR Compliance

The European Union's (EU) General Product Safety Regulation (GPSR) is a set of rules that requires consumer products to be safe and our obligations to ensure this.

If you have any concerns about our products, you can contact us on ProductSafety@springernature.com

In case Publisher is established outside the EU, the EU authorized representative is:

Springer Nature Customer Service Center GmbH
Europaplatz 3
69115 Heidelberg, Germany

The manufacturer's authorised representative in the EU is Springer Nature Customer Service Centre GmbH, Europaplatz 3, 69115 Heidelberg, Germany. If you have any concerns regarding our products, please contact ProductSafety@springernature.com

Printed and bound by CPI Group (UK) Ltd, Croydon, CR0 4YY
30/09/2026
02225561-0007